Curvature and Homology

Curvature and Homology

SAMUEL I. GOLDBERG

Department of Mathematics
University of Illinois, Urbana, Illinois

DOVER PUBLICATIONS, INC., NEW YORK

Published in Canada by General Publishing Company, Ltd., 30 Lesmill Road, Don Mills, Toronto, Ontario.
Published in the United Kingdom by Constable and Company, Ltd.

This Dover edition, first published in 1982, is an unabridged and further corrected republication of the second printing (1970) of the work published in 1962 by Academic Press, N.Y., as Volume 11 in the series *Pure and Applied Mathematics*. Twelve additional titles have been added to the References.

Manufactured in the United States of America
Dover Publications, Inc.
180 Varick Street
New York, N.Y. 10014

Library of Congress Cataloging in Publication Data

Goldberg, Samuel I.
Curvature and homology.

Reprint. Originally published: New York : Academic Press, 1962 (1970 printing) (Pure and applied mathematics ; 11)
Bibliography: p.
Includes indexes.
1. Curvature. 2. Homology theory. 3. Geometry, Riemannian. I. Title. II. Series: Pure and applied mathematics (Academic Press) ; 11.
QA645.G6 1982 516.3'62 81-19459
ISBN 0-486-64314-X AACR2

To my parents and my wife

PREFACE

The purpose of this book is to give a systematic and "self-contained" account along modern lines of the subject with which the title deals, as well as to discuss problems of current interest in the field. With this statement the author wishes to recall another book, "Curvature and Betti Numbers," by K. Yano and S. Bochner; this tract is aimed at those already familiar with differential geometry, and has served admirably as a useful reference during the nine years since its appearance. In the present volume, a coordinate-free treatment is presented wherever it is considered feasible and desirable. On the other hand, the index notation for tensors is employed whenever it seems to be more adequate.

The book is intended for the reader who has taken the standard courses in linear algebra, real and complex variables, differential equations, and point-set topology. Should he lack an elementary knowledge of algebraic topology, he may accept the results of Chapter II and proceed from there. In Appendix C he will find that some knowledge of Hilbert space methods is required. This book is also intended for the more seasoned mathematician, who seeks familiarity with the developments in this branch of differential geometry in the large. For him to feel at home a knowledge of the elements of Riemannian geometry, Lie groups, and algebraic topology is desirable.

The exercises are intended, for the most part, to supplement and to clarify the material wherever necessary. This has the advantage of maintaining emphasis on the subject under consideration. Several might well have been explained in the main body of the text, but were omitted in order to focus attention on the main ideas. The exercises are also devoted to miscellaneous results on the homology properties of rather special spaces, in particular, δ-pinched manifolds, locally convex hypersurfaces, and minimal varieties. The inexperienced reader should not be discouraged if the exercises appear difficult. Rather, should he be interested, he is referred to the literature for clarification.

References are enclosed in square brackets. Proper credit is almost always given except where a reference to a later article is either more informative or otherwise appropriate. Cross references appear as (6.8.2) referring to Chapter VI, Section 8, Formula 2 and also as (VI.A.3) referring to Chapter VI, Exercise A, Problem 3.

The author owes thanks to several colleagues who read various parts of the manuscript. He is particularly indebted to Professor M. Obata, whose advice and diligent care has led to many improvements. Professor R. Bishop suggested some exercises and further additions. Gratitude is also extended to Professors R. G. Bartle and A. Heller for their critical reading of Appendices A and C as well as to Dr. L. McCulloh and Mr. R. Vogt for assisting with the proofs. For the privilege of attending his lectures on Harmonic Integrals at Harvard University, which led to the inclusion of Appendix A, thanks are extended to Professor L. Ahlfors. Finally, the author expresses his appreciation to Harvard University for the opportunity of conducting a seminar on this subject.

It is a pleasure to acknowledge the invaluable assistance received in the form of partial financial support from the Air Force Office of Scientific Research.

S. I. GOLDBERG

Urbana, Illinois
February, 1962

CONTENTS

NOTATION INDEX

The symbols used have gained general acceptance with some exceptions. In particular, R and C are the fields of real and complex numbers, respectively. (In § 7.1, the same letter C is employed as an operator and should cause no confusion.) The commonly used symbols $\in$, $\cup$, $\cap$, $\cong$, sup, inf, are not listed. The exterior or Grassman algebra of a vector space V (over R or C) is written as $\wedge(V)$. By $\wedge^p(V)$ is meant the vector space of its elements of degree p and $\wedge$ denotes multiplication in $\wedge(V)$. The elements of $\wedge(V)$ are designated by Greek letters. The symbol M is reserved for a topological manifold, T_P its tangent space at a point $P \in M$ (in case M is a differentiable manifold) and T_P^* the dual space (of covectors). The space of tangent vector fields is denoted by T and its dual by T^*. The Lie bracket of tangent vectors X and Y is written as $[X, Y]$. Tensors are generally denoted by Latin letters. For example, the metric tensor of a Riemannian manifold will usually be denoted by g. The covariant form of X (with respect to g) is designated by the corresponding Greek symbol ξ. The notation for composition of functions (maps) employed is flexible. It is sometimes written as $g \cdot f$ and at other times the dot is not present. The dot is also used to denote the (local) scalar product of vectors (relative to g). However, no confusion should arise.

$O(n) = O(n, R)$: The subgroup of $GL(n, R)$ consisting of those matrices a for which ${}^t a = a^{-1}$ where a^{-1} is the inverse of a and ${}^t a$ denotes its transpose : ${}^t(a_j^i) = (a_i^j)$.

$U(n) = \{a \in GL(n, C) \mid \bar{a} = {}^t a^{-1}\}$, where $\bar{a} = (\overline{a_j^i})$.

$SU(n) = \{a \in U(n) \mid \det(a) = 1\}$.

INTRODUCTION

The most important aspect of differential geometry is perhaps that which deals with the relationship between the curvature properties of a Riemannian manifold and its topological structure. One of the beautiful results in this connection is the generalized Gauss-Bonnet theorem which for orientable surfaces has long been known. In recent years there has been a considerable increase in activity in global differential geometry thanks to the celebrated work of W. V. D. Hodge and the applications of it made by S. Bochner, A. Lichnerowicz, and K. Yano. In the decade since the appearance of Bochner's first papers in this field many fruitful investigations on the subject matter of "curvature and betti numbers" have been inaugurated. The applications are, to some extent, based on a theorem in differential equations due to E. Hopf. The Laplace-Beltrami operator Δ is elliptic and when applied to a function f of class 2 defined on a compact Riemannian manifold M yields the Bochner lemma: "If $\Delta f \geqq 0$ everywhere on M, then f is a constant and Δf vanishes identically." Many diverse applications to the relationship between the curvature properties of a Riemannian manifold and its homology structure have been made as a consequence of this "observation." Of equal importance, however, a "dual" set of results on groups of motions is realized.

The existence of harmonic tensor fields over compact orientable Riemannian manifolds depends largely on the signature of a certain quadratic form. The operator Δ introduces curvature, and these properties of the manifold determine to some extent the global structure via Hodge's theorem relating harmonic forms with betti numbers. In Chapter II, therefore, the theory of harmonic integrals is developed to the extent necessary for our purposes. A proof of the existence theorem of Hodge is given (modulo the fundamental differentiability lemma C.1 of Appendix C), and the essential material and information necessary for the treatment and presentation of the subject of curvature and homology is presented. The idea of the proof of the existence theorem is to show that $\bar{\Delta}^{-1}$—the inverse of the closure of Δ-is a completely continuous operator. The reader is referred to de Rham's book "Variétés Différentiables" for an excellent exposition of this result.

The spaces studied in this book are important in various branches of mathematics. Locally they are those of classical Riemannian geometry, and from a global standpoint they are compact orientable manifolds. Chapter I is concerned with the local structure, that is, the geometry of the space over which the harmonic forms are defined. The properties necessary for an understanding of later chapters are those relating to the differential geometry of the space, and those which are topological properties. The topology of a differentiable manifold is therefore discussed in Chapter II. Since these subjects have been given essentially complete and detailed treatments elsewhere, and since a thorough discussion given here would reduce the emphasis intended, only a brief survey of the bare essentials is outlined. Families of Riemannian manifolds are described in Chapter III, each including the n-sphere and retaining its betti numbers. In particular, a 4-dimensional δ-pinched manifold is a homology sphere provided $\delta > \frac{1}{4}$. More generally, the second betti number of a δ-pinched even-dimensional manifold is zero if $\delta > \frac{1}{4}$.

The theory of harmonic integrals has its origin in an attempt to generalize the well-known existence theorem of Riemann to everywhere finite integrals over a Riemann surface. As it turns out in the generalization a $2n$-dimensional Riemannian manifold plays the part of the Riemann surface in the classical 2-dimensional case although a Riemannian manifold of 2 dimensions is not the same as a Riemann surface. The essential difference lies in the geometry which in the latter case is conformal. In higher dimensions, the concept of a complex analytic manifold is the natural generalization of that of a Riemann surface in the abstract sense. In this generalization concepts such as holomorphic function have an invariant meaning with respect to the given complex structure. Algebraic varieties in a complex projective space P_n have a natural complex structure and are therefore complex manifolds provided there are no "singularities." There exist, on the other hand, examples of complex manifolds which cannot be imbedded in a P_n. A complex manifold is therefore more general than a projective variety. This approach is in keeping with the modern developments due principally to A. Weil.

It is well-known that all orientable surfaces admit complex structures. However, for higher even-dimensional orientable manifolds this is not the case. It is not possible, for example, to define a complex structure on the 4-dimensional sphere. (In fact, it was recently shown that not every topological manifold possesses a differentiable structure.) For a given complex manifold M not much is known about the complex structure itself; all consequences are derived from assumptions which are weaker—the "almost-complex" structure, or stronger—the existence of a "Kaehler metric." The former is an assumption concerning the tangent bundle of M and therefore suitable for fibre space methods, whereas the latter is an assumption on the Riemannian geometry of M, which can be investigated by the theory of harmonic forms. The material of Chapter V is partially concerned with a development of hermitian geometry, in particular, Kaehler geometry along the lines proposed by S. Chern. Its influence on the homology structure of the manifold is discussed in Chapters V and VI. Whereas the homology properties described in Chapter III are similar to those of the ordinary sphere (insofar as betti numbers are concerned), the corresponding properties in Chapter VI are possessed by P_n itself. Families of hermitian manifolds are described, each including P_n and retaining its betti numbers. One of the most important applications of the effect of curvature on homology is to be found in the vanishing theorems due to K. Kodaira. They are essential in the applications of sheaf theory to complex manifolds.

A conformal transformation of a compact Riemann surface is a holomorphic homeomorphism. For compact Kaehler manifolds of higher dimension, an element of the connected component of the identity of the group of conformal transformations is an isometry, and consequently a holomorphic homeomorphism. More generally, an infinitesimal conformal map of a compact Riemannian manifold admitting a harmonic form of constant length is an infinitesimal isometry. Thus, if a compact homogeneous Riemannian manifold admits an infinitesimal non-isometric conformal transformation, it is a homology sphere. Indeed, it is then isometric with a sphere. The conformal transformation group is studied in Chapter III, and in Chapter VII groups of holomorphic as well as conformal homeomorphisms of Kaehler manifolds are investigated.

In Appendix A, a proof of de Rham's theorems based on the concept of a sheaf is given although this notion is not defined. Indeed, the proof is but an adaptation from the general theory of sheaves and a knowledge of the subject is not required.

CHAPTER I

RIEMANNIAN MANIFOLDS

In seeking to generalize the well-known theorem of Riemann on the existence of holomorphic integrals over a Riemann surface, W. V. D. Hodge [*39*] considers an n-dimensional Riemannian manifold as the space over which a certain class of integrals is defined. Now, a Riemannian manifold of two dimensions is not a Riemann surface, for the geometry of the former is Riemannian geometry whereas that of a Riemann surface is conformal geometry. However, in a certain sense a 2-dimensional Riemannian manifold may be thought of as a Riemann surface. Moreover, conformally homeomorphic Riemannian manifolds of two dimensions define equivalent Riemann surfaces. Conversely, a Riemann surface determines an infinite set of conformally homeomorphic 2-dimensional Riemannian manifolds. Since the underlying structure of a Riemannian manifold is a differentiable structure, we discuss in this chapter the concept of a differentiable manifold, and then construct over the manifold the integrals, tensor fields and differential forms which are basically the objects of study in the remainder of this book.

1.1. Differentiable manifolds

The differential calculus is the main tool used in the study of the geometrical properties of curves and surfaces in ordinary Euclidean space E^3. The concept of a curve or surface is not a simple one, so that in many treatises on differential geometry a rigorous definition is lacking. The discussions on surfaces are further complicated since one is interested in those properties which remain invariant under the group of motions in E^3. This group is itself a 6-dimensional manifold. The purpose of this section is to develop the fundamental concepts of differentiable manifolds necessary for a rigorous treatment of differential geometry.

Given a topological space, one can decide whether a given function defined over it is continuous or not. A discussion of the properties of the classical surfaces in differential geometry requires more than continuity, however, for the functions considered. By a *regular closed surface* S in E^3 is meant an ordered pair $\{S_0, X\}$ consisting of a topological space S_0 and a differentiable map X of S_0 into E^3. As a topological space, S_0 is to be a separable, Hausdorff space with the further properties:

(i) S_0 is compact (that is $X(S_0)$ is closed and bounded);

(ii) S_0 is connected (a topological space is said to be *connected* if it cannot be expressed as the union of two non-empty disjoint open subsets);

(iii) Each point of S_0 has an open neighborhood homeomorphic with E^2: The map $X : P \to (x(P), y(P), z(P))$, $P \in S_0$ where $x(P)$, $y(P)$ and $z(P)$ are differentiable functions is to have rank 2 at each point $P \in S_0$, that is the matrix

$$\begin{pmatrix} x_u & y_u & z_u \\ x_v & y_v & z_v \end{pmatrix}$$

of partial derivatives must be of rank 2 where u, v are local parameters at P. Let U and V be any two open neighborhoods of P homeomorphic with E^2 and with non-empty intersection. Then, their local parameters or coordinates (cf. definition given below of a differentiable structure) must be related by differentiable functions with non-vanishing Jacobian. It follows that the rank of X is invariant with respect to a change of coordinates.

That a certain amount of differentiability is necessary is clear from several points of view. In the first place, the condition on the rank of X implies the existence of a tangent plane at each point of the surface. Moreover, only those local parameters are "allowed" which are related by differentiable functions.

A regular closed surface is but a special case of a more general concept which we proceed to define.

Roughly speaking, a differentiable manifold is a topological space in which the concept of derivative has a meaning. Locally, the space is to behave like Euclidean space. But first, a topological space M is said to be *separable* if it contains a countable basis for its topology. It is called a *Hausdorff space* if to any two points of M there are disjoint open sets each containing exactly one of the points.

A separable Hausdorff space M of dimension n is said to have a *differentiable structure* of class $k > 0$ if it has the following properties:

(i) Each point of M has an open neighborhood homeomorphic with an open subset in R^n the (number) space of n real variables, that is, there is a finite or countable open covering $\{U_\alpha\}$ and, for each α a homeomorphism $u_\alpha : U_\alpha \to R^n$ of U_α onto an open subset in R^n;

(ii) For any two open sets U_α and U_β with non-empty intersection the map $u_\beta u_\alpha^{-1} : u_\alpha(U_\alpha \cap U_\beta) \to R^n$ is of class k (that is, it possesses continuous derivatives of order k) with non-vanishing Jacobian.

The functions defining u_α are called *local coordinates in* U_α. Clearly, one may also speak of structures of class ∞ (that is, structures of class k for every positive integer k) and *analytic* structures (that is, every map $u_\beta u_\alpha^{-1}$ is expressible as a convergent power series in the n variables). The local coordinates constitute an essential tool in the study of M. However, the geometrical properties should be independent of the choice of local coordinates.

The space M with the property (i) will be called a *topological manifold.* We shall generally assume that the spaces considered are connected although many of the results are independent of this hypothesis.

Examples: 1. The Euclidean space E^n is perhaps the simplest example of a topological manifold with a differentiable structure. The identity map I in E^n together with the *unit covering* (R^n, I) is its natural differentiable structure: $(U_1, u_1) = (R^n, I)$.

2. The $(n - 1)$-dimensional sphere in E^n defined by the equation

$$\sum_{i=1}^{n} (x^i)^2 = 1: \tag{1.1.1}$$

It can be covered by $2n$ coordinate neighborhoods defined by $x^i > 0$ and $x^i < 0$ $(i = 1, ..., n)$.

3. The general linear group: Let V be a vector space over R (the real numbers) of dimension n and let $\{e_1, ..., e_n\}$ be a basis of V. The group of all linear automorphisms a of V may be expressed as the group of all non-singular matrices (a^i_j);

$$ae_i = a^j_i\, e_j, \qquad i, j = 1, ..., n \tag{1.1.2}$$

called the *general linear group* and denoted by $GL(n, R)$. We shall also denote it by $GL(V)$ when dealing with more than one vector space. (The Einstein summation convention where repeated indices implies addition has been employed in formula (1.1.2) and, in the sequel we shall adhere to this notation.) The multiplication law is

$$(ab)^i_j = a^i_k\, b^k_j\,.$$

$GL(n, R)$ may be considered as an open set [and hence as an open submanifold (cf. §1.5)] of E^{n^2}. With this structure (as an analytic manifold), $GL(n, R)$ is a Lie group (cf. §3.6).

Let f be a real-valued continuous function defined in an open subset S of M. Let P be a point of S and U_α a coordinate neighborhood containing P. Then, in $S \cap U_\alpha$, f can be expressed as a function of the local coordinates $u^1, \ldots, u^n$ in U_α. (If $x^1, \ldots, x^n$ are the n coordinate functions on R^n, then $u^i(P) = x^i(u_\alpha(P))$, $i = 1, \ldots, n$ and we may write $u^i = x^i \cdot u_\alpha$). The function f is said to be *differentiable* at P if $f(u^1, \ldots, u^n)$ possesses all first partial derivatives at P. The *partial derivative* of f with respect to u^i at P is defined as

$$\left(\frac{\partial f}{\partial u^i}\right)_P = \left(\frac{\partial (f u_\alpha^{-1})}{\partial x^i}\right)_{u_\alpha(P)}.$$

This property is evidently independent of the choice of U_α. The function f is called *differentiable* in S, if it is differentiable at every point of S. Moreover, f is of the form $g \cdot u_\alpha$ if the domain is restricted to $S \cap U_\alpha$ where g is a continuous function in $u_\alpha(S \cap U_\alpha) \subset R^n$. Two differentiable structures are said to be *equivalent* if they give rise to the same family of differentiable functions over open subsets of M. This is an equivalence relation. The family of functions of class k determines the differentiable structures in the equivalence class.

A topological manifold M together with an equivalence class of differentiable structures on M is called a *differentiable manifold.* It has recently been shown that not every topological manifold can be given a differentiable structure [*44*]. On the other hand, a topological manifold may carry differentiable structures belonging to distinct equivalence classes. Indeed, the 7-dimensional sphere possesses several inequivalent differentiable structures [60].

A differentiable mapping f of an open subset S of R^n into R^n is called *sense-preserving* if the Jacobian of the map is positive in S. If, for any pair of coordinate neighborhoods with non-empty intersection, the mapping $u_\beta u_\alpha^{-1}$ is sense-preserving, the differentiable structure is said to be *oriented* and, in this case, the differentiable manifold is called *orientable.* Thus, if $f_{\beta\alpha}(x)$ denotes the Jacobian of the map $u_\beta u_\alpha^{-1}$ at $x^i(u_\alpha(P))$, $i = 1, \ldots, n$, then $f_{\gamma\beta}(x) f_{\beta\alpha}(x) = f_{\gamma\alpha}(x)$, $P \in U_\alpha \cap U_\beta \cap U_\gamma$.

The 2-sphere in E^3 is an orientable manifold whereas the real projective plane (the set of lines through the origin in E^3) is not (cf. I.B. 2).

Let M be a differentiable manifold of class k and S an open subset of M. By restricting the functions (of class k) on M to S, the differentiable structure so obtained on S is called an *induced structure* of class k. In particular, on every open subset of E^1 there is an induced structure

1.2. Tensors

To every point P of a regular surface S there is associated the tangent plane at P consisting of the tangent vectors to the curves on S through P. A tangent vector t may be expressed as a linear combination of the tangent vectors X_u and X_v "defining" the tangent plane:

$$t = \xi^1 X_u + \xi^2 X_v, \qquad \xi^i \in R, \qquad i = 1, 2. \tag{1.2.1}$$

At this point, we make a slight change in our notation: We put $u^1 = u$, $u^2 = v$, $X_1 = X_u$ and $X_2 = X_v$, so that (1.2.1) becomes

$$t = \xi^i X_i. \tag{1.2.2}$$

Now, in the coordinates $\bar{u}^1$, $\bar{u}^2$ where the $\bar{u}^i$ are related to the u^j by means of differentiable functions with non-vanishing Jacobian

$$t = \xi^i \frac{\partial \bar{u}^j}{\partial u^i} \bar{X}_j \tag{1.2.3}$$

where $\bar{X} = X(u^1(\bar{u}^1, \bar{u}^2), u^2(\bar{u}^1, \bar{u}^2))$. If we put

$$\bar{\xi}^j = \frac{\partial \bar{u}^j}{\partial u^i} \xi^i \tag{1.2.4}$$

equation (1.2.3) becomes

$$t = \bar{\xi}^j \bar{X}_j. \tag{1.2.5}$$

In classical differential geometry the vector t is called a contravariant vector, the equations of transformation (1.2.4) determining its character.

Guided by this example we proceed to define the notion of contravariant vector for a differentiable manifold M of dimension n. Consider the triple (P, U_α, ξ^i) consisting of a point $P \in M$, a coordinate neighborhood U_α containing P and a set of n real numbers ξ^i. An equivalence relation is defined if we agree that the triples (P, U_α, ξ^i) and $(\bar{P}, U_\beta, \bar{\xi}^i)$ are *equivalent* if $P = \bar{P}$ and

$$\bar{\xi}^j = \left(\frac{\partial \bar{u}^j}{\partial u^i}\right)_{u_\alpha(P)} \xi^i, \tag{1.2.6}$$

where the u^i are the coordinates of $u_\alpha(P)$ and $\bar{u}^i$ those of $u_\beta(P)$, $P \in U_\alpha \cap U_\beta$. An equivalence class of such triples is called a *contravariant vector* at P. When there is no danger of confusion we simply speak of the contravariant vector by choosing a particular set of representatives ξ^i

($i = 1, ..., n$). That the contravariant vectors form a linear space over R is clear. In analogy with surface theory this linear space is called the *tangent space* at P and will be denoted by T_P. (For a rather sophisticated definition of tangent vector the reader is referred to §3.4.)

Let f be a differentiable function defined in a neighborhood of $P \in U_\alpha \cap U_\beta$. Then,

$$\left(\frac{\partial(fu_\alpha^{-1})}{\partial x^i}\right)_{u_\alpha(P)} = \left(\frac{\partial(fu_\beta^{-1})}{\partial x^j}\right)_{u_\beta(P)} \left(\frac{\partial \bar{u}^j}{\partial u^i}\right)_{u_\alpha(P)}. \tag{1.2.7}$$

Now, applying (1.2.6) we obtain

$$\left(\frac{\partial(fu_\alpha^{-1})}{\partial x^i}\right)_{u_\alpha(P)} \xi^i = \left(\frac{\partial(fu_\beta^{-1})}{\partial x^j}\right)_{u_\beta(P)} \bar{\xi}^j. \tag{1.2.8}$$

The equivalence class of "functions" of which the left hand member of (1.2.8) is a representative is commonly called the *directional derivative* of f along the contravariant vector ξ^i. In particular, if the components $\xi^i(i = 1, ..., n)$ all vanish except the kth which is 1, the directional derivative is the partial derivative with respect to u^k and the corresponding contravariant vector is denoted by $\partial/\partial u^k$. Evidently, these vectors for all $k = 1, ..., n$ form a base of T_P called the *natural base*. On the other hand, the partial derivatives of f in (1.2.8) are representatives of a vector (which we denote by df) in the dual space T_P^* of T_P. The elements of T_P^* are called *covariant vectors* or, simply, *covectors*. In the sequel, when we speak of a covariant vector at P, we will occasionally employ a set of representatives. Hence, if η_i is a covariant vector and ξ^i a contravariant vector the expression $\eta_i\xi^i$ is a *scalar invariant* or, simply *scalar*, that is

$$\bar{\eta}_i\bar{\xi}^i = \eta_i\xi^i, \tag{1.2.9}$$

and so,

$$\bar{\eta}_i = \frac{\partial u^j}{\partial \bar{u}^i}\eta_j \tag{1.2.10}$$

are the equations of transformation defining a covariant vector. We define the *inner product* of a contravariant vector $v = \xi^i$ and a covariant vector $w^* = \eta_i$ by the formula

$$\langle v, w^* \rangle = \eta_i\xi^i. \tag{1.2.11}$$

That the inner product is bilinear is clear. Now, from (1.2.10) we obtain

$$\bar{\eta}_i\, d\bar{u}^i = \eta_i\, du^i \tag{1.2.12}$$

where the du^i ($i = 1, ..., n$) are the differentials of the functions $u^1, ..., u^n$.

The invariant expression $\eta_i du^i$ is called a *linear (differential) form* or *1-form*. Conversely, when a linear (differential) form is given, its coefficients define an element of T_P^*. If we agree to identify T_P^* with the space of 1-forms at P, the du^i at P form a base of T_P^* dual to the base $\partial/\partial u^i$ $(i = 1, ..., n)$ of tangent vectors at P:

$$\left\langle \frac{\partial}{\partial u^i}, du^j \right\rangle = \delta_i^j, \qquad i, j = 1, ..., n \tag{1.2.13}$$

where δ_i^j is the 'Kronecker delta', that is, $\delta_i^j = 1$ if $i = j$ and $\delta_i^j = 0$ if $i \neq j$.

We proceed to generalize the notions of contravariant and covariant vectors at a point $P \in M$. To this end we proceed in analogy with the definitions of contravariant and covariant vector. Consider the triples $(P, U_\alpha, \xi^{i_1 \dots i_r}{}_{j_1 \dots j_s})$ and $(\bar{P}, U_\beta, \bar{\xi}^{i_1 \dots i_r}{}_{j_1 \dots j_s})$. They are said to be equivalent if $\bar{P} = P$ and if the n^{r+s} constants $\bar{\xi}^{i_1 \dots i_r}{}_{j_1 \dots j_s}$ are related to the n^{r+s} constants $\xi^{i_1 \dots i_r}{}_{j_1 \dots j_s}$ by the formulae

$$\bar{\xi}^{i_1 \dots i_r}{}_{j_1 \dots j_s} = \left(\frac{\partial \bar{u}^{i_1}}{\partial u^{k_1}}\right)_{u_\alpha(P)} \cdots \left(\frac{\partial \bar{u}^{i_r}}{\partial u^{k_r}}\right)_{u_\alpha(P)} \left(\frac{\partial u^{l_1}}{\partial \bar{u}^{j_1}}\right)_{u_\beta(P)} \cdots \left(\frac{\partial u^{l_s}}{\partial \bar{u}^{j_s}}\right)_{u_\beta(P)} \xi^{k_1 \dots k_r}{}_{l_1 \dots l_s}. \tag{1.2.14}$$

An equivalence class of triples $(P, U_\alpha, \xi^{i_1 \dots i_r}{}_{j_1 \dots j_s})$ is called a *tensor of type* (r, s) *over* T_P *contravariant of order* r *and covariant of order* s. A tensor of type $(r, 0)$ is called a *contravariant tensor* and one of type $(0, s)$ a *covariant tensor*. Clearly, the tensors of type (r, s) form a linear space—the *tensor space* of tensors of type (r, s). By convention a scalar is a *tensor of type* $(0, 0)$.

If the *components* $\xi^{i_1 \dots i_r}{}_{j_1 \dots j_s}$ of a tensor are all zero in one local coordinate system they are zero in any other local coordinate system. This tensor is then called a *zero tensor*. Again, if $\xi^{i_1 \dots i_r}{}_{j_1 \dots j_s}$ is symmetric or skew-symmetric in i_p, i_q (or in j_p, j_q), $\bar{\xi}^{i_1 \dots i_r}{}_{j_1 \dots j_s}$ has the same property. These properties are therefore characteristic of tensors. The tensor $\xi^{i_1 \dots i_r}$ (or $\xi_{j_1 \dots j_s}$) is said to be *symmetric (skew-symmetric)* if it is symmetric (skew-symmetric) in every pair of indices.

The *product* of two tensors $(P, U_\alpha, \xi^{i_1 \dots i_r}{}_{j_1 \dots j_s})$ and $(P, U_\alpha, \eta^{i_1 \dots i_{r'}}{}_{j_1 \dots j_{s'}})$ one of type (r, s) the other of type (r', s') is the tensor $(P, U_\alpha, \xi^{i_1 \dots i_r}{}_{j_1 \dots j_s} \eta^{i_{r+1} \dots i_{r+r'}}{}_{j_{s+1} \dots j_{s+s'}})$ of type $(r + r', s + s')$. In fact,

$$\bar{\xi}^{i_1 \dots i_r}{}_{j_1 \dots j_s} \bar{\eta}^{i_{r+1} \dots i_{r+r'}}{}_{j_{s+1} \dots j_{s+s'}} =$$

$$\left(\frac{\partial \bar{u}_{i_1}}{\partial u^{k_1}}\right)_{u_\alpha(P)} \cdots \left(\frac{\partial u^{l_s}}{\partial \bar{u}^{j_s}}\right)_{u_\beta(P)} \left(\frac{\partial \bar{u}^{i_{r+1}}}{\partial u^{k_{r+1}}}\right)_{u_\alpha(P)} \cdots \left(\frac{\partial u^{l_{s+s'}}}{\partial \bar{u}^{j_{s+s'}}}\right)_{u_\beta(P)} \xi^{k_1 \dots k_r}{}_{l_1 \dots l_s} \eta^{k_{r+1} \dots k_{r+r'}}{}_{l_{s+1} \dots l_{s+s'}}.$$

It is also possible to form new tensors from a given tensor. In fact, let $(P, U_\alpha, \xi^{i_1 \ldots i_r}{}_{j_1 \ldots j_s})$ be a tensor of type (r, s). The triple $(P, U_\alpha, \xi^{i_1 \ldots i_p \ldots i_r}{}_{j_1 \ldots j_q \ldots j_s})$ where the indices i_p and j_q are equal (recall that repeated indices indicate summation from 1 to n) is a representative of a tensor of type $(r-1, s-1)$. For,

$$\xi^{i_1 \ldots i_p \ldots i_r}{}_{j_1 \ldots j_{q-1} i_p j_{q+1} \ldots j_s} = \left(\frac{\partial \bar{u}^{i_1}}{\partial u^{k_1}}\right)_{u_\alpha(P)} \cdots \left(\frac{\partial \bar{u}^{i_p}}{\partial u^{k_p}}\right)_{u_\alpha(P)} \cdots \left(\frac{\partial \bar{u}^{i_r}}{\partial u^{k_r}}\right)_{u_\alpha(P)}$$

$$\cdot \left(\frac{\partial u^{l_1}}{\partial \bar{u}^{j_1}}\right)_{u_\beta(P)} \cdots \left(\frac{\partial u^{l_p}}{\partial \bar{u}^{i_p}}\right)_{u_\beta(P)} \cdots \left(\frac{\partial u^{l_s}}{\partial \bar{u}^{j_s}}\right)_{u_\beta(P)} \xi^{k_1 \ldots k_p \ldots k_r}{}_{l_1 \ldots l_{q-1} l_p l_{q+1} \ldots l_s}$$

$$= \left(\frac{\partial \bar{u}^{i_1}}{\partial u^{k_1}}\right)_{u_\alpha(P)} \cdots \left(\frac{\partial \bar{u}^{i_{p-1}}}{\partial u^{k_{p-1}}}\right)_{u_\alpha(P)} \left(\frac{\partial \bar{u}^{i_{p+1}}}{\partial u^{k_{p+1}}}\right)_{u_\alpha(P)} \cdots \left(\frac{\partial \bar{u}^{i_r}}{\partial u^{k_r}}\right)_{u_\alpha(P)}$$

$$\cdot \left(\frac{\partial u^{l_1}}{\partial \bar{u}^{j_1}}\right)_{u_\beta(P)} \cdots \left(\frac{\partial u^{l_{q-1}}}{\partial \bar{u}^{j_{q-1}}}\right)_{u_\beta(P)} \left(\frac{\partial u^{l_{q+1}}}{\partial \bar{u}^{j_{q+1}}}\right)_{u_\beta(P)} \cdots \left(\frac{\partial u^{l_s}}{\partial \bar{u}^{j_s}}\right)_{u_\beta(P)} \xi^{k_1 \ldots k_p \ldots k_r}{}_{l_1 \ldots l_{q-1} k_p l_{q+1} \ldots l_s}$$

since

$$\frac{\partial \bar{u}^{i_p}}{\partial u^{k_p}} \frac{\partial u^{l_p}}{\partial \bar{u}^{i_p}} = \delta^{l_p}_{k_p} .$$

This operation is known as *contraction* and the tensor so obtained is called the *contracted tensor*.

These operations may obviously be combined to yield other tensors. A particularly important case occurs when the tensor ξ_{ij} is a symmetric covariant tensor of order 2. If η^i is a contravariant vector, the quadratic form $\xi_{ij}\, \eta^i\, \eta^j$ is a scalar. The property that this quadratic form be positive definite is a property of the tensor ξ_{ij} and, in this case, we call the tensor *positive definite*.

Our definition of a tensor of type (r, s) is rather artificial and is actually the one given in classical differential geometry. An intrinsic definition is given in the next section. But first, let V be a vector space of dimension n over R and let V^* be the dual space of V. A *tensor of type* (r, s) *over* V, contravariant of order r and covariant of order s, is defined to be a multilinear map of the direct product $V \times \ldots \times V \times V^* \times \ldots \times V^*$ (V:s times, V^*:r times) into R. All tensors of type (r, s) form a linear space over R with respect to the usual addition and scalar multiplication for multilinear maps. This space will be denoted by T^r_s. In particular, tensors of type (1,0) may be identified with elements of V and those of type (0,1) with elements of V^* by taking into account the duality between V and V^*. Hence $T_0^1 \cong V$ and $T_1^0 \cong V^*$.

The tensor space T_r^1 may be considered as the vector space of all multilinear maps of $V\times ...\times V$ (r times) into V. In fact, given $f \in T_r^1$, a multilinear map $t: V\times ...\times V \to V$ is uniquely determined by the relation

$$\langle t(v_1, ..., v_r), v^* \rangle = f(v_1, ..., v_r, v^*) \in R \tag{1.2.15}$$

for all $v_1, ..., v_r \in V$ and $v^* \in V^*$, where, as before, $\langle\, ,\, \rangle$ denotes the value which v^* takes on $t(v_1, ..., v_r)$. Clearly, this establishes a canonical-isomorphism of T_r^1 with the linear space of all multilinear maps of $V\times ...\times V$ into V. In particular, T_1^1 may be identified with the space of all linear endomorphisms of V.

Let $\{e_i\}$ and $\{e^{*k}\}$ be dual bases in V and V^*, respectively:

$$\langle e_i, e^{*k} \rangle = \delta_i^k. \tag{1.2.16}$$

These bases give rise to a base in T_s^r whose elements we write as $e_{i_1...i_r}{}^{k_1...k_s} = e_{i_1} \otimes ... \otimes e_{i_r} \otimes e^{*k_1} \otimes ... \otimes e^{*k_s}$ (cf. I. A for a definition of the tensor product). A tensor $t \in T_s^r$ may then be represented in the form

$$t = \xi^{i_1...i_r}{}_{k_1...k_s} e_{i_1...i_r}{}^{k_1...k_s}, \tag{1.2.17}$$

that is, as a linear combination of the basis elements of T_s^r. The coefficients $\xi^{i_1...i_r}{}_{k_1...k_s}$ then define t in relation to the bases $\{e_i\}$ and $\{e^{*k}\}$.

1.3. Tensor bundles

In differential geometry one is not interested in tensors but rather in tensor fields which we now proceed to define. The definition given is but one consequence of a general theory (cf. I. J) having other applications to differential geometry which will be considered in § 1.4 and § 1.7. Let $T_s^r(P)$ denote the tensor space of tensors of type (r, s) over T_P and put

$$\mathcal{T}_s^r = \bigcup_{P \in M} T_s^r(P).$$

We wish to show that $\mathcal{T}_s^r$ actually defines a differentiable manifold and that a 'tensor field' of type (r, s) is a certain map from M into $\mathcal{T}_s^r$, that is a rule which assigns to every $P \in M$ a tensor of type (r, s) on the tangent space T_P. Let $\tilde{V}$ be a vector space of dimension n over R and $\tilde{T}_s^r$

the corresponding space of tensors of type (r, s). If we fix a base in $\tilde{V}$, a base of $\tilde{T}^r_s$ is determined. Let U be a coordinate neighborhood and u the corresponding homeomorphism from U to E^n. The local coordinates of a point $P \in U$ will be denoted by $(u^i(P))$; they determine a base $\{du^i(P)\}$ in T^*_P and a dual base $\{e_i(P)\}$ in T_P. These bases give rise to a well-defined base in $T^r_s(P)$. Consider the map

$$\varphi_U : U \times \tilde{T}^r_s \rightarrow \mathscr{T}^r_s$$

where $\varphi_U(P, t)$, $P \in U$, $t \in \tilde{T}^r_s$ belongs to $T^r_s(P)$ and has the same components $\xi^{i_1 \dots i_r}{}_{j_1 \dots j_s}$ relative to the (natural) base of $T^r_s(P)$ as t has in $\tilde{T}^r_s$. That φ_U is 1-1 is clear. Now, let V be a second coordinate neighborhood such that $U \cap V \neq \Box$ (the empty set), and consider the map

$$\varphi_{U,P} : \tilde{T}^r_s \rightarrow T^r_s(P)$$

defined by

$$\varphi_{U,P}(t) = \varphi_U(P,t), \qquad t \in \tilde{T}^r_s. \tag{1.3.1}$$

Then,

$$g_{UV}(P) = \varphi^{-1}_{U,P} \cdot \varphi_{V,P} \tag{1.3.2}$$

is a 1-1 map of $\tilde{T}^r_s$ onto itself. Let $(v^i(P))$ denote the local coordinates of P in V. They determine a base $\{dv^i(P)\}$ in T^*_P and a dual base $\{f_i(P)\}$ in T_P. If we set

$$\bar{t} = g_{UV}(P)t, \tag{1.3.3}$$

it follows that

$$\varphi_U(P,\bar{t}) = \varphi_V(P,t). \tag{1.3.4}$$

Since

$$\varphi_U(P,\bar{t}) = \bar{\xi}^{i_1 \dots i_r}{}_{j_1 \dots j_s}\, e_{i_1 \dots i_r}{}^{j_1 \dots j_s}(P)$$

and

$$\varphi_V(P,t) = \xi^{i_1 \dots i_r}{}_{j_1 \dots j_s}\, f_{i_1 \dots i_r}{}^{j_1 \dots j_s}(P) \tag{1.3.5, 1.3.6}$$

where $\{e_{i_1 \dots i_r}{}^{j_1 \dots j_s}(P)\}$ and $\{f_{i_1 \dots i_r}{}^{j_1 \dots j_s}(P)\}$ are the induced bases in $T^r_s(P)$,

$$\bar{\xi}^{i_1 \dots i_r}{}_{j_1 \dots j_s} = \left(\frac{\partial u^{i_1}}{\partial v^{k_1}}\right)_{v(P)} \cdots \left(\frac{\partial v^{l_s}}{\partial u^{j_s}}\right)_{u(P)} \xi^{k_1 \dots k_r}{}_{l_1 \dots l_s}. \tag{1.3.7}$$

These are the equations defining $g_{UV}(P)$. Hence $g_{UV}(P)$ is a linear automorphism of $\tilde{T}^r_s$. If we give to $\tilde{T}^r_s$ the topology and differentiable structure derived from the Euclidean space of the components of its elements it becomes a differentiable manifold. Now, a topology is defined in $\mathscr{T}^r_s$ by the requirement that for each U, φ_U maps open sets of $U \times \tilde{T}^r_s$ into open sets of $\mathscr{T}^r_s$. In this way, it can be shown that $\mathscr{T}^r_s$ is a separable Hausdorff space. In fact, $\mathscr{T}^r_s$ is a differentiable manifold of class $k - 1$ as one sees from the equations (1.3.7).

The map $g_{UV}: U \cap V \rightarrow GL(\tilde{T}^r_s)$ is continuous since M is of class $k \geqq 1$. Let P be a point in the overlap of the three coordinate neighborhoods, U,V,W: $U \cap V \cap W \neq \square$. Then,

$$g_{UV}(P) g_{VW}(P) = g_{UW}(P) \tag{1.3.8}$$

and since

$$g_{VU}(P) = g^{-1}_{UV}(P), \tag{1.3.9}$$

these maps form a topological subgroup of $GL(\tilde{T}^r_s)$. The family of maps g_{UV} for $U \cap V \neq \square$ where $U, V, \ldots$ is a covering of M is called the set of *transition functions* corresponding to the given covering.

Now, let

$$\pi : \mathscr{T}^r_s \rightarrow M$$

be the projection map defined by $\pi(T^r_s(P)) = P$. For $l < k$, a map $f: M \rightarrow \mathscr{T}^r_s$ of class l satisfying $\pi \cdot f =$ identity is called a *tensor field* of type (r, s) and class l. In particular, a tensor field of type (1,0) is called a *vector field* or an *infinitesimal transformation*. The manifold $\mathscr{T}^r_s$ is called the *tensor bundle* over the *base space* M with *structural group* GL (n^{r+s}, R) and *fibre* T^r_s. In the general theory of fibre bundles, the map f is called a *cross-section*. Hence, a tensor field of type (r, s) and class $l < k$ is a cross-section of class l in the tensor bundle $\mathscr{T}^r_s$ over M.

The bundle $\mathscr{T}^1_0$ is usually called the *tangent bundle*.

Since a tensor field is an assignment of a tensor over T_P for each point $P \in M$, the components $\partial f / \partial u^i$ $(i = 1, \ldots, n)$ in (1.2.8) define a covariant vector field (that is, there is a local cross-section) called the *gradient of f*. We may ask whether differentiation of vector fields gives rise to tensor fields, that is given a covariant vector field ξ_i, for example (the ξ_i are the components of a tensor field of type (0,1)), do the n^2 functions $\partial \xi_i / \partial u^j$ define a tensor field (of type (0,2)) over U? We see from (1.2.12) that the presence of the term $(\partial^2 u^j / \partial \bar{u}^k \partial \bar{u}^i) \xi_j$ in

$$\bar{\eta}_{ik} = \frac{\partial \bar{\xi}_i}{\partial \bar{u}^k} = \frac{\partial u^j}{\partial \bar{u}^i} \frac{\partial u^l}{\partial \bar{u}^k} \eta_{jl} + \frac{\partial^2 u^j}{\partial \bar{u}^k \, \partial \bar{u}^i} \xi_j \tag{1.3.10}$$

yields a negative reply. However, because of the symmetry of i and k in the second term on the right the components $\bar{\eta}_{ik} - \bar{\eta}_{ki}$ define a skew-symmetric tensor field called the *curl* of the vector ξ_i. If the ξ_i define a gradient vector field, that is, if there exists a real-valued function f defined on an open subset of M such that $\xi_i = (\partial f/du^i)$, the curl must vanish. Conversely, if the curl of a (covariant) vector field vanishes, the vector field is necessarily a (local) gradient field.

1.4. Differential forms

Let M be a differentiable manifold of dimension n. Associated to each point $P \in M$, there is the dual space T_P^* of the tangent space T_P at P. We have seen that T_P^* can be identified with the space of linear differential forms at P. Hence, to a 1-dimensional subspace of the tangent space there corresponds a linear differential form. We proceed to show that to a p-dimensional subspace of T_P corresponds a skew-symmetric covariant tensor of type $(0, p)$, in fact, a 'differential form of degree p'. To this end, we construct an algebra over T_P^* called the Grassman or exterior algebra:

An associative algebra $\wedge(V)$ (with addition denoted by $+$ and multiplication by $\wedge$) over R containing the vector space V over R is called a *Grassman* or *exterior algebra* if

(i) $\wedge(V)$ contains the unit element 1 of R,

(ii) $\wedge(V)$ is generated by 1 and the elements of V,

(iii) If $x \in V$, $x \wedge x = 0$,

(iv) The dimension of $\wedge(V)$ (as a vector space) is 2^n, $n = \dim V$.

Property (ii) means that any element of $\wedge(V)$ can be written as a linear combination of $1 \in R$ and of products of elements of V, that is $\wedge(V)$ is generated from V and 1 by the three operations of the algebra. Property (iii) implies that $x \wedge y = -y \wedge x$ for any two elements x, $y \in V$. Select any basis $\{e_1, ..., e_n\}$ of V. Then, $\wedge(V)$ contains all products of the e_i $(i = 1, ..., n)$. By using the rules

$$e_i \wedge e_i = 0, \qquad e_i \wedge e_j = -e_j \wedge e_i, \qquad i, j = 1, ..., n, \tag{1.4.1}$$

we can arrange any product of the e_i so that it is of the form

$$e_{i_1} \wedge ... \wedge e_{i_p}, \qquad i_1 < ... < i_p$$

or else, zero. The latter case arises when the original product contains a repeated factor. It follows that we can compute any product of two or more vectors $a_1e_1 + ... + a_ne_n$ of V as a linear combination of the *decomposable p-vectors* $e_{i_1} \wedge \ldots \wedge e_{i_p}$. Since, by assumption, $\wedge(V)$ is spanned by 1 and such products, it follows that $\wedge(V)$ is spanned by the elements $e_{i_1} \wedge \ldots \wedge e_{i_p}$ where $(i_1, \ldots, i_p)$ is a subset of the set $(1, \ldots, n)$ arranged in increasing order. But there are exactly 2^n subsets of $(1, \ldots, n)$, while by assumption $\dim \wedge(V) = 2^n$. These elements must therefore be linearly independent. Hence, any element of $\wedge(V)$ can be uniquely represented as a linear combination

$$\sum_{p=0}^{n} \sum_{(i_1 \ldots i_p)} a_{i_1 \ldots i_p} e_{i_1} \wedge \ldots \wedge e_{i_p}, \quad a_{i_1 \ldots i_p} \in R, \tag{1.4.2}$$

where now and in the sequel $(i_1 \ldots i_p)$ implies $i_1 < \ldots < i_p$. An element of the first sum is called *homogeneous* of degree p.

It may be shown that any two Grassman algebras over the same vector space are isomorphic. For a realization of $\wedge(V)$ in terms of the 'tensor algebra' over V the reader is referred to (I.C.2).

The elements $x_1, \ldots, x_q$ in V are linearly independent, if and only if, their product $x_1 \wedge \ldots \wedge x_q$ in $\wedge(V)$ is not zero. The proof is an easy exercise in linear algebra. In particular, for the basis elements $e_1, \ldots, e_n$ of V, $e_1 \wedge \ldots \wedge e_n \neq 0$. However, any product of $n + 1$ elements of V must vanish.

All the elements

$$e_{i_1} \wedge \ldots \wedge e_{i_p}, \qquad i_1 < \ldots < i_p$$

for a fixed p span a linear subspace of $\wedge(V)$ which we denote by $\wedge^p(V)$. This subspace is evidently independent of the choice of base. An element of $\wedge^p(V)$ is called an *exterior p-vector* or, simply a *p-vector*. Clearly, $\wedge^1(V) = V$. We define $\wedge^0(V) = R$. As a vector space, $\wedge(V)$ is then the direct sum of the subspaces $\wedge^p(V)$, $0 \leqq p \leqq n$.

Let W be the subspace of V spanned by $y_1, \ldots, y_p \in V$. This gives rise to a p-vector $\eta = y_1 \wedge \ldots \wedge y_p$ which is unique up to a constant factor as one sees from the theory of linear equations. Moreover, any vector $y \in W$ has the property that $y \wedge \eta$ vanishes. The subspace W also determines its orthogonal complement (relative to an inner product) in V, and this subspace in turn defines a 'unique' $(n - p)$-vector. Note that for each p, the spaces $\wedge^p(V)$ and $\wedge^{n-p}(V)$ have the same dimensions. Any p-vector ξ and any $(n - p)$-vector η determine an n-vector $\xi \wedge \eta$ which in terms of the basis $e = e_1 \wedge \ldots \wedge e_n$ of $\wedge^n(V)$ may be expressed as

$$\xi \wedge \eta = (\xi, \eta)\, e \tag{1.4.3}$$

where $(\xi, \eta) \in R$. It can be shown that this 'pairing' defines an isomorphism of $\wedge^p(V)$ with $(\wedge^{n-p}(V))^*$ (cf. 1.5.1 and II.A).

Let V^* denote the dual space of V and consider the Grassman algebra $\wedge (V^*)$ over V^*. It can be shown that the spaces $\wedge^p(V^*)$ are canonically isomorphic with the spaces $(\wedge^p(V))^*$ dual to $\wedge^p(V)$. The linear space $\wedge^p(V^*)$ is called the space of *exterior p-forms* over V; its elements are called *p-forms*. The isomorphism between $\wedge^p(V^*)$ and $\wedge^{n-p}(V^*)$ will be considered in Chapter II, § 2.7 as well as in II.A.

We return to the vector space T_P^* of covariant vectors at a point P of the differentiable manifold M of class k and let U be a coordinate neighborhood containing P with the local coordinates $u^1, ..., u^n$ and natural base $du^1, ..., du^n$ for the space T_P^*. An element $\alpha(P) \in \wedge^p(T_P^*)$ then has the following representation in U:

$$\alpha(P) = a_{(i_1...i_p)}(P)\, du^{i_1}(P) \wedge ... \wedge du^{i_p}(P). \tag{1.4.4}$$

If to each point $P \in U$ we assign an element $\alpha(P) \in \wedge^p(T_P^*)$ in such a way that the coefficients $a_{i_1\cdots i_p}$ are of class $l \geqq 1 (l < k)$ then α is said to be a *differential form* of *degree p and class l*. More precisely, an *exterior differential polynomial* of class $l \leqq k - 1$ is a *cross-section* α of class l of the bundle

$$\wedge^*(M) \equiv \wedge(T^*) = \bigcup_{P \in M} \wedge (T_P^*),$$

that is, if π is the projection map:

$$\pi : \wedge^*(M) \to M$$

defined by $\pi(\wedge(T_P^*)) = P$, then $\alpha : M \to \wedge^*(M)$ must satisfy $\pi\alpha(P) = P$ for all $P \in M$ (cf. § 1.3 and I.J). If, for every $P \in M$, $\alpha(P) \in \wedge^p(T_P^*)$ for some (fixed) p, the exterior polynomial is called an *exterior differential form* of *degree p*, or simply a *p-form*. In this case, we shall simply write $\alpha \in \wedge^p(T^*)$. (When reference to a given point is unnecessary we shall usually write T and T^* for T_P and T_P^* respectively).

Let M be a differentiable manifold of class $k \geqq 2$. Then, there is a map

$$d : \wedge (T^*) \to \wedge (T^*)$$

sending exterior polynomials of class l into exterior polynomials of class $l - 1$ with the properties:

(i) For $p = 0$ (differentiable functions f), df is a covector (the differential of f),

(ii) d is a linear map such that $d(\wedge^p(T^*)) \subset \wedge^{p+1}(T^*)$,

(iii) For $\alpha \in \wedge^p(T^*)$, $\beta \in \wedge^q(T^*)$,

$$d(\alpha \wedge \beta) = d\alpha \wedge \beta + (-1)^p \alpha \wedge d\beta,$$

(iv) $d(df) = 0$.

To see this, we need only define

$$d\alpha = da_{(i_1 \ldots i_p)}\, du^{i_1} \wedge \ldots \wedge du^{i_p}$$
$$= \frac{\partial a_{(i_1 \ldots i_p)}}{\partial u^i}\, du^i \wedge du^{i_1} \wedge \ldots \wedge du^{i_p} \tag{1.4.5}$$

where

$$\alpha = a_{(i_1 \ldots i_p)}\, du^{i_1} \wedge \ldots \wedge du^{i_p}.$$

In fact, the operator d is uniquely determined by these properties: Let d^* be another operator with the properties (i)-(iv). Since it is linear, we need only consider its effect upon $\beta = f du^{i_1} \wedge \ldots \wedge du^{i_p}$. By property (iii), $d^*\beta = d^*f \wedge du^{i_1} \wedge \ldots \wedge du^{i_p} + f d^*(du^{i_1} \wedge \ldots \wedge du^{i_p})$. Applying (iii) inductively, then (i) followed by (iv) we obtain the desired conclusion.

It follows easily from property (iv) and (1.4.5) that $d(d\alpha) = 0$ for any exterior polynomial α of class $\geqq 2$.

The operator d is a *local operator*, that is if α and β are forms which coincide on an open subset S of M, then $d\alpha = d\beta$ on S.

The elements $\wedge_c^p(T^*)$ of the kernel of d: $\wedge^p(T^*) \to \wedge^{p+1}(T^*)$ are called *closed p-forms* and the images $\wedge_e^p(T^*)$ of $\wedge^{p-1}(T^*)$ under d are called *exact p-forms*. They are clearly linear spaces (over R). The quotient space of the closed forms of degree p by the subspace of exact p-forms will be denoted by $D^p(M)$ and called the *p-dimensional cohomology group of M obtained using differential forms*. Since the exterior product defines a multiplication of elements (cohomology classes) in $D^p(M)$ and $D^q(M)$ with values in $D^{p+q}(M)$ for all p and q, the direct sum

$$D(M) = \sum_{p=0}^{n} D^p(M) \tag{1.4.6}$$

becomes a ring (over R) called the *cohomology ring of M obtained using differential forms*. In fact, from property (iii) we may write

$$\begin{aligned} &\text{closed form} \wedge \text{closed form} = \text{closed form},\\ &\text{closed form} \wedge \text{exact form} = \text{exact form},\\ &\text{exact form} \wedge \text{closed form} = \text{exact form}. \end{aligned} \tag{1.4.7}$$

Examples: Let M be a 3-dimensional manifold and consider the coordinate neighborhood with the local coordinates x, y, z. The linear differential form

$$\alpha = p\,dx + q\,dy + r\,dz \tag{1.4.8}$$

where p, q, and r are functions of class 2 (at least) of x, y, and z has for its differential the 2-form

$$2d\alpha = (q_x - p_y)\, dx \wedge dy + (r_y - q_z)\, dy \wedge dz + (p_z - r_x)\, dz \wedge dx.$$

Moreover, the 2-form

$$\beta = p\, dy \wedge dz + q\, dz \wedge dx + r\, dx \wedge dy \tag{1.4.9}$$

has the differential

$$d\beta = (p_x + q_y + r_z)\, dx \wedge dy \wedge dz.$$

In more familiar language, $d\alpha$ is the curl of α and $d\beta$ its divergence. That $dd\alpha = 0$ is expressed by the identity

$$\text{div curl}\ \alpha = 0.$$

We now show that the coefficients $a_{i_1 \ldots i_p}$ of a differential form α can be considered as the components of a skew-symmetric tensor field of type $(0, p)$. Indeed, the $a_{i_1 \ldots i_p}$ are defined for $i_1 < \ldots < i_p$. They may be defined for all values of the indices by taking account of the anti-commutativity of the covectors du^i, that is we may write

$$\alpha = \frac{1}{p!}\, a_{i_1 \ldots i_p}\, du^{i_1} \wedge \ldots \wedge du^{i_p}.$$

That the $a_{i_1 \ldots i_p}$ are the components of a tensor field is easy to show. In the sequel, we will absorb the factor $1/p!$ in the expression of a p-form except when its presence is important.

In order to express the exterior product of two forms and the differential of a form (cf. (1.4.5)) in a canonical fashion the *Kronecker symbol*

$$\delta^{j_1 \ldots j_p}_{i_1 \ldots i_p} = \begin{vmatrix} \delta^{j_1}_{i_1} & \cdots & \delta^{j_1}_{i_p} \\ & & \\ \delta^{j_p}_{i_1} & \cdots & \delta^{j_p}_{i_p} \end{vmatrix}$$

will be useful. The important properties of this symbol are:

(i) $\delta^{j_1 \ldots j_p}_{i_1 \ldots i_p}$ is skew-symmetric in the i_k and j_k,

(ii) $\delta^{(j_1 \ldots j_p)}_{(i_1 \ldots i_p)} = \delta^{j_1}_{i_1} \cdots \delta^{j_p}_{i_p}$.

This condition is equivalent to

(ii)′ For every system of $\binom{n}{p}$ numbers $a_{(i_1 \ldots i_p)}$,

$$a_{(i_1 \ldots i_p)} = \delta^{j_1 \ldots j_p}_{i_1 \ldots i_p}\, a_{(j_1 \ldots j_p)}$$

and (ii)′ is equivalent to

$$\text{(ii)}'' \quad a_{i_1\ldots i_p} = \frac{1}{p!}\,\delta^{j_1\ldots j_p}_{i_1\ldots i_p}\,a_{j_1\ldots j_p}$$

where $a_{i_1\ldots i_p}$ is a p-vector.

The condition (ii)″ shows that the Kronecker symbol is actually a tensor of type (p, p).

Now, let

$$\alpha = a_{(i_1\ldots i_p)}\,du^{i_1}\wedge\ldots\wedge du^{i_p}$$

and

$$\beta = b_{(i_1\ldots i_q)}\,du^{i_1}\wedge\ldots\wedge du^{i_q}.$$

Then,

$$\alpha\wedge\beta = c_{i_1\ldots i_{p+q}}\,du^{i_1}\wedge\ldots\wedge du^{i_{p+q}} \tag{1.4.10}$$

where

$$(p+q)!\,c_{i_1\ldots i_{p+q}} = \delta^{(j_1\ldots j_p)(k_1\ldots k_q)}_{i_1\ldots i_{p+q}}\,a_{j_1\ldots j_p}\,b_{k_1\ldots k_q}$$

and

$$(p+1)!\,d\alpha = \delta^{j(i_1\ldots i_p)}_{k_1\ldots k_{p+1}}\,\frac{\partial a_{(i_1\ldots i_p)}}{\partial u^j}\,du^{k_1}\wedge\ldots\wedge du^{k_{p+1}}. \tag{1.4.11}$$

From (1.4.10) we deduce

$$\alpha\wedge\beta = (-1)^{pq}\,\beta\wedge\alpha. \tag{1.4.12}$$

1.5. Submanifolds

The set of differentiable functions F (of class k) in a differentiable manifold M (of class k) forms an algebra over R with the usual rules of addition, multiplication and scalar multiplication by elements of R. Given two differentiable manifolds M and M', a map ϕ of M into M' is called differentiable, if $f'\cdot\phi$ is a differentiable function in M for every such function f' in M'. This may be expressed in terms of local coordinates in the following manner: Let $u^1, \ldots, u^n$ be local coordinates at $P \in M$ and $v^1, \ldots, v^m$ local coordinates at $\phi(P) \in M'$. Then ϕ is a differentiable map, if and only if, the $v^i(\phi(u^1, \ldots, u^n)) \equiv v^i(u^1, \ldots, u^n)$ are differentiable functions of $u^1, \ldots, u^n$. The map ϕ induces a (linear) differentiable map ϕ_* of the tangent space T_P at $P \in M$ into the tangent space $T_{\phi(P)}$ at $P' = \phi(P) \in M'$. Let $X \in T_P$ and consider a differentiable function f' in the algebra F' of differentiable functions in M'. The directional derivative of $f'\cdot\phi$ along X is given by

$$\xi^i\,\frac{\partial v^j}{\partial u^i}\,\frac{\partial f'}{\partial v^j}\,,\; i = 1, \ldots, n;\quad j = 1, \ldots, m$$

where the ξ^i are the (contravariant) components of X in the local coordinates $u^1, \ldots, u^n$. This, in turn is equal to the directional derivative of f' along the contravariant vector

$$X' = \xi^i \frac{\partial v^j}{\partial u^i} \frac{\partial}{\partial v^j}$$

at $\phi(P)$. By mapping X in T_P into X' in $T_{P'}$, we get a linear map of T_P into $T_{\phi(P)}$. This is the *induced map* ϕ_*. The map ϕ is said to be *regular* (at P) if the induced map ϕ_* is 1-1.

A subset M' of M is called a *submanifold* of M if it is itself a differentiable manifold, and if the injection ϕ' of M' into M is a regular differentiable map. When necessary we shall denote M' by (ϕ', M'). Obviously, we have $\dim M' \leqq \dim M$. The topology of M' need not coincide with that induced by M on M'. If M' is an open subset of M, then it possesses a naturally induced differentiable structure. In this case, M' is called an *open submanifold* of M.

Recalling the definition of regular surface we see that the above univalence condition is equivalent to the condition that the Jacobian of ϕ is of rank n.

By a *closed submanifold* of dimension r is meant a submanifold M' with the properties: (i) $\phi'(M')$ is closed in M and (ii) every point $P \in \phi'(M')$ belongs to a coordinate neighborhood U with the local coordinates $u^1, \ldots, u^n$ such that the set $\phi'(M') \cap U$ is defined by the equations $u^{r+1} = 0, \ldots, u^n = 0$. The definition of a regular closed surface given in § 1.1 may be included in the definition of closed submanifold.

We shall require the following notion: A *parametrized curve* in M is a differentiable map of class k of a connected open interval of R into M.

The differentiable map $\phi : M \to M'$ induces a map ϕ^* called the *dual of* ϕ_* defined as follows:

$$\phi^* : T^*_{\phi(P)} \to T^*_P$$

and

$$\langle v, \phi^*(w^*)\rangle = \langle \phi_*(v), w^*\rangle', \quad v \in T_P,\ w^* \in T^*_{\phi(P)}.$$

The map ϕ^* may be extended to a map which we again denote by ϕ^*

$$\phi^* : \wedge\, (T^*_{\phi(P)}) \to \wedge\, (T^*_P)$$

as follows: Consider the pairing $\langle v_1 \wedge \ldots \wedge v_p, w_1^* \wedge \ldots \wedge w_p^*\rangle$ defined by

$$\langle v_1 \wedge \ldots \wedge v_p,\ w_1^* \wedge \ldots \wedge w_p^*\rangle = p!\det\,(\langle v_i, w_j^*\rangle) \qquad (1.5.1)$$

and put

$$\langle v_1 \wedge \dots \wedge v_p,\ \phi^*(w_1^* \wedge \dots \wedge w_p^*)\rangle = \det(\langle v_i,\ \phi^*(w_j^*)\rangle).$$

Clearly, ϕ^* is a ring homomorphism. Moreover,

$$\phi^*(d\alpha) = d(\phi^*\alpha),$$

that is, the exterior differential operator d commutes with the induced dual map of a differentiable map from one differentiable manifold into another.

1.6. Integration of differential forms

It is our intention in this section to sketch a proof of the formula of Stokes not merely because of its fundamental importance in the theory of harmonic integrals but because of the applications we make of it in later chapters. However, a satisfactory integration theory of differential forms over a differentiable manifold must first be developed.

The classical definition of a p-fold integral

$$\int_D f\, du^1 \dots du^p$$

of a continuous function $f = f(u^1, \dots, u^p)$ of p variables defined over a domain D of the space of the variables $u^1, \dots, u^p$ as given, for example, by Goursat does not take explicit account of the orientation of D. The definition of an orientable differentiable manifold M given in § 1.1 together with the isomorphism which exists between $\wedge^p(T_P^*)$ and $\wedge^{n-p}(T_P^*)$ at each point P of M (cf. § 2.7) results in the following equivalent definition:

A differentiable manifold M of dimension n is said to be *orientable* if there exists over M a continuous differential form of degree n which is nowhere zero (cf. I.B).

Let α and β define orientations of M. These orientations are the same if $\beta = f\alpha$ where the function f is always positive. An orientable manifold therefore has exactly two orientations. The manifold is called *oriented* if such a form $\alpha \neq 0$ is given. The form α induces an orientation in the tangent space at each point $P \in M$. Any other form of degree n can then be written as $f(P)\alpha$ and is be said to be > 0, < 0 or $= 0$ at P provided that $f(P) > 0$, < 0 or $= 0$. This depends only on the orientation of M and not on the choice of the differential form defining the orientation.

The *carrier*, carr (α) of a differential form α is the closure of the set of points outside of which α is equal to zero. The following theorem due to J. Dieudonné is of crucial importance. (Its proof is given in Appendix D.)

To a locally finite open covering $\{U_i\}$ of a differentiable manifold of class $k \geqq 1$ there is associated a set of functions $\{g_j\}$ with the properties

(i) Each g_j is of class k and satisfies the inequalities

$$0 \leqq g_j \leqq 1$$

everywhere. Moreover, its carrier is compact and is contained in one of the open sets U_i,

(ii) $$\sum_j g_j = 1,$$

(iii) Every point of M has a neighborhood met by only a finite number of the carriers of g_j.

The g_j are said to form a *partition of unity subordinated* to $\{U_i\}$ that is, a partition of the function 1 into non-negative functions with small carriers. Property (iii) states that the partition of unity is *locally finite*, that is, each point $P \in M$ has a neighborhood met by only a finite number of the carriers of g_j. If M is compact, there can be a finite number of g_j; in any case, the g_j form a countable set. With these preparations we can now prove the following theorem:

Let M be an oriented differentiable manifold of dimension n. Then, there is a unique functional which associates to a continuous differential form α of degree n with compact carrier a real number denoted by $\int_M \alpha$ and called the *integral of* α. This functional has the properties:

(i) $\int_M (\alpha + \beta) = \int_M \alpha + \int_M \beta$,

(ii) If the carrier of α is contained in a coordinate neighborhood U with the local coordinates $u^1, \ldots, u^n$ such that $du^1 \wedge \ldots \wedge du^n > 0$ (in U) and $\alpha = a_{1\ldots n}\, du^1 \wedge \ldots \wedge du^n$ where $a_{1\ldots n}$ is a function of $u^1, \ldots, u^n$, then

$$\int_M \alpha = \int_U a_{1\ldots n}\, du^1 \ldots du^n \tag{1.6.1}$$

where the n-fold integral on the right is a Riemann integral.

Since carr $(\alpha) \subset U$ we can extend the definition of the function $a_{1\ldots n}$ to the whole of E^n, so that (1.6.1) becomes the the n-fold integral

$$\int_M \alpha = \int_{-\infty}^{+\infty} \ldots \int_{-\infty}^{+\infty} a_{1\ldots n}\, du^1 \ldots du^n. \tag{1.6.2}$$

In order to define the integral of an n-form α with compact carrier S we take a locally finite open covering $\{U_i\}$ of M by coordinate neighborhoods and a partition of unity $\{g_j\}$ subordinated to $\{U_i\}$. Since every point $P \in S$ has a neighborhood met by only a finite number of the carriers of the g_j, these neighborhoods for all $P \in S$ form a covering of S. Since S is compact, it has a finite sub-covering, and so there is at most a finite number of g_j different from zero. Since $\int g_j\alpha$ is defined, we put

$$\int_M \alpha = \sum_j \int g_j\alpha. \tag{1.6.3}$$

That the integral of α over M so defined is independent of the choice of the neighborhood containing the carrier of g_j as well as the covering $\{U_i\}$ and its corresponding partition of unity is not difficult to show. Moreover, it is convergent and satisfies the properties (i) and (ii). The uniqueness is obvious.

Suppose now that M is a compact orientable manifold and let β be an $(n-1)$-form defined over M. Then,

$$\int_M d\beta = 0. \tag{1.6.4}$$

To prove this, we take a partition of unity $\{g_i\}$ and replace β by $\Sigma_i g_i\beta$. This result is also immediate from the theorem of Stokes which we now proceed to establish.

Stokes' theorem expresses a relation between an integral over a domain and one over its boundary. Its applications in mathematical physics are many but by no means outstrip its usefulness in the theory of harmonic integrals.

Let M be a differentiable manifold of dimension n. A domain D with *regular boundary* is a point set of M whose points may be classified as either interior or boundary points. A point P of D is an *interior point* if it has a neighborhood in D. P is a *boundary point* if there is a coordinate neighborhood U of P such that $U \cap D$ consists of those points $Q \in U$ satisfying $u^n(Q) \geqq u^n(P)$, that is, D lies on only one side of its boundary. That these point sets are mutually exclusive is clear. (Consider, as an example, the upper hemisphere including the rim. On the other hand, a closed triangle has singularities). The *boundary* ∂D of D is the set of all its boundary points. The following theorem is stated without proof:

The boundary of a domain with regular boundary is a closed sub-manifold of M. Moreover, if M is orientable, so is ∂D whose orientation is canonically induced by that of D.

Now, let D be a compact domain with regular boundary and let h be a real-valued function on M with the property that $h(P) = 1$ if $P \in D$ and is otherwise zero. Then, the integral of an n-form α may be defined over D by the formula

$$\int_D \alpha = \int_M h\alpha. \tag{1.6.5}$$

Although the form $h\alpha$ is not continuous the right side is meaningful as one sees by taking a partition of unity.

Let α be a differential form of degree $n-1$ and class $k \geqq 1$ in M. Then

$$\int_{\partial D} i^*\alpha = \int_D d\alpha \tag{1.6.6}$$

where the map i sending ∂D into M is the identity and ∂D has the orientation canonically induced by that of D. This is the *theorem of Stokes*. In order to prove it, we select a countable open covering of M by coordinate neighborhoods $\{U_i\}$ in such a way that either U_i does not meet ∂D, or it has the property of the neighborhood U in the definition of boundary point. Let $\{g_j\}$ be a partition of unity subordinated to this covering. Since D and its boundary are both compact, each of them meets only a finite number of the carriers of g_j. Hence,

$$\int_{\partial D} i^*\alpha = \sum_j \int_{\partial D} g_j\alpha$$

and

$$\int_D d\alpha = \sum_j \int_D d(g_j\alpha).$$

These sums being finite, it is only necessary to establish that

$$\int_{\partial D} g_i\alpha = \int_D d(g_i\alpha)$$

for each i, the integrals being evaluated by formula (1.6.1). To complete the proof then, choose a local coordinate system $u^1, \ldots, u^n$ for the coordinate neighborhood U_i in such a way that $du^1 \wedge \ldots \wedge du^n > 0$ and put

$$\alpha = \sum_{k=1}^{n} (-1)^{k-1} a_k du^1 \wedge \ldots \wedge du^{k-1} \wedge du^{k+1} \wedge \ldots \wedge du^n$$

where the functions a_k are of class $\geqq 1$. Then,

$$d\alpha = \sum_{k=1}^{n} \frac{\partial a_k}{\partial u^k} du^1 \wedge \ldots \wedge du^n.$$

Compare with (1.4.9). The remainder of the proof is left as an exercise.

1.7. Affine connections

We have seen that the partial derivatives of a function with respect to a given system of local coordinates are the components of a covariant vector field or, stated in an invariant manner, the differential of a function is a covector. That this case is unique has already been shown (cf. equation 1.3.10). A similar computation for the contravariant vector field $X = \xi^i(\partial/\partial u^i)$ results in

$$\frac{\partial \bar{\xi}^i}{\partial \bar{u}^k} = \frac{\partial \bar{u}^i}{\partial u^j}\frac{\partial u^l}{\partial \bar{u}^k}\eta^j_{\ l} + \frac{\partial^2 \bar{u}^i}{\partial u^j \partial u^l}\frac{\partial u^l}{\partial \bar{u}^k}\xi^j \tag{1.7.1}$$

where

$$\bar{\xi}^i = \frac{\partial \bar{u}^i}{\partial u^j}\xi^j \tag{1.7.2}$$

in $U \cap \bar{U}$. Again, the presence of the second term on the right indicates that the derivative of a contravariant vector field does not have tensor character. Differentiation may be given an invariant meaning on a manifold by introducing a set of n^2 linear differential forms $\omega^i_j = \Gamma^i_{jk}\,du^k$ in each coordinate neighborhood, so that in the overlap $U \cap \bar{U}$ of two coordinate neighborhoods

$$\frac{\partial \bar{u}^k}{\partial u^j}\bar{\omega}^i_{\ k} = \frac{\partial \bar{u}^i}{\partial u^k}\omega^k_{\ j} - \frac{\partial^2 \bar{u}^i}{\partial u^l \partial u^j}\,du^l. \tag{1.7.3}$$

A direct computation shows that in the intersection of three coordinate neighborhoods one of the relations (1.7.3) is a consequence of the others. In terms of the n^3 coefficients Γ^i_{jk}, equations (1.7.3) may be written in the form

$$\Gamma^i_{jk} = \frac{\partial^2 \bar{u}^l}{\partial u^j \partial u^k}\frac{\partial u^i}{\partial \bar{u}^l} + \frac{\partial \bar{u}^r}{\partial u^j}\frac{\partial \bar{u}^s}{\partial u^k}\frac{\partial u^i}{\partial \bar{u}^l}\bar{\Gamma}^l_{rs}. \tag{1.7.4}$$

These equations are the classical equations of transformation of an affine connection. With these preliminaries we arrive at the notion we are seeking. We shall see that the ω^i_j permit us to define an invariant type of differentiation over a differentiable manifold.

An *affine connection* on a differentiable manifold M is defined by prescribing a set of n^2 linear differential forms ω^i_j in each coordinate neighborhood of M in such a way that in the overlap of two coordinate neighborhoods

$$dp^i_j + p^k_j\,\bar{\omega}^i_k = p^i_k\,\omega^k_j, \quad p^i_j = \frac{\partial \bar{u}^i}{\partial u^j}. \tag{1.7.5}$$

A manifold with an affine connection is called an *affinely connected manifold.*

The existence of an affine connection on a differentiable manifold will be shown in § 1.9. In the sequel, we shall assume that M is an affinely connected manifold. Now, from the equations of transformation of a contravariant vector field $X = \xi^i(\partial/\partial u^i)$ we obtain by virtue of the equations (1.7.5)

$$\begin{aligned} d\bar{\xi}^i &= dp^i_j\, \xi^j + p^i_j\, d\xi^j \\ &= (\omega^k_j\, p^i_k - \bar{\omega}^i_k\, p^k_j)\, \xi^j + p^i_j\, d\xi^j. \end{aligned} \tag{1.7.6}$$

By rewriting these equations in the symmetrical form

$$d\bar{\xi}^i + \bar{\omega}^i_k\, \bar{\xi}^k = p^i_j(d\xi^j + \omega^j_k\, \xi^k) \tag{1.7.7}$$

we see that the quantity in brackets transforms like a contravariant vector field. We call this quantity the *covariant differential* of X and denote it by DX: Its jth component $d\xi^j + \omega^j_k\, \xi^k$ will be denoted by $(DX)^j$. In terms of the natural base for covectors, (1.7.7) becomes

$$\left(\frac{\partial \bar{\xi}^i}{\partial \bar{u}^j} + \bar{\xi}^k\, \bar{\Gamma}^i_{kj}\right) d\bar{u}^j = p^i_m\left(\frac{\partial \xi^m}{\partial u^l} + \xi^k\, \Gamma^m_{kl}\right) du^l. \tag{1.7.8}$$

We set

$$D_l\, \xi^j = \frac{\partial \xi^j}{\partial u^l} + \xi^k\, \Gamma^j_{kl} \tag{1.7.9}$$

and call it the *covariant derivative* of X with respect to u^l. That the components $D_l\, \xi^j$ transform like a tensor field of type (1,1) is clear. In fact, it follows from (1.7.8) that

$$\bar{D}_j\, \bar{\xi}^i = \frac{\partial \bar{u}^i}{\partial u^m}\frac{\partial u^l}{\partial \bar{u}^j} D_l\, \xi^m \tag{1.7.10}$$

where the l.h.s. denotes the covariant derivative of X with respect to $\bar{u}^j$.

A similar discussion in the case of the covariant vector field ξ_i permits us to define the covariant derivative of ξ_i as the tensor field $D_j\xi_i$ of type (0,2) where

$$D_j\, \xi_i = \frac{\partial \xi_i}{\partial u^j} - \xi_k\, \Gamma^k_{ij}. \tag{1.7.11}$$

The extension of the above argument to tensor fields of type (r, s) is straightforward—the covariant derivative of the tensor field $\xi^{i_1\cdots i_r}{}_{j_1\ldots j_s}$ with respect to u^k being

$$\begin{aligned} D_k\, \xi^{i_1\ldots i_r}{}_{j_1\ldots j_s} = \frac{\partial}{\partial u^k}\, \xi^{i_1\ldots i_r}{}_{j_1\ldots j_s} + \xi^{l i_2\ldots i_r}{}_{j_1\ldots j_s}\, \Gamma^{i_1}_{lk} + \ldots + \xi^{i_1\ldots i_{r-1} l}{}_{j_1\ldots j_s}\, \Gamma^{i_r}_{lk} \\ - \xi^{i_1\ldots i_r}{}_{l j_2\ldots j_s}\, \Gamma^l_{j_1 k} - \ldots - \xi^{i_1\ldots i_r}{}_{j_1\ldots j_{s-1} l}\, \Gamma^l_{j_s k}\,. \end{aligned} \tag{1.7.12}$$

The covariant derivative of a tensor field being itself a tensor field, we may speak of second covariant derivatives, etc., the result again being a tensor field.

Since Euclidean space E^n, considered as a differentiable manifold, is covered by one coordinate neighborhood, it is not essential from our point of view to introduce the concept of covariant derivative. In fact, the affine connection is defined by setting the Γ^i_{jk} equal to zero. The underlying affine space A^n is the ordinary n-dimensional vector space—the tangent space at each point P of E^n coinciding with A^n. Indeed, the linear map sending the tangent vector $\partial/\partial u^i$ to the vector (0, ..., 0, 1, 0, ..., 0) (1 in the ith place) identifies the tangent space T_P with A^n itself. Let P and Q be two points of A^n. A tangent vector at P and one at Q are said to be *parallel* if they may be identified with the same vector of A^n. Clearly, the concept of parallelism (of tangent vectors) in A^n is independent of the curve joining them. However, in general, this is not the case as one readily sees from the differential geometry of surfaces in E^3. We therefore make the following definition:

Let $C = C(t)$ be a piecewise differentiable curve in M. The tangent vectors

$$X(t) = \xi^i(t)\frac{\partial}{\partial u^i} \tag{1.7.13}$$

are said to be *parallel along* C if the covariant derivative $DX(t)$ of $X(t)$ vanishes in the direction of C, that is, if

$$\frac{d\xi^i}{dt} + \Gamma^i_{jk}\frac{du^k}{dt}\xi^j = 0. \tag{1.7.14}$$

A piecewise differentiable curve is called an *auto-parallel curve*, if its tangent vectors are parallel along the curve itself.

The equations (1.7.14) are a system of n first order differential equations, and so corresponding to the initial value $X = X(t_0)$ at $t = t_0$ there is a unique solution. Geometrically, we say that the vector $X(t_0)$ has been given a *parallel displacement* along C. Algebraically, the parallel displacement along C is a linear isomorphism of the tangent spaces at the points of C. By definition, the auto-parallel curves are the integral curves of the system

$$\frac{d^2u^i}{dt^2} + \Gamma^i_{jk}\frac{du^j}{dt}\frac{du^k}{dt} = 0. \tag{1.7.15}$$

Hence, corresponding to given initial values, there is a unique auto-parallel curve through a given point tangent to a given vector. Note that the auto-parallel curves in A^n are straight lines.

Affine space has the further property that functions defined in it have symmetric second covariant derivatives. This is, however, not the case in an arbitrary differentiable manifold. For, let f be a function expressed in the local coordinates (u^i). Then

$$D_i f = \frac{\partial f}{\partial u^i},$$

$$D_j D_i f = D_j(D_i f) = \frac{\partial^2 f}{\partial u^i \, \partial u^j} - \frac{\partial f}{\partial u^k} \Gamma^k_{ij}, \tag{1.7.16}$$

from which

$$D_j D_i f - D_i D_j f = \frac{\partial f}{\partial u^k} (\Gamma^k_{ji} - \Gamma^k_{ij}). \tag{1.7.17}$$

If we put

$$T_{jk}{}^i = \Gamma^i_{jk} - \Gamma^i_{kj}, \tag{1.7.18}$$

it follows that the $T_{jk}{}^i$ are the components of a tensor field of type (1,2) called the *torsion tensor* of the affine connection Γ^i_{jk}. We remark at this point, that if $\tilde{\omega}^i_j = \tilde{\Gamma}^i_{jk}\, du^k$ are a set of n^2 linear differential forms in each coordinate neighborhood defining another affine connection on M, then it follows from the equations (1.7.4) that $\Gamma^i_{jk} - \tilde{\Gamma}^i_{jk}$ is a tensor field. In particular, if we put $\tilde{\Gamma}^i_{jk} = \Gamma^i_{kj}$, that is, if $\tilde{\omega}^i_j = \Gamma^i_{kj} du^k$, $\Gamma^i_{jk} - \Gamma^i_{kj}$ is a tensor field. When we come to discuss the geometry of a Riemannian manifold we shall see that there is an affine connection whose torsion tensor vanishes. However, even in this case, it is not true that covariant differentiation is symmetric although for (scalar) functions this is certainly the case. In fact, a computation shows that

$$D_k D_j \, \xi^i - D_j D_k \, \xi^i = \xi^l \, R^i{}_{ljk} - D_l \, \xi^i \, T_{jk}{}^l \tag{1.7.19}$$

where

$$R^i_{jkl} = \frac{\partial \Gamma^i_{jk}}{\partial u^l} - \frac{\partial \Gamma^i_{jl}}{\partial u^k} + \Gamma^i_{sl} \, \Gamma^s_{jk} - \Gamma^i_{sk} \, \Gamma^s_{jl}. \tag{1.7.20}$$

(In the case under consideration the components $T_{jk}{}^l$ are zero). Clearly, $R^i{}_{jkl}$ is a tensor field of type (1,3) which is skew-symmetric in its last two indices. It is called the *curvature tensor* and depends only on the affine connection, that is, the functions $R^i{}_{jkl}$ are functions of the Γ^i_{jk} only. More generally, for a tensor of type (r, s)

$$
\begin{aligned}
D_l D_k\, \xi^{i_1 \dots i_r}{}_{j_1 \dots j_s} - D_k D_l\, \xi^{i_1 \dots i_r}{}_{j_1 \dots j_s} \\
= \sum_{\rho=1}^{r} \xi^{i_1 \dots i_{\rho-1} i i_{\rho+1} \dots i_r}{}_{j_1 \dots j_s}\, R^{i_\rho}{}_{ikl} \\
- \sum_{\sigma=1}^{s} \xi^{i_1 \dots i_r}{}_{j_1 \dots j_{\sigma-1} j j_{\sigma+1} \dots j_s}\, R^{j}{}_{j_\sigma kl} \\
- D_i\, \xi^{i_1 \dots i_r}{}_{j_1 \dots j_s}\, T_{kl}{}^{i}.
\end{aligned}
\qquad (1.7.21)
$$

Now, if both the torsion and curvature tensors vanish, covariant differentiation is symmetric. It does not follow, however, that the Γ^i_{jk} vanish, that is, the space is not necessarily affine space.

An affinely connected manifold is said to be *locally affine* or *locally flat* if a coordinate system exists relative to which the coefficients of connection vanish. Under the circumstances, both the torsion and curvature tensors vanish. Conversely, if the torsion and curvature are zero it can be shown that the manifold is locally flat (cf. I.E).

1.8. Bundle of frames

The necessity of the concept of an affine connection on a differentiable manifold has been clearly established from an analytical point of view. A geometrical interpretation of this notion is desirable. Hence, in this section a realization of this very important concept will be given in terms of the bundle of frames over M.

By a *frame* x at the point $P \in M$ is meant a set $\{X_1, \dots, X_n\}$ of linearly independent tangent vectors at P. Let B be the set of all frames x at all points P of M. Every element $a \in GL(n, R)$ acts on B to the right, that is, if a denotes the matrix (a^i_j) and $x = \{X_1, \dots, X_n\}$, then $x \cdot a = \{a^j_1 X_j, \dots, a^j_n X_j\} \in B$ is another frame at P. The map $\pi : B \to M$ of B onto M defined by $\pi(x) = P$ assigns to each frame x its point of origin. In terms of a system of local coordinates $u^1, \dots, u^n$ in M the local coordinates in B are given by $(u^j, \xi^k_{(i)})$—the n^2 functions $\xi^k_{(i)}$ being defined by the n vectors X_i of the frame:

$$X_i = \xi^k_{(i)} \frac{\partial}{\partial u^k}, \qquad i = 1, \dots, n. \qquad (1.8.1)$$

Clearly, the $\xi^k_{(i)}$, $i, k = 1, \ldots, n$ are the elements of a non-singular matrix $(\xi^k_{(i)})$. Conversely, every non-singular matrix defines a frame expressed in the above form. The set of all frames at all points of M can be given a topology, and in fact, a differentiable structure by taking $u^1, \ldots, u^n$ and $(\xi^k_{(i)})$ as local coordinates in $\pi^{-1}(U)$. The differentiable manifold B is called the *bundle of frames* or *bases* over M with *structural group* $GL(n, R)$.

Let $(\xi^{(i)}_k)$ denote the inverse matrix of $(\xi^k_{(i)})$. In the overlap of two coordinate neighborhoods, $(u^i, \xi^k_{(i)})$ and $(\bar{u}^i, \bar{\xi}^k_{(i)})$ are related by

$$\bar{\xi}^k_{(i)} = \frac{\partial \bar{u}^k}{\partial u^j} \xi^j_{(i)}. \tag{1.8.2}$$

It follows that

$$\bar{\xi}^{(i)}_k = \frac{\partial u^j}{\partial \bar{u}^k} \xi^{(i)}_j$$

from which

$$\bar{\xi}^{(i)}_k \, d\bar{u}^k = \xi^{(i)}_j \, du^j. \tag{1.8.3}$$

Hence, for each i, the function $\xi^{(i)}_k$ assigns to every point x of $\pi^{-1}(U)$ a 1-form $\alpha^i = \xi^{(i)}_j du^j$ at $\pi(x)$ in U. Defining $\theta^i = \pi^*\alpha^i$, $i = 1, \ldots, n$ we obtain n linearly independent 1-forms θ^i on the whole of B. Now, we take the covariant differential of each of the vectors X_i. From (1.7.7) and (1.7.8) we obtain

$$\overline{(DX_{(i)})}^j = p^j_m (DX_{(i)})^m \tag{1.8.4}$$

where

$$(DX_{(i)})^k = d\xi^k_{(i)} + \omega^k_j \, \xi^j_{(i)}, \tag{1.8.5}$$

and so from (1.8.3)

$$\bar{\xi}^{(k)}_j \, \overline{(DX_{(i)})}^j = \xi^{(k)}_m \, (DX_{(i)})^m. \tag{1.8.6}$$

Denoting the common expression in (1.8.6) by α^k_i we see that the α^k_i define n^2 linear differential forms $\theta^k_i = \pi^*\alpha^k_i$ on the whole of the bundle B.

The $n^2 + n$ forms θ^i, θ^i_j in B are vector-valued differential forms in B. To see this, identify B with the collection of vector space isomorphisms $x : R^n \to T_P$; namely, if x is the frame $\{X_1, \ldots, X_n\}$ at P, then $x(a^1, \ldots, a^n) = a^i X_i$. Now, for each $t \in T_x$, define θ to be an R^n-valued 1-form by

$$\theta(t) = x^{-1}(\pi_*(t)).$$

As an exercise we leave to the reader the verification of the formulae for the exterior derivatives of the θ^j and θ^j_i:

$$d\theta^j - \theta^k \wedge \theta^j_k = \Theta^j, \tag{1.8.7}$$

$$d\theta^j_i - \theta^l_i \wedge \theta^j_l = \Theta^j_i, \tag{1.8.8}$$

where

$$\Theta^j = \tfrac{1}{2}P_{lm}{}^j\,\theta^l \wedge \theta^m, \quad \Theta^j_i = \tfrac{1}{2}S^j{}_{ilm}\,\theta^l \wedge \theta^m, \tag{1.8.9}$$

and

$$P_{lm}{}^j = -\,\xi^{(j)}_i\,\xi^p_{(l)}\,\xi^q_{(m)}\,T_{pq}{}^i \cdot \pi, \tag{1.8.10}$$

$$S^j{}_{ilm} = -\,\xi^{(j)}_k\,\xi^p_{(i)}\,\xi^q_{(l)}\,\xi^r_{(m)}\,R^k{}_{pqr} \cdot \pi \tag{1.8.11}$$

—the $P_{lm}{}^j$ and $S^j{}_{ilm}$ being functions on B whereas the torsion and curvature tensors are defined in M. Equations (1.8.7) - (1.8.9) are called the *equations of structure*. They are independent of the particular choice of frames, so that if we consider only those frames for which

$$\xi^k_{(i)} = \xi^{(k)}_i = \delta^k_i,$$

$$du^k \wedge \omega^j_k = \tfrac{1}{2}T_{lm}{}^j\,du^l \wedge du^m,$$

and

$$d\omega^j_i - \omega^l_i \wedge \omega^j_l = -\,\tfrac{1}{2}R^j{}_{ilm}\,du^l \wedge du^m. \tag{1.8.12}$$

Differentiating equations (1.8.7) and (1.8.8) we obtain the *Bianchi identities*:

$$d\Theta^j = \theta^k \wedge \Theta^j_k - \Theta^k \wedge \theta^j_k, \tag{1.8.13}$$

$$d\Theta^j_i = \theta^k_i \wedge \Theta^j_k - \Theta^k_i \wedge \theta^j_k. \tag{1.8.14}$$

We have seen that an affine connection on M gives rise to a *complete parallelisability* of the bundle of frames B over M, that is the affine connection determines $n^2 + n$ linearly independent linear differential forms in B. Conversely, if n^2 linear differential forms θ^j_i are given in B which together with the n-forms θ^j satisfy the equations of structure, they define an affine connection. The proof of this important fact is omitted.

Let a be an element of the structural group $GL(n, R)$ of the bundle of frames B over M. It induces a linear isomorphism of the tangent space T_x at $x \in B$ onto the tangent space $T_{x.a}$. This, in turn gives rise to an isomorphism of $T^*_{x.a}$ onto T^*_x. On the other hand, the projection map π induces a map π^* of T^*_P (the space of covectors at $P \in M$). An affine connection on M may then be described as follows:

(i) T^*_x is the direct sum of W^*_x and $\pi^*(T^*_P)$ where W^*_x is a linear subspace at $x \in B$ and $\pi(x) = P$;

(ii) For every $a \in GL(n, R)$ and $x \in B$, W^*_x is the image of $W^*_{x.a}$ by the induced map on the space of covectors.

In other words, *an affine connection on M is a choice of a subspace W^*_x in T^*_x at each point x of B subject to the conditions* (*i*) *and* (*ii*). Note that the dimension of W^*_x is n^2. Hence, it can be defined by prescribing n^2 linearly independent differential forms which together with the θ^i span T^*_x.

1.9. Riemannian geometry

Unless otherwise indicated, we shall assume in the sequel that we are given a differentiable manifold M of dimension n and class ∞.

A *Riemannian metric* on M is a tensor field g of type (0,2) on M subject to the conditions:

(i) g is a symmetric tensor field, and

(ii) g is positive definite.

This tensor field is called the *fundamental tensor field.* When a Riemannian metric is given on M the manifold is called a *Riemannian manifold.* Geometry based upon a Riemannian metric is called *Riemannian geometry*. A Riemannian metric gives rise to an inner (scalar) product on each tangent space T_P at $P \in M$: the *scalar product* of the contravariant vector fields $X = \xi^i(\partial/\partial u^i)$ and $Y = \eta^i(\partial/\partial u^i)$ at the point P is defined to be the scalar

$$X \cdot Y = g_{jk}\, \xi^j\, \eta^k, \quad g_{jk} = g\left(\frac{\partial}{\partial u^j}, \frac{\partial}{\partial u^k}\right). \tag{1.9.1}$$

The positive square root of $X \cdot X$ is called the *length* of the vector X. Since the Riemannian metric is a tensor field, the quadratic differential form

$$ds^2 = g_{jk}\, du^j\, du^k \tag{1.9.2}$$

(where we have written $du^j\, du^k$ in place of $du^j \otimes du^k$ for convenience) is independent of the choice of local coordinates u^i. In this way, if we are given a parametrized curve $C(t)$, the integral

$$s = \int_{t_0}^{t_1} \sqrt{X(t) \cdot X(t)}\, dt \tag{1.9.3}$$

where $X(t)$ is the tangent vector to $C(t)$ defines the length s of the arc joining the points $(u^i(t_0))$ and $(u^i(t_1))$.

Now, every differentiable manifold M (of class k) possesses a Riemannian metric. Indeed, we take an open covering $\{U_\alpha\}$ of M by coordinate neighborhoods and a partition of unity $\{g_\alpha\}$ subordinated to U_α. Let $ds_\alpha^2(= \sum_{i=1}^n du^i\, du^i)$ be a positive definite quadratic differential form defined in each U_α and let the carrier of g_α be contained in U_α. Then, $\sum_\alpha g_\alpha ds_\alpha^2$ defines a Riemannian metric on M.

Since the $du^i\, du^i$ have coefficients of class $k - 1$ in any other coordinate system and the g_α can be taken to be of class k the manifold M possesses a Riemannian metric of class $k - 1$.

It is now shown that there exists an affine connection on a differentiable manifold. In fact, *we prove that there is a unique connection with the properties*: (*a*) *the torsion tensor is zero and* (*b*) *the scalar product* (*relative to some metric*) *is preserved during parallel displacement.* To show this, assume that we have a connection Γ^i_{jk} satisfying conditions (a) and (b). We will obtain a formula for the coefficients Γ^i_{jk} in terms of the metric tensor g of (b). Let $X(t) = \xi^i(t)(\partial/\partial u^i)$ and $Y(t) = \eta^i(t)(\partial/\partial u^i)$ be tangent vectors at the point $(u^i(t))$ on the parametrized curve $C(t)$. The condition that these vectors be parallel along $C(t)$ are

$$\frac{d\xi^i}{dt} + \Gamma^i_{jk} \frac{du^k}{dt} \xi^j = 0 \tag{1.9.4}$$

and

$$\frac{d\eta^i}{dt} + \Gamma^i_{jk} \frac{du^k}{dt} \eta^j = 0. \tag{1.9.5}$$

By condition (b),

$$\frac{d}{dt}(g_{ij}\, \xi^i\, \eta^j) = 0, \tag{1.9.6}$$

that is

$$\left(\frac{dg_{ij}}{dt} - g_{lj}\, \Gamma^l_{ik} \frac{du^k}{dt} - g_{il}\, \Gamma^l_{jk} \frac{du^k}{dt}\right) \xi^i\, \eta^j = 0. \tag{1.9.7}$$

Since (1.9.6) holds for any pair of vectors X and Y and any parametrized curve $C(t)$,

$$\frac{\partial g_{ij}}{\partial u^k} = g_{lj}\, \Gamma^l_{ik} + g_{il}\, \Gamma^l_{jk}. \tag{1.9.8}$$

By permuting the indices i, j, and k, two further equations are obtained:

$$\frac{\partial g_{jk}}{\partial u^i} = g_{lk}\,\Gamma^l_{ji} + g_{jl}\,\Gamma^l_{ki}, \tag{1.9.9}$$

$$\frac{\partial g_{ki}}{\partial u^j} = g_{li}\,\Gamma^l_{kj} + g_{kl}\,\Gamma^l_{ij}. \tag{1.9.10}$$

We define the contravariant tensor field g^{jk} by means of the equations

$$g_{ij}\,g^{jk} = \delta^k_i. \tag{1.9.11}$$

Adding (1.9.8) to (1.9.9) and subtracting (1.9.10), one obtains after multiplying the result by $\frac{1}{2}\,g^{jm}$ and contracting

$$\Gamma^m_{ki} = \left\{{m \atop ki}\right\} + \tfrac{1}{2}\,(T_{ki}{}^m - T^m{}_{ki} - T^m{}_{ik}), \tag{1.9.12}$$

where

$$\left\{{m \atop ki}\right\} = \tfrac{1}{2}g^{mj}\left(\frac{\partial g_{ij}}{\partial u^k} + \frac{\partial g_{jk}}{\partial u^i} - \frac{\partial g_{ki}}{\partial u^j}\right) \tag{1.9.13}$$

and

$$T^m{}_{ki} = g^{mr}\,g_{is}\,T_{rk}{}^s. \tag{1.9.14}$$

(Although the torsion tensor vanishes, it will be convenient in § 5.3 to have the formula (1.9.12)). Hence, since the torsion tensor vanishes (condition (a)), the connection Γ^i_{jk} is given explicitly in terms of the metric by formula (1.9.13). That the $\left\{{i \atop jk}\right\}$ transform as they should is an easy exercise. This is the *connection of Levi Civita*. We remark that condition (b) says that parallel displacement is an isometry. This follows since parallel displacement is an isomorphic linear map between tangent spaces.

A Riemannian metric gives rise to a submanifold $\tilde{B}$ of the bundle of frames over M. This is the bundle of all orthonormal frames over M. An orthonormal frame at a point P of M is a set of n mutually perpendicular unit vectors in the tangent space at P. In this case, the structural group of the bundle is the orthogonal group. *A connection defined by a parallelization of $\tilde{B}$ gives a parallel displacement which is an isometry—the Levi Civita connection being the only one which is torsion free.* If we denote by θ_i, θ_{ij}, Θ_{ij}, S_{jikl} the restrictions of θ^i, θ^j_i, Θ^j_i, $S^i{}_{jkl}$ to the orthonormal frames (cf. § 1.8), then by 'developing' the frames along a parametrized curve C into affine space A^n (see the following paragraph), it can be shown that

$$\theta_{ij} + \theta_{ji} = 0, \qquad \Theta_{ij} + \Theta_{ji} = 0 \tag{1.9.15}$$

(cf. I.G.5),

$$d\theta_i = \sum_j \theta_j \wedge \theta_{ji}, \tag{1.9.16}$$

$$d\theta_{ij} - \sum_k \theta_{ik} \wedge \theta_{kj} = \Theta_{ij}, \tag{1.9.17}$$

$$d\Theta_{ij} = \sum_k \theta_{ik} \wedge \Theta_{kj} - \Theta_{ik} \wedge \theta_{kj} \tag{1.9.18}$$

where the forms θ_i and θ_{ij} $(i<j)$ are linearly independent; moreover, the functions S_{ijkl} (cf. (1.8.9)) have the symmetry properties

$$S_{ijkl} = -S_{jikl} = -S_{ijlk}, \tag{1.9.19}$$

$$S_{ijkl} = S_{klij}, \tag{1.9.20}$$

$$S_{ijkl} + S_{iklj} + S_{iljk} = 0. \tag{1.9.21}$$

Equations (1.9.16) and (1.9.17) are the restrictions to $\tilde{B}$ of the corresponding equations (1.8.7) and (1.8.8).

Consider the bundle of frames over $C(t)$ and denote once again the restrictions of θ_i, θ_{ij} to the submanifold over $C(t)$ by the same symbols. To describe this bundle we choose a family of orthonormal frames $\{A_1(t), ..., A_n(t)\}$ along $C(t)$—one for each value of t. Then, for a given value of t the vectors $X_1(t), ..., X_n(t)$ of a general frame are given by

$$X_i(t) = x_i^j A_j(t), \quad (x_i^j) \in O(n,R).$$

The frames $\{X_1(t), ..., X_n(t)\}$ can be mapped into frames in the bundle $\tilde{A}^n$ of frames over A^n so that their relative positions remain unchanged. In particular, frames with the same origin along $C(t)$ are mapped into frames with the same origin in $\tilde{A}^n$. This follows from the fact that under the mapping the θ_i and θ_{ij} are the dual images of corresponding differential forms in $\tilde{A}^n$ (cf. I.F.1).

Let $C(t_1)$ and $C(t_2)$ be any two points of $C(t)$. A vector of $T_{C(t)}$ is given by $x^i A_i(t)$. Consider the map which associates with a vector $x^i A_i(t_1) \in T_{C(t_1)}$ the vector $\tilde{x}^i A_i(t_2) \in T_{C(t_2)}$ defined by

$$C'(t_1) + x^i A_i'(t_1) = C'(t_2) + \tilde{x}^i A_i'(t_2) \in A^n \tag{1.9.22}$$

where the prime denotes the image in A^n of the corresponding vector with origin on $C(t)$ and $C'(t)$ is the image of $C(t)$. In this way, the various tangent spaces along $C(t)$ can be 'compared'. This situation may be geometrically described by saying that the tangent spaces along $C(t)$ are *developed* into A^n and compared by means of the development.

An element of $\tilde{B}$ over $P \in M$ is a set of n mutually perpendicular unit vectors $X_1, \ldots, X_n$ in the tangent space at P. The frames along C are developed into affine space A^n and, as before, the images are denoted by a prime, so that $P \to P'$ and $X_i \to X'_i$ $(i = 1, 2, \ldots, n)$. In this way, a scalar product may be defined in A^n by identifying A^n with one of its tangent spaces and putting

$$X' \cdot Y' = X \cdot Y.$$

Since the Levi Civita parallelism is an isometric linear map f_* between tangent spaces, the scalar product defined in A^n has an invariant meaning; for, $f_*X \cdot f_*Y = X \cdot Y$.

Since the vectors of a frame are contravariant vectors, they determine a set of n linearly independent vectors in the space of covectors at the same point P, and since this latter space may be identified with $\wedge^1(T_P^*)$ a frame at P defines a set of independent 1-forms θ_i at that point. We make a change in our notation at this stage: Since we deal with a development of the tangent spaces along C into the vector space A^n we shall denote by $P,\{e_1, \ldots, e_n\}$ a typical frame in $\tilde{B}$ over P so that the image frame $P',\{e'_1, \ldots, e'_n\}(P \to P')$ in A^n is a 'fixed' basis for the frames in A^n. Now, consider the vectorial 1-form $\sum_{i=1}^n \theta_i e'_i$ in A^n (cf. I.A.6) which we denote by the 'displacement vector' dP'. Since A^n may be covered by one coordinate neighborhood R^n with local coordinates $u^1, \ldots, u^n$, we may look upon dP' as the vector whose components are the differentials $du^1, \ldots, du^n$. Moreover, the e'_i are the natural basis vectors $\partial/\partial u^i$ $(i = 1, \ldots, n)$. Now, in affine space it is not necessary to introduce the concept of covariant differential, and so the differential de'_i is a vectorial 1-form for each i, and we may write

$$de'_i = \sum_{j=1}^n \theta_{ij}\, e'_j. \tag{1.9.23}$$

Differentiating the equations

$$e_i \cdot e_k = \delta_{ik}$$

and applying (1.9.23) we obtain the first of equations (1.9.15) (cf. *I.G*). The remaining formulae follow from those in § 1.8 as well as (1.9.15).

We remark that the tensor $R_{ijkl} = g_{im}R^m{}_{jkl}$ satisfies the relations (1.9.19) - (1.9.21).

The forms θ_i and θ_{ij} are determined by the Riemannian metric of the manifold. If we are given two such metrics ds^2 and $d\bar{s}^2$ in the local coordinates (u^i) and $(\bar{u}^i)$, respectively, then it can be shown that if f is a local differentiable homeomorphism $f: U \to \bar{U}$ such that $f^*(d\bar{s}^2) = ds^2$, then $f^*\bar{\theta}_i = \theta_i$ and $f^*\bar{\theta}_{ij} = \theta_{ij}$, and conversely, if we write $\theta_i^2 = \theta_i \otimes \theta_i$, $i = 1, ..., n$ where $\otimes$ denotes the tensor product of covectors (cf. I.A)

$$f^*(\bar{\theta}_1^2 + \cdots + \bar{\theta}_n^2) = \theta_1^2 + \cdots + \theta_n^2$$

where f^* is the induced dual map. (The forms θ_i, $i = 1, ..., n$ are vectors determined by duality from the vectors e_i by means of the metric). Therefore, f induces a homeomorphism of the bundles $\tilde{B}_U$ and $\tilde{B}_{\bar{U}}$ of orthonormal frames over U and $\bar{U}$, respectively.

It follows that the forms θ_i and θ_{ij} are intrinsically associated with the Riemannian metric in the sense that the dual of the homeomorphism $\tilde{B}_U \to \tilde{B}_{\bar{U}}$ maps the $\bar{\theta}_i$ into the θ_i and the $\bar{\theta}_{ij}$ into the θ_{ij}, and for this reason they account for the important properties of Riemannian geometry.

1.10. Sectional curvature

In a 2-dimensional Riemannian manifold the only non-vanishing functions S_{ijkl} are $S_{1212} = -S_{1221} = -S_{2112} = S_{2121}$. We remark that the S_{ijkl} are not the components of a tensor but are, in any case, functions defined on the bundle $\tilde{B}$ of orthonormal frames. Moreover, the quantity $-S_{1212}$ is the Gaussian curvature of the manifold. We proceed to show that the value of the function $-S_{1212}$ at a point P in an n-dimensional Riemannian manifold M is the Gaussian curvature at P of some surface (2-dimensional submanifold) through P. To this end, consider the family $\mathscr{F}$ of orthonormal frames $\{e_1, ..., e_n\}$ at a point P of M with the property that the 'first' two vectors of each of these frames lie in the same plane π through P. Let S be a 2-dimensional submanifold through P whose tangent plane at P is π. The surface S is said to be *geodesic at* P if the geodesics (cf. § 1.11) through P tangent to π all lie on S. We seek the condition that S be geodesic at P. Let C be a parametrized curve on S through P tangent to the vector $\sum_{\alpha=1}^2 x_\alpha e_\alpha$ at P and develop the frames along C into E^n. If we denote the image of a frame $\{e_1, ..., e_n\}$ by $\{e_1', ..., e_n'\}$, we have

$$d\Big(\sum_{\alpha=1}^{2} x_\alpha e_\alpha'\Big) = \sum_{\alpha=1}^{2}\sum_{r=3}^{n} x_\alpha\, \theta_{\alpha r}\, e_r' + \sum_{\alpha,\beta=1}^{2} x_\alpha\, \theta_{\alpha\beta}\, e_\beta'.$$

In order that C be a geodesic, $\Sigma_{\alpha=1}^2 x_\alpha\, \theta_{\alpha r}$ must vanish, and since this holds for arbitrary initial values of the x_α, the forms $\theta_{\alpha r}$ $(1 \leqq \alpha \leqq 2,$

$3 \leqq r \leqq n$) are equal to zero at P. Conversely, if the $\theta_{\alpha r}$ vanish at P, then from (1.9.16) and (1.9.17)

$$
\begin{aligned}
d\theta_1 &= \theta_2 \wedge \theta_{21} + \sum_r \theta_r \wedge \theta_{r1} = \theta_2 \wedge \theta_{21}, \\
d\theta_2 &= \theta_1 \wedge \theta_{12} + \sum_r \theta_r \wedge \theta_{r2} = \theta_1 \wedge \theta_{12}, \\
d\theta_{12} &= \sum_r \theta_{1r} \wedge \theta_{r2} + S_{1212}\, \theta_1 \wedge \theta_2 = S_{1212}\, \theta_1 \wedge \theta_2.
\end{aligned}
\tag{1.10.1}
$$

These are the equations which hold on S. Hence, the quantity $-S_{1212}$ at a point P of a Riemannian manifold is equal to the Gaussian curvature at P of the surface tangent to the plane spanned by the first two vectors and which is geodesic at P.

The Gaussian curvature at a point P of the surface geodesic at P and tangent to a plane π in the tangent space at P is called the *sectional curvature* at (P, π) and is denoted by $R(P, \pi)$. If ξ^i, η^i are two orthonormal vectors which span π, it follows from (1.8.11) that

$$R(P,\pi) = - R_{ijkl}\, \xi^i \eta^j \xi^k \eta^l, \tag{1.10.2}$$

since $R_{ijkl} = g_{im} R^m{}_{jkl}$.

Let ξ^{*i}, η^{*i} be any two vectors spanning π. Then,

$$\xi^i = a\xi^{*i} + b\eta^{*i}, \quad \eta^i = c\xi^{*i} + d\eta^{*i}$$

where $ad - bc \neq 0$. In terms of the vectors ξ^{*i}, η^{*i},

$$R(P,\pi) = -(ad-bc)^2 R_{ijkl}\, \xi^{*i} \eta^{*j} \xi^{*k} \eta^{*l},$$

where $1/ad - bc$ is the oriented area of the parallelogram with ξ^{*i}, η^{*i} as adjacent sides:

$$\frac{1}{(ad-bc)^2} = (g_{ik} g_{jl} - g_{il} g_{jk})\, \xi^{*i} \eta^{*j} \xi^{*k} \eta^{*l}.$$

If we drop the asterisks, we obtain the following formula for the sectional curvature at (P, π):

$$R(P,\pi) = \frac{R_{ijkl}\, \xi^i \eta^j \xi^k \eta^l}{(g_{jk} g_{il} - g_{jl} g_{ik})\, \xi^i \eta^j \xi^k \eta^l}\,. \tag{1.10.3}$$

Now, assume that $R(P, \pi)$ is independent of π, that is, suppose that the sectional curvature at (P, π) does not depend on the two-dimensional section passing through this point. Then, from (1.10.3), we obtain

$$R_{ijkl} = K(g_{jk} g_{il} - g_{jl} g_{ik}) \tag{1.10.4}$$

where K denotes the common value of $R(P, \pi)$ for all planes π. By (1.8.11)

$$\begin{aligned} S_{ijkl} &= K\xi^p_{(i)}\,\xi^q_{(j)}\,\xi^r_{(k)}\,\xi^s_{(l)}\,(g_{qr}\,g_{ps} - g_{qs}\,g_{pr}) \\ &= K(\delta_{jk}\,\delta_{il} - \delta_{jl}\,\delta_{ik}) \end{aligned} \tag{1.10.5}$$

since the frames are orthonormal. Equation (1.10.5) may be rewritten by virtue of the second of equations (1.8.9) as

$$\Theta_{ij} = -K\theta_i \wedge \theta_j. \tag{1.10.6}$$

If we assume that at every point $P \in M$, $R(P, \pi)$ is independent of the plane section π, then, by substituting (1.10.6) into (1.9.18) and applying (1.9.16) we get

$$dK \wedge \theta_i \wedge \theta_j = 0.$$

Hence, dK must be a linear combination of θ_i and θ_j from which $dK = 0$ if $n \geqq 3$. This result is due to F. Schur: *If the sectional curvature at every point of a Riemannian manifold does not depend on the two-dimensional section passing through the point, then it is constant over the manifold.* Such a Riemannian manifold is said to be of *constant curvature.*

Assume that the constant sectional curvature K vanishes. We may conclude then that the tensor $R^i{}_{jkl}$ vanishes, and so the manifold is locally flat. This means that there is a coordinate system with the property that relative to it the components $\{^{\,i}_{jk}\}$ of the Levi Civita connection vanish. For, the equations

$$\frac{\partial^2 \bar{u}^i}{\partial u^j\,\partial u^k} = \Gamma^l_{jk}\,\frac{\partial \bar{u}^i}{\partial u^l}$$

obtained from (1.7.4) by putting $\bar{\Gamma}^l_{rs} = 0$ are completely integrable. Hence, there is a coordinate system in which the $\bar{\Gamma}^l_{rs}$ vanish. It follows that the components g_{jk} of the fundamental tensor are constants. Thus, we have a local isometry from the manifold to E^n. Conversely, if such a map exists, then clearly $R^i{}_{jkl}$ vanishes.

Let $X_i = \xi^k_{(i)}(\partial/\partial u^k)$ $(i = 1, \ldots, n)$ denote n mutually orthogonal unit vectors at a point in a Riemannian manifold with the local coordinates $u^1, \ldots, u^n$. Then from (1.9.1)

$$g_{ij}\,\xi^i_{(r)}\,\xi^j_{(s)} = \delta_{rs}. \tag{1.10.7}$$

It follows from the equations (1.9.11) that

$$g^{ij} = \sum_{r=1}^{n} \xi^i_{(r)}\,\xi^j_{(r)}. \tag{1.10.8}$$

The sectional curvature K_{rs} determined by the vectors X_r and X_s is given by

$$K_{rs} = -R_{ijkl}\,\xi^i_{(r)}\,\xi^j_{(s)}\,\xi^k_{(r)}\,\xi^l_{(s)}. \tag{1.10.9}$$

Taking the sum of both sides of this equation from $s = 1$ to $s = n$ we obtain

$$\sum_{s=1}^{n} K_{rs} = R_{ik}\,\xi^i_{(r)}\,\xi^k_{(r)} \tag{1.10.10}$$

where we have put $R_{ik} = -g^{jl}R_{ijkl}$, that is

$$R_{jk} = g^{il}\,R_{ijkl}. \tag{1.10.11}$$

The tensor R_{jk} is called the *Ricci curvature tensor* or simply the *Ricci tensor*. Again,

$$\sum_{r=1}^{n}\sum_{s=1}^{n} K_{rs} = R \tag{1.10.12}$$

where we have put

$$R = g^{ik}\,R_{ik}. \tag{1.10.13}$$

The scalar $R_{ik}\,\xi^i_{(r)}\,\xi^k_{(r)}$ is called the *Ricci curvature* with respect to the unit tangent vector X_r. The scalar R determined by equation (1.10.12) is independent of the choice of orthonormal frame used to define it. It is called the *Ricci scalar curvature* or simply the *scalar curvature.* The Ricci curvature κ in the direction of the tangent vector ξ^i is defined by

$$\kappa = \frac{R_{jk}\,\xi^j\,\xi^k}{g_{jk}\,\xi^j\,\xi^k}. \tag{1.10.14}$$

It follows that

$$(R_{jk} - \kappa g_{jk})\,\xi^j\,\xi^k = 0. \tag{1.10.15}$$

The directions which give the extrema of κ are given by

$$(R_{jk} - \kappa g_{jk})\,\xi^j = 0. \tag{1.10.16}$$

In general, there are n solutions $\xi^j_{(1)}, \ldots, \xi^j_{(n)}$ of this equation which are mutually orthogonal. These directions are called *Ricci directions.* A manifold for which the Ricci directions are indeterminate is called an *Einstein manifold.* In this case, the Ricci curvature is given by

$$R_{jk} = \kappa g_{jk}. \tag{1.10.17}$$

If we multiply both sides of this equation by g^{jk}, we obtain

$$R = n\kappa. \tag{1.10.18}$$

(In the sequel, the operation of multiplying the components of a tensor by the components of the metric tensor and contracting will be called *transvection.*) It follows that

$$R_{jk} = \frac{R}{n} g_{jk}. \tag{1.10.19}$$

Now, the Bianchi identity (1.8.14), or rather (1.9.18) can be expressed as

$$D_m R_{ijkl} + D_k R_{ijlm} + D_l R_{ijmk} = 0 \tag{1.10.20}$$

where D_j denotes covariant differentiation in terms of the Levi Civita connection. Transvecting this identity with g^{im} we obtain

$$D_s R^s{}_{jkl} = D_l R_{jk} - D_k R_{jl} \tag{1.10.21}$$

which upon transvection with g^{jk} results in

$$2D_s R^s{}_l = D_l R. \tag{1.10.22}$$

Substituting (1.10.19) into (1.10.22) and noting that

$$D_l g_{jk} = 0, \tag{1.10.23}$$

we see that for $n > 2$, the scalar curvature is a constant. Hence, *in an Einstein manifold the scalar curvature is constant* ($n > 2$).

It should be remarked that the tensor R_{jk} is symmetric. In fact, from equations (1.8.11) and (1.9.21) we obtain

$$R^i{}_{jkl} + R^i{}_{klj} + R^i{}_{ljk} = 0. \tag{1.10.24}$$

Contracting (1.10.24) with respect to i and l gives

$$R_{jk} - R_{kj} = 0$$

by virtue of the symmetry relations (1.9.19) and the definition (1.10.11). Hence, *the Ricci curvature tensor is symmetric.*

1.11. Geodesic coordinates

In this section we digress to define a rather special system of local coordinates at an arbitrary point P_0 of a Riemannian manifold M of dimension n and metric g. But first, we have seen that the differential equations of the auto-parallel curves $u^i = u^i(t)$, $i = 1, ..., n$ of an affine connection $\omega_j^i = \Gamma^i_{jk}\, du^k$ are given by

$$\frac{d^2u^i}{dt^2} + \Gamma^i_{jk} \frac{du^j}{dt} \frac{du^k}{dt} = 0, \quad i = 1, ..., n \tag{1.11.1}$$

and that any integral curve of (1.11.1) is determined by a point P_0 and a direction at P_0. If the affine connection is the Levi Civita connection, a *geodesic curve* (or, simply, *geodesic*) is defined as a solution of (1.11.1) where the parameter t denotes arc length.

We define a local coordinate system $(\bar{u}^i)$ at P_0 as follows: At the *pole* P_0 the partial derivatives of the components $\bar{g}_{ij}$ of the metric tensor vanish, that is

$$\left(\frac{\partial \bar{g}_{ij}}{\partial \bar{u}^k}\right)_{P_0} = 0, \quad i, j, k = 1, ..., n. \tag{1.11.2}$$

Hence, the coefficients $\bar{\Gamma}^i_{jk}$ of the canonical connection also vanish at P_0:

$$(\bar{\Gamma}^i_{jk})_{P_0} = 0. \tag{1.11.3}$$

Such a system of local coordinates is called a *geodesic coordinate system.* Thus, at the pole of a geodesic coordinate system, covariant differentiation is identical with ordinary differentiation. On the other hand, from (1.11.1)

$$\left(\frac{d^2\bar{u}^i}{dt^2}\right)_{P_0} = 0$$

—a property enjoyed by the geodesics of E^n relative to a system of cartesian coordinates. These are the reasons for exhibiting such coordinates at a point of a Riemannian manifold. Indeed, in a given computation substantial simplifications may result.

The existence of geodesic coordinates is easily established. For, if we write the equations of transformation (1.7.4) of an affine connection in the form

$$-\bar{\Gamma}^l_{rs} \frac{\partial \bar{u}^r}{\partial u^j} \frac{\partial \bar{u}^s}{\partial u^k} = \frac{\partial^2 \bar{u}^l}{\partial u^j\, \partial u^k} - \Gamma^i_{jk} \frac{\partial \bar{u}^l}{\partial u^i} \tag{1.11.4}$$

and define the n functions $\bar{u}^1, \ldots, \bar{u}^n$ by

$$\bar{u}^i = a^i_k(u^k - u^k(P_0)) + \tfrac{1}{2}a^i_l \Gamma^l_{jk}(u^j - u^j(P_0))\,(u^k - u^k(P_0))$$

where the a^i_k are n^2 constants with non-vanishing determinant, then

$$\left(\frac{\partial \bar{u}^i}{\partial u^k}\right)_{P_0} = a^i_k, \quad \left(\frac{\partial^2 \bar{u}^i}{\partial u^j\, \partial u^k}\right)_{P_0} = a^i_l\,(\Gamma^l_{jk})_{P_0}.$$

It follows that the right side of (1.11.4) vanishes at P_0. Consequently, by (1.9.8) the equations (1.11.2) are satisfied.

Incidentally, there exists a geodesic coordinate system in terms of which $(g_{ij})_{P_0} = \delta^i_j$. For, we can find real linear transformations of the $(\bar{u}^i)$, $i = 1, \ldots, n$ with constant coefficients so that the fundamental quadratic form may be expressed as a sum of squares.

EXERCISES

A. The tensor product

Let V and W be vector spaces of dimension n over the field F and denote by V^* and W^* the dual spaces of V and W, respectively. Let $L(V^*, W^*; F)$ denote the space of bilinear maps of $V^* \times W^*$ into F. This vector space is defined to be the tensor product of V and W and is denoted by $V \otimes W$.

1. Define the map $u : V \times W \to V \otimes W$ as follows:

$u(v, w)\,(v^*, w^*) = \langle v, v^*\rangle\,\langle w, w^*\rangle$. Then, u is bilinear and $u(V \times W)$ generates $V \otimes W$. Denote $u(v, w)$ by $v \otimes w$ and call u the natural map. u is onto but not 1-1.

Hint: To prove that u is onto choose a basis $e_1, \ldots, e_n$ for V and a basis $f_1, \ldots, f_n$ for W.

2. Let Z be a vector space over F and $\theta : V \times W \to Z$ a bilinear map. Then, there is a unique linear map $\bar{\theta} : V \otimes W \to Z$ such that $\bar{\theta} \cdot u = \theta$.

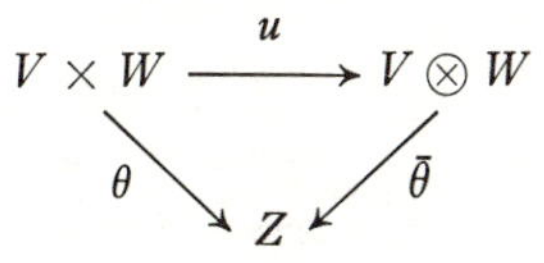

This property characterizes the tensor product as is shown in the following exercise.

3. If P is a vector space over F, $\bar{u} : V \times W \to P$ is a bilinear map onto P, and if for any vector space Z, $\theta : V \times W \to Z$ (θ, bilinear), there is a unique linear map $\bar{\theta} : P \to Z$ with $\bar{\theta} \cdot \bar{u} = \theta$,

$$\begin{array}{ccc} V \times W & \xrightarrow{\bar{u}} & P \\ & {\scriptstyle\theta}\searrow \quad \swarrow{\scriptstyle\bar{\theta}} & \\ & Z & \end{array}$$

then P and $V \otimes W$ are canonically isomorphic.

We are now able to give an important alternate construction of the tensor product. The importance of this construction rests in the fact that it is a typical example of a more general process, viz., dividing free algebras by relations.

4. Let $F_{V \times W}$ be the free vector space generated by $V \times W$ and consider $V \times W$ as a subset of $F_{V \times W}$ with the obvious imbedding. Let K be the subspace of $F_{V \times W}$ generated by elements of the form

$$(\alpha x + \beta y, z) - \alpha(x,z) - \beta(y,z),$$
$$(x, \alpha z + \beta w) - \alpha(x,z) - \beta(x,w).$$

Then, $(F_{V \times W})/K$ together with the projection map $u : V \times W \to (F_{V \times W})/K$ satisfies the characteristic property for the tensor product of V and W. In particular, u is bilinear. It follows that $(F_{V \times W})/K$ is canonically isomorphic with $V \otimes W$.

In the following exercise we discuss the concept of a tensorial form.

5. By a *tensorial p-form of type* (r, s) at a point P of a differentiable manifold M we shall mean an element of the tensor product of the vector space $T^r_s(P)$ of tensors of type (r, s) at P with the vector space $\wedge^p(T_P)$ of p-forms at P. A *tensorial p-form of type* (r, s) is a map $M \to T^r_s \otimes \wedge^p(T)$ assigning to each $P \in M$ an element of the tensor space $T^r_s(P) \otimes \wedge^p(T_P)$. A tensorial p-form of type $(0, 0)$ is simply a p-form and a tensorial 1-form of type $(1, 0)$ or *vectorial 1-form* may be considered as a 1-form with values in T.

Show that a tensorial p-form of type (r,s) may be expressed as a p-form whose coefficients are tensors of type (r, s) or as a tensor field of type (r, s) with p-forms as coefficients.

6. The notation of the latter part of § 1.9 is employed in this exercise. We shall use the symbol P' to denote the position vector OP' relative to some fixed point $O \in A^n$. Then, the vectors e'_i may be expressed as

$$e'_i = \frac{\partial P'}{\partial u^i}, \quad i = 1, \dots, n. \qquad (*)$$

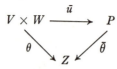

If P' moves along the curve $C'(t)$, we have

$$\frac{dP'}{dt} = \frac{\partial P'}{\partial u^i}\frac{du^i}{dt} = e'_i\frac{du^i}{dt},$$

that is,

$$dP' = e'_i du^i = (du^i)\, e'_i.$$

Thus, dP' is a vectorial 1-form. Show that dP' may be considered as that vectorial 1-form giving the identity map of A^n into itself.

Differentiating the relations $(*)$ with respect to u^j we obtain

$$\frac{\partial e'_i}{\partial u^j} = \frac{\partial^2 P'}{\partial u^j \partial u^i}.$$

Again, since e_i is a function of the parameter t along $C'(t)$,

$$\frac{de'_i}{dt} = \frac{\partial^2 P'}{\partial u^j \partial u^i}\frac{du^j}{dt} = \frac{\partial e'_i}{\partial u^j}\frac{du^j}{dt},$$

that is,

$$de'_i = \frac{\partial e'_i}{\partial u^j}\, du^j.$$

The de'_i $(i = 1, ..., n)$ are vectorial 1-forms. Hence, in terms of the basis $\{e'_i \otimes du^j\}$,

$$\frac{\partial e'_i}{\partial u^j} = \Gamma^k_{ij} e'_k$$

where the Γ^k_{ij} are the components of de'_i relative to this basis. Put

$$\theta^k_i = \Gamma^k_{ij}\, du^j, \quad i, k = 1, ..., n.$$

Then,

$$de'_i = \theta^k_i\, e'_k.$$

Show that the matrix (θ^k_i) defines a map of the tangent space at $P' + dP'$ onto the tangent space at P'. Consequently, the functions Γ^k_{ij} are the coefficients of connection relative to the natural basis.

B. Orientation

1. Show the equivalence of the two definitions of an orientation for a differentiable manifold. Assume that the form α of § 1.6 is differentiable.

Hint: Use a partition of unity.

2. If θ, ϕ denote polar coordinates on a sphere in E^3 the manifold can be covered by the neighborhoods

$$U : 0 \leqq \theta < \frac{\pi}{2} + \delta,$$
$$U' : \frac{\pi}{2} - \delta < \theta \leqq \pi$$

with coordinates

$$u^1 = \tan \frac{\theta}{2} \cos \phi, \quad u^2 = - \tan \frac{\theta}{2} \sin \phi.$$

and

$$u'^1 = \cot \frac{\theta}{2} \cos \phi, \quad u'^2 = \cot \frac{\theta}{2} \sin \phi,$$

respectively. Show that the sphere is orientable.

On the other hand, the real projective plane P^2 is not an orientable manifold. For, denoting by x, y, z rectangular cartesian coordinates in E^3, P^2 can be covered by the neighborhoods:

$$U \ : \left|\frac{y}{x}\right| < 2, \left|\frac{z}{x}\right| < 2,$$
$$U' \ : \left|\frac{z}{y}\right| < 2, \left|\frac{x}{y}\right| < 2,$$
$$U'' : \left|\frac{x}{z}\right| < 2, \left|\frac{y}{z}\right| < 2$$

with the corresponding coordinates

$$u^1 = \frac{y}{x}, \quad u^2 = \frac{z}{x},$$
$$u'^1 = \frac{z}{y}, \quad u'^2 = \frac{x}{y},$$

and

$$u''^1 = \frac{x}{z}, \quad u''^2 = \frac{y}{z}.$$

Incidentally, the compact surfaces can be classified as spheres or projective planes with various numbers of handles attached.

C. Grassman algebra

1. Let E be an associative algebra over the reals R with the properties:

1) E is a graded algebra (cf. § 3.3), that is $E = E_0 \oplus E_1 \oplus \ldots \oplus E_n \oplus \ldots$, where the operation $\oplus$ denotes the direct sum; each E_i is a subspace of E and for $e_i \in E_i$, $e_j \in E_j$, $e_i \wedge e_j \in E_{i+j}$ where $\wedge$ denotes multiplication in E;

2) $E_1 = V$ where V is a real n-dimensional vector space and $E_0 = R$;

3) E_1 together with the identity $1 \in R$ generates E;

4) $x \wedge x = 0, \quad x \in E_1$;

5) $\rho x_1 \wedge \ldots \wedge x_n = 0$, $x_1 \wedge \ldots \wedge x_n \neq 0$, $x_1, \ldots, x_n \in E_1$ implies $\rho = 0$.

Then E is isomorphic to $\wedge (V)$.

2. The algebra E can be realized as $T(V)/I_e$, where $T(V)$ is the tensor algebra over V and I_e is the ideal generated by the elements of the form $x \otimes x$, $x \in V$.

D. Frobenius' theorem [23]

The ensuing discussion is purely local. To begin with, we operate in a neighborhood of the origin O in R^n. Let θ be a 1-form which is not zero at O. The problem considered is to find conditions for the existence of functions f and g such that

$$\theta = fdg,$$

that is, an integrating factor for the differential equation

$$\theta = 0$$

is required. If $\theta = fdg$, then $f(O) \neq 0$. Thus, $d\theta = df \wedge dg = df \wedge \theta/f$ or

$$d\theta = \omega \wedge \theta \quad \text{where} \quad \omega = \frac{df}{f}.$$

Hence,

$$\theta \wedge d\theta = 0.$$

Observe that if $\theta = fdg$, the equation $\theta = 0$ implies $dg = 0$ and conversely. Consequently, the solutions or integral surfaces of $\theta = 0$ are the hypersurfaces $g =$ const.

As an example, let $n = 3$ and consider the 1-form

$$\theta = yz\,dx + xz\,dy + dz$$

where (x, y, z) are rectangular coordinates of a point in R^3. Then, $d\theta = y\,dz \wedge dx + x\,dz \wedge dy$. It follows that $d\theta = dz/z \wedge \theta$. However, $\omega = dz/z$ is singular along the z-axis. To avoid this, we may take $\omega = -y\,dx - x\,dy$. The function g may be determined. by employing the fact that the integral surfaces $g =$ const. are cut by the plane $x = at$, $y = bt$ in the solution z of $g(0, 0, z) =$ const. On this plane, the equation $\theta = 0$ becomes

$$dz + 2abzt\,dt = 0.$$

The solution of this ordinary differential equation with the initial condition $z(0) = c$ is

$$z = ce^{-abt^2}.$$

Since $abt^2 = xy$, these curves span a surface

$$z = ce^{-xy}.$$

If we think of a, b, c as variables and make the transformation $x = a$, $y = b$, $z = ce^{-ab}$, it is seen that the integral surfaces are

$$ze^{xy} = \text{const.}$$

Apply the above procedure to the form

$$\theta = dz - y\,dx - dy$$

and show that on the planes $x = at$, $y = bt$ the surfaces $z = \frac{1}{2}xy + y + c$ are obtained whereas on the parabolic cylinders $x = at$, $y = bt^2$, the surfaces obtained are $z = \frac{1}{3}xy + y + c$. (This is not the case in the first example.) Show that the reason integral surfaces are not obtained is given by $\theta \wedge d\theta \neq 0$.

1. Let P be a point of the n-dimensional differentiable manifold M of class k and V_r an r-dimensional subspace of the tangent space T_P at P. Put $q = n - r$. Let $x(r, P)$ be a frame at P whose last r vectors $e_A(A, B, \ldots = q + 1, \ldots, n)$ are in V_r. Then, V_r may be defined in terms of the vectors $\theta^1, \ldots, \theta^r$ of the dual space T_P^*, that is by the system of equations

$$\theta^1 = 0, \ldots, \theta^q = 0.$$

The vectors of any other frame $\bar{x}(r, P)$ satisfying these conditions may be expressed in terms of the vectors of $x(r, P)$ as follows:

$$\bar{e}_A = a_A^B\, e_B, \quad \bar{e}_i = a_i^\alpha e_\alpha, \quad \alpha, \beta, \ldots = 1, \ldots, n.$$

It follows that $a_A^i = 0$ for $i = 1, \ldots, q$ and $A = q + 1, \ldots, n$. Hence, the corresponding coframes (cf. D. 2) are given by

$$\theta^i = a_j^i\, \bar{\theta}^j, \quad \theta^A = a_\alpha^A\, \bar{\theta}^\alpha$$

where the matrix $(a_j^i) \in GL(q, R)$.

2. Conversely, let $\theta^1, \ldots, \theta^q$ be q linearly independent (over R) pfaffian forms at P. Let (θ^A), $A = q + 1, \ldots, n$ be r pfaffian forms given in such a way that the (θ^α), $\alpha = 1, \ldots, n$ define a coframe (that is, the dual vectors form a frame). The system of equations $\theta^1 = 0, \ldots, \theta^q = 0$ then determines uniquely an

r-dimensional subspace V_r of T_P. In order that the systems (θ^i), $(\bar{\theta}^i)$ give rise to the same r-dimensional subspace it is necessary and sufficient that there exist a matrix $(a^i_j) \in GL(q, R)$ satisfying

$$\theta^i = a^i_j \bar{\theta}^j.$$

3. Let D be a domain of M. A *pfaffian system* of *rank* q and *class* $l(2 \leqq l \leqq k-1)$ is defined, if, for every covering of D by coordinate neighborhoods $\{U\}$ and every point P of U a system of q linearly independent pfaffian forms is given such that for $P \in U \cap \bar{U}$

$$\theta^i = a^i_j \bar{\theta}^j$$

where the matrix $(a^i_j) \in GL(q, R)$ is of class l.

A pfaffian system of rank $q(= n - r)$ on D defines an r-dimensional subspace of the tangent space T_P at each point $P \in D$, that is, a *field of r-planes* of class l. A manifold may not possess pfaffian systems of a given rank. For example, the existence of a pfaffian system of rank $n - 1$ is equivalent to the existence of a field of directions. This is not possible on an even-dimensional sphere.

4. Suppose a pfaffian system of rank q and class l is defined on the coordinate neighborhood U by the 1-forms θ^i, $i = 1, \ldots, q$. This system is said to be *completely integrable* if there are q functions f^i of class $l + 1$ such that

$$\theta^i = a^i_j \, df^j, \quad (a^i_j) \in GL(q,R).$$

The pfaffian system may then be defined by the q differentials df^i. Under the circumstances the functions f^i form a *first integral* of the system.

The following result is due to Frobenius:

In order that a pfaffian system (θ^i) *be completely integrable it is necessary and sufficient that* $d\theta^i \wedge \theta^1 \wedge \ldots \wedge \theta^q = 0$ *for every* $i = 1, \ldots, q$.

The necessity is clear. The sufficiency may be proved by employing a result on the existence of a 'canonical pfaffian system' in R^n and then proceeding by induction on r [23]. Since a pfaffian system of rank q on U defines and can be defined by a non-zero decomposable form Θ of degree q determined up to a non-zero factor this result may be stated as follows:

If a pfaffian system of rank q has the property that at every point $P \in M$ *there is a local coordinate system such that the form* Θ *can be chosen to involve only q of these coordinates, the system is completely integrable.*

5. If $\Theta = \theta^1 \wedge \ldots \wedge \theta^q$, the condition

$$d\Theta \wedge \theta^i = 0$$

is equivalent to the condition

$$d\Theta = \omega \wedge \Theta$$

for some 1-form ω.

6. The linear subspaces of dimension r of T_P are in 1-1 correspondence with the classes of non-zero decomposable r-vectors—each class consisting of r-vectors differing from one another by a scalar factor. The set of r-vectors can be given a topology by means of the components relative to some basis. This defines a topology and, in fact, a differentiable structure in the set of subspaces denoted by $G^r(T_P)$ of dimension r of T_P. The manifold so obtained is called the *Grassman manifold* over T_P. The Grassman manifold $G^r(T_P^*)$ over the dual space may be similarly defined. There is a 1-1 correspondence

$$G^r(T_P) \to G^q(T_P^*).$$

This map is independent of the choice of a basis in $\wedge^n(T_P^*)$. Evidently then, it is a homeomorphism.

Define the fibre bundle

$$G^r(M) = \bigcup_{P \in M} G^r(T_P)$$

over M and show that it can be given a topology and a differentiable structure of class $k - 1$.

7. A cross section

$$F : M \to G^r(M)$$

of this bundle is a pfaffian system of rank q sometimes called a *differential system of dimension* r or *r-distribution*. A differential system of dimension r therefore associates with every point P of M a linear subspace of dimension r of T_P. By means of the correspondence $G^r(T_P) \to G^q(T_P^*)$, F defines (up to a non-zero factor) a decomposable form of degree q.

8. A submanifold (φ, M') is called an *integral manifold* of F if, for every $P' \in M'$,

$$\varphi_* : T_{P'} \to F(\varphi(P')).$$

The dimension of an integral manifold is therefore $\leq r$. Show that F is completely integrable if every $P \in M$ has a coordinate neighborhood with the local coordinates $u^1, ..., u^n$ such that the 'coordinate slices'

$$u^1 = \text{const.}, ..., u^q = \text{const.}$$

are integral manifolds of F.

Consider a completely integrable pfaffian system. The manifold (φ, M') is an integral manifold, if on every neighborhood U of M such that $U \cap M' \neq \square$ the pfaffian forms $\theta^1, \ldots, \theta^q$ vanish. If $P \in M'$, the tangent space to M' at P is the r-plane defined by the pfaffian system.

9. The Frobenius theorem is a generalization of well-known theorems on total differential equations. Consider, for example, the case $n = 3$, $r = 2$ with the form Θ considered above given in the local coordinates x, y, z by

$$\Theta = P(x, y, z)dx + Q(x, y, z)dy + R(x, y, z)dz.$$

By Frobenius' theorem, a necessary and sufficient condition for complete integrability is given by

$$d\Theta \wedge \Theta = 0,$$

that is

$$P(R_y - Q_z) + Q(P_z - R_x) + R(Q_x - P_y) = 0.$$

E. Local flatness [23]

1. If the curvature and torsion of an affinely connected manifold M are both zero, show that the manifold is locally flat.

Hint: By means of the equations (1.7.5) it suffices to show the existence of a local coordinate system $(\bar{u}^i)$ such that

$$d\bar{u}^i = p^i_j \, du^j$$

and

$$dp^i_j = p^i_k \, \omega^k_j .$$

Use Frobenius' theorem.

This may also be seen as follows: From the structural equations it is seen that zero curvature implies that the distribution of horizontal planes in B given by $\theta^i_j = 0$ is completely integrable. An integral manifold is thus a covering of M. Since the torsion is also zero the other structural equation gives $d\theta^i = 0$, $i = 1, \ldots, n$ on the integral manifold. Consequently, $\theta^i = du^i$, where $(u^1, \ldots, u^n)$ is a flat coordinate system.

F. Development of frames along a parametrized curve into A^n [23]

1. In the notation of § 1.9 show that the frames $\{X_1(t), \ldots, X_n(t)\}$ can be mapped into A^n in such a way that the pfaffian forms θ_i, θ_{ij} are dual images of corresponding forms in A^n:

Let $X_1'(t), \ldots, X_n'(t)$ denote the images of the frame vectors under the mapping. In the notation of § 1.9 a typical frame along C is denoted by $P, \{e_1, \ldots, e_n\}$ and its image vectors in A^n by $P', \{e_1', \ldots, e_n'\}$. If the θ_i and θ_{ij} are the dual images of corresponding forms in A^n the position vector P' together with the vectors e'_i satisfy the pfaffian system

$$dP' = \sum_{i=1}^{n} \theta_i e'_i,$$

$$de'_i = \sum_{j=1}^{n} \theta_{ij} e'_j \qquad (*)$$

(cf. equations (1.9.23)). The variables of this system are t, x^i_j and the components of the vectors P', $e'_1, ..., e'_n$. Since the curvature forms Θ_{ij} are quadratic in the differentials of the local coordinates, they vanish along a parametrized curve. It follows that there exists a local differentiable homeomorphism f from the bundle of frames over the submanifold $C(t)$ to the bundle of frames over $C'(t)$—the submanifold defined by the image of $C(t)$ in A^n, such that

$$f^*\bar{\theta}_i = \theta_i, \quad f^*\bar{\theta}_{ij} = \theta_{ij}$$

where $\bar{\theta}_i$, $\bar{\theta}_{ij}$ denote the forms in A^n corresponding to θ_i, θ_{ij}. Show that the conditions in Frobenius' theorem are satisfied by this system and hence that it is completely integrable. As a consequence of this, show that there is exactly one set of vectors P', $e'_1, ..., e'_n$ satisfying (*) and taking arbitrary initial values for $t = t_0$ and $x^i_j = \delta^i_j$. If $e'_1, ..., e'_n$ are linearly independent for $t = t_0$ show that they are independent for all values of t, that is, for all $t, \{e'_1, ..., e'_n\}$ is a frame on $C'(t)$.

G. Holonomy [23]

1. Denote the affine transformation defined by equation (1.9.22) by $T_{t_2 t_1}$

$$T_{t_2 t_1}: T_{P(t_1)} \to T_{P(t_2)}\,.$$

Show that $T_{t_2 t_1}$ is not, in general, a linear map. Define the linear map

$$T'_{t_2 t_1}: T_{P(t_1)} \to T_{P(t_2)}$$

sending the vector $x^i A_i(t_1) \in T_{P(t_1)}$ into the vector $\tilde{x}^i A_i(t_2) \in T_{P(t_2)}$ by means of the equation

$$x^i A'_i(t_1) = \tilde{x}^i A'_i(t_2).$$

Show that $T_{t_2 t_1}$ is independent of (a) the choice of initial frame $x^j_i = \delta^j_i$ for $t = t_0$ and (b) the choice of the family $\{A_1(t), ..., A_n(t)\}$ of frames along $C(t)$.

2. Let O be an arbitrary point of M and $\{\gamma\}$ the family of closed parametrized curves on M with O as origin. The map

$$\gamma \longrightarrow T_\gamma$$

associates with each $\gamma \in \{\gamma\}$ an affine transformation T_γ of the tangent space at O. These transformations form a group denoted by H_o-called the *holonomy group* at O. The *restricted holonomy group* H_o' consisting of the affine linear maps T_γ' is similarly defined. Show that the group H_o when considered as an abstract group is independent of the choice of O.

Hint: M is arcwise connected.

3. An affine connection is called a *metrical connection* if its restricted holonomy group leaves invariant a positive definite quadratic form. Let M be an affinely connected manifold with a metrical connection and assume that the scalar product of two vectors is defined at some point O of M. Show that the scalar product may be defined everywhere on M.

Hint: Let P be an arbitrary point of M, C a parametrized curve joining O and P and T_C' the affine linear map from T_O to T_P along C. Define the scalar product at P by

$$X_P \cdot Y_P = T_C^{-1} X_P \cdot T_C^{-1} Y_P$$

and show that this definition is independent of the choice of C.

4. Show that the Levi Civita connection is a metrical connection.

5. Establish the equations (1.9.15).

One may proceed as follows: Develop the frames along C into affine space A^n. Let $X(t_0)$ and $Y(t_0)$ be two vectors at $C(t_0)$ and $X'(t_0)$, $Y'(t_0)$ the corresponding vectors at $C'(t_0)$. Define a scalar product at $C'(t_0)$ by

$$X'(t_0) \cdot Y'(t_0) = X(t_0) \cdot Y(t_0).$$

By identifying A^n with one of its tangent spaces, a scalar product is defined in A^n. From $G.3$, this scalar product is independent of the choice of t_0. In this way, it follows that the orthonormal frames along C can be developed into A^n in such a way that

$$dP' = \sum_{i=1}^{n} \theta_i\, e_i', \quad de_i' = \sum_{j=1}^{n} \theta_{ij}\, e_j'$$

where

$$e_i' \cdot e_j' = \delta_{ij}.$$

The equations (1.9.15) follow by differentiating the last relation and applying $I.F.$

The idea of translating, wherever possible, problems of Riemannian geometry to problems of Euclidean geometry is due to E. Cartan [Leçons sur la géométrie des espaces de Riemann, Gauthier-Villars (1928; 2nd edition, 1946)].

H. Geodesic coordinates

1. Show that at the pole of geodesic coordinates (u^i) the Riemannian curvature tensor has the components

$$R_{ijkl} = \frac{1}{2}\left(\frac{\partial^2 g_{ik}}{\partial u^j\, \partial u^l} - \frac{\partial^2 g_{il}}{\partial u^j\, \partial u^k} - \frac{\partial^2 g_{jk}}{\partial u^i\, \partial u^l} + \frac{\partial^2 g_{jl}}{\partial u^i\, \partial u^k}\right).$$

Hence, the curvature tensor has the symmetry property (1.9.20).

I. The curvature tensor

1. The curvature tensor (which we now denote by L) of a Riemannian manifold with metric tensor g is completely determined by the sectional curvatures.

To see this, consider L as a transformation

$$L : T \times T \times T \to T$$

(cf. 1.2.15); then, the symmetry relations (1.9.19)-(1.9.21) become

(a) $L(X,Y,Z) = -L(Y,X,Z)$,

(b) $g(L(X,Y,Z),W) = -g(L(X,Y,W),Z)$,

(c) $g(L(X,Y,Z),\, W) = g(L(Z,W,X),Y)$,

(d) $g(L(X,Y,Z),W) + g(L(X,Z,W),Y) + g(L(X,W,Y),Z) = 0$.

The relation (a) says that as a function of the first two variables L depends only on $X \wedge Y$. Thus, we may write

$$L(X \wedge Y,Z) = L(X,Y,Z).$$

The metric tensor g may be extended to an inner product on $\wedge^2(T)$ as follows:

$$g(X_{11} \wedge X_{12}, X_{21} \wedge X_{22}) = \det g(X_{ij}, X_{i^*j^*})$$

for any vectors $X_{11}, X_{12}, X_{21}, X_{22} \in T$ where $i,j = 1,2;\ 1^* = 2,\ 2^* = 1$. Then, (b) says that $g(L(X \wedge Y,Z),W)$ is a function of $Z \wedge W$ only. Hence, there is a unique $\tilde{L}(X \wedge Y) \in \wedge^2(T)$ such that

$$g(L(X \wedge Y,Z),W) = g(\tilde{L}(X \wedge Y),Z \wedge W).$$

The relation (c) says that $\tilde{L}$ is a symmetric transformation of $\wedge^2(T)$.

By the usual 'polarization trick':
$(2g(X,Y) = g(X + Y, X + Y) - g(X,X) - g(Y,Y))$, a symmetric linear transformation is determined by the quadratic form corresponding to it. Hence $\tilde{L}$ is determined by

$$g(\tilde{L}(\xi), \xi)$$

where the bivector ξ runs through $\wedge^2(T)$. It is sufficient to consider only decomposable ξ. Consequently, $\tilde{L}$ is determined by the sectional curvatures

$$K(X,Y) = -\frac{g(\tilde{L}(X \wedge Y), X \wedge Y)}{g(X \wedge Y, X \wedge Y)}$$

of the planes spanned by X and Y for all $X,Y \in T$.

2. Put

$$R(X,Y)Z = L(X \wedge Y,Z)$$

and show that $R(X,Y)$ is a tensor of type (1,1). The sectional curvature determined by the vectors X and Y may then be written as

$$K(X,Y) = -\frac{g(R(X,Y)X,Y)}{g(X \wedge Y,X \wedge Y)}\cdot$$

For any set $\{X_i, X_j, X_k, X_l\}$ of orthonormal vectors, show that

$$R_{ijkl} = g(R(X_i,X_j)X_k,X_l).$$

3. Show that the curve C in the orthogonal group of T_P given by the matrix $(C(t)^i_j)$ defining the parallel translation of T_P around the coordinate square with corners (*a*) $u_i = u_i(P)$, $u_j = u_j(P)$, (b) $u_i = u_i(P) + \sqrt{t}$, $u_j = u_j(P)$, (c) $u_i = u_i(P) + \sqrt{t}$, $u_j = u_j(P) + \sqrt{t}$, (d) $u_i = u_i(P)$, $u_j = u_j(P) + \sqrt{t}$, all other u's constant has derivative

$$(C(t)^i_j)_{kl} = R^i{}_{jkl}.$$

J. Principal fibre bundles

1. Given a differentiable manifold M and Lie group G we define a new differentiable manifold $B = B(M,G)$ called a *principal fibre bundle* with *base space* M and *structural group* G as follows:

(i) The group G acts differentiably on B without fixed points, that is the map $(x,g) \to xg$, $x \in B$, $g \in G$ from $B \times G \to B$ is differentiable;

(ii) The manifold M is the quotient space of B by the equivalence relation defined by G;

(iii) The canonical projection $\pi : B \to M$ is differentiable;

(iv) Each point $P \in M$ has a neighborhood U such that $\pi^{-1}(U)$ is isomorphic with $U \times G$, that is, if $x \in \pi^{-1}(U)$ the map $x \to (\pi(x), \phi(x))$ from $\pi^{-1}(U) \to U \times G$ is a differentiable isomorphism with $\phi(xg) = \phi(x)g$, $g \in G$.

Show that $M \times G$ is a principal fibre bundle by allowing G to act on $M \times G$ as follows: $(P,g)h = (P,gh)$, $P \in M$, g, $h \in G$.

2. The submanifold $\pi^{-1}(P)$ associated with each $P \in M$ is a closed submanifold of $B(M,G)$ differentiably isomorphic with G. It is called the *fibre* over P. If M' is an open submanifold of M, show that $\pi^{-1}(M')$ is a principal fibre bundle with base space M' and structural group G.

3. Let $\{U_\alpha\}$ be an open covering of M. Show that the map $\pi^{-1}(U_\alpha \cap U_\beta) \to G$ defined by

$$\phi_\beta(xg)\,(\phi_\alpha(xg))^{-1} = \phi_\beta(x)\,(\phi_\alpha(x))^{-1}, \quad x \in \pi^{-1}(U_\alpha \cap U_\beta)$$

is constant on each fibre. Denote the induced maps of $U_\alpha \cap U_\beta \to G$ by $f_{\beta\alpha}$. For $U_\alpha \cap U_\beta \neq \square$ the $f_{\beta\alpha}$ are called the *transition functions* corresponding to the covering $\{U_\alpha\}$. They have the property

$$f_{\gamma\alpha}(P) = f_{\gamma\beta}(P) f_{\beta\alpha}(P), \quad P \in U_\alpha \cap U_\beta \cap U_\gamma.$$

4. Let $\{U_\alpha\}$ be an open covering of M and $f_{\beta\alpha}: U_\alpha \cap U_\beta \to G$, $U_\alpha \cap U_\beta \neq \square$ a family of differentiable maps satisfying the above relation. Construct a principal fibre bundle $B(M,G)$ whose transition functions are the $f_{\beta\alpha}$.

Hint: Define $N_\alpha = U_\alpha \times G$ for each open set U_α of the covering $\{U_\alpha\}$ and put $N = \bigcup_\alpha N_\alpha$. If we take as open sets in N the open sets of the N_α, N becomes a differentiable manifold. Define an equivalence relation $\sim$ in N in the following way: $(P,g) \sim (P,h)$, if and only if $h = f_{\beta\alpha}(P)g$. Finally, define B as the quotient space of N by this equivalence relation. Let $\pi^{-1}(U_\alpha)$ be an open submanifold of B differentiably homeomorphic with $U_\alpha \times G$. In this way, B becomes a differentiable manifold and one may now check conditions (i) - (iv) above.

5. Show that the homogeneous space G/H of the Lie group G by the closed subgroup H defines a principal fibre bundle $G(G/H,H)$ with base space G/H and structural group H (cf. VI. E. 1).

6. Show that the bundle of frames with group $G = GL(n,R)$ is a principal fibre bundle.

7. Consider the principal fibre bundle $B(M,G)$ and let F be a differentiable manifold on which G acts differentiably, that is the map $(g,v) \to g\,v$ from $G \times F \to F$ is differentiable. The group G can be made to act differentiably on $B \times F$ in the following manner: $(x,v) \to (x,v)g = (xg, g^{-1}v)$. Denote by E the quotient space $(B \times F)/G$; the points of E are the classes $[(x,v)]$, $x \in B$, $v \in F$. Denote by π_B the canonical projection of B onto M. A projection π_E of E onto M is defined by $\pi_E[(x,v)] = \pi_B(x)$. For each $P \in M$, the *fibre*

$\pi_E^{-1}(P) \subset E$ of E is the set of points represented by the class $[(x,v)]$ where x is an arbitrary point of B satisfying $\pi_B(x) = P$ and v is an arbitrary point of F. Show that E is a differentiable manifold by considering $\pi_E^{-1}(U)$ as an open submanifold of E which may be identified with $U \times F$. In terms of the differentiable structure given to E the map π_E is differentiable. The manifold E is known as the *associated fibre bundle* of B with *base space* M, *standard fibre* F and *structural group* G. Note that E and B have the same base spaces and structural groups.

8. Let F be an n-dimensional vector space with the fixed basis $(v_1, \ldots, v_n)$. The group $G = GL(n,R)$ acts on F by $g\, v_i = g_i^j\, v_j$. The *tangent bundle* is the associated fibre bundle of B with F as standard fibre. Show that the tangent bundle is the bundle of contravariant vectors of § 1.3.

It is surprising indeed that a manifold structure can be defined on the set of all tangent vectors, for there is no *a priori* relation between tangent spaces defined abstractly. Moreover, the idea of a vector varying continuously in a vector space which itself varies is *a priori* remarkable.

9. Let M be a (connected) differentiable manifold and B its universal covering space. By considering the action of the fundamental group $\pi_1(M)$ on B, show that B is a principal fibre bundle with base space M and structural group $\pi_1(M)$. Show also that any covering space is an associated fibre bundle of B with discrete standard fibre.

K. Riemannian metrics

1. It has been shown that a (connected) differentiable manifold M admits a Riemannian metric (cf. § 1.9). With respect to a Riemannian metric, a natural metric d may be defined as follows: $d(P, Q)$ is the greatest lower bound of the lengths of all piecewise differentiable curves joining P and Q. A Riemannian manifold is therefore a metric space. It is a complete metric space if the metric d is complete (cf. § 7.7). In this case the Riemannian metric is said to be *complete.* Every differentiable manifold carries a complete Riemannian metric. If every Riemannian metric carried by M is complete, M is compact [*86*]. A Riemannian manifold is said to be *complete* if its metric is complete.

CHAPTER II

TOPOLOGY OF DIFFERENTIABLE MANIFOLDS

In Chapter I we studied the local geometry of a Riemannian manifold M. In the sequel, we will be interested in how the local properties of M affect its global behaviour. The Grassman algebra of exterior forms is a structure defined at each point of a differentiable manifold. In the theory of multiple integrals we consider rather the Grassman bundle which, as we have seen, is the union of these algebras taken over the manifold. It is the purpose of this chapter to describe a class of differential forms (the harmonic forms) which have important topological implications. To this end, we describe the topology of M insofar as it is necessary to define certain algebraic characters, namely the cohomology groups of M. These groups are, in fact, topological invariants of the manifold. The procedure followed is similar to that of Chapter I where the Grassman algebra was first defined over an 'arbitrary' vector space and then associated with a differentiable manifold via the tangent space at each point of the manifold. We begin then by defining an abstract complex K over which an algebraic structure will be defined. We will then associate K with a related construction K' on M. The corresponding algebra over K' will yield the topological invariants we seek. The chapter is concluded with a theorem relating the class of forms referred to above with these invariants.

2.1. Complexes

A *closure finite abstract complex* K is a countable collection of objects $\{S_i^p\}$, $i = 1, 2, \cdots$ called *simplexes* satisfying the following properties.

(i) To each simplex S_i^p there is associated an integer $p \geqq 0$ called its *dimension*;

(ii) To the simplexes S_i^p and S_j^{p-1} is associated an integer denoted by $[S_i^p : S_j^{p-1}]$, called their *incidence number*;

(iii) There are only a finite number of simplexes S_j^{p-1} such that $[S_i^p : S_j^{p-1}] \neq 0$;

(iv) For every pair of simplexes S_i^{p+1}, S_j^{p-1} whose dimensions differ by two

$$\sum_k [S_i^{p+1} : S_k^p]\,[S_k^p : S_j^{p-1}] = 0.$$

We associate with K an integer *dim* K (which may be infinite) called its dimension which is defined as the maximum dimension of the simplexes of K.

An algebraic structure is imposed on K as follows: The p-simplexes are taken as free generators of an abelian group. A (formal) finite sum

$$C_p = \sum_i g_i S_i^p, \quad g_i \in G$$

where G is an abelian group is called a *p-dimensional chain* or, simply a *p-chain*. Two p-chains may be added, their sum being defined in the obvious manner:

$$C_p + C_p' = \sum_i (g_i + g_i')S_i^p, \quad g_i, g_i' \in G.$$

In this way, the p-chains form an abelian group which is denoted by $C_p(K, G)$. This group can be shown to be isomorphic with $C_p(K, Z) \otimes G$ where Z denotes the ring of integers, that is, the tensor product (see below) of the free abelian group generated by K with the abelian group G.

Let Λ be a ring with unity 1. A *Λ-module* is an (additive) abelian group A together with a map $(\lambda, a) \to \lambda a$ of $\Lambda \times A \to A$ satisfying

(i) $\lambda(a_1 + a_2) = \lambda a_1 + \lambda a_2$,

(ii) $(\lambda_1 + \lambda_2)a = \lambda_1 a + \lambda_2 a$,

(iii) $(\lambda_1\lambda_2)a = \lambda_1(\lambda_2 a)$,

(iv) $1a = a$.

Since the ring Λ operates on the group A on the left such a module is called a *left Λ-module*. A *right Λ-module* is defined similarly; indeed, one need only replace λa by $a\lambda$ and (iii) becomes

(iii)′ $a(\lambda_1\lambda_2) = (a\lambda_1)\lambda_2$.

For commutative rings no distinction is made between left and right Λ-modules.

Note that a Z-module is simply an abelian group and that for every integer n

$$na = a + \cdots + a \text{ (n times).}$$

Let A be a right Λ-module and B a left Λ-module. Denote by $F_{A\times B}$ the free abelian group having as basis the set $A \times B$ of pairs (a, b), $a \in A$, $b \in B$ and by Γ the subgroup of $F_{A\times B}$ generated by the elements of the form

$$(a_1 + a_2, b) - (a_1, b) - (a_2, b),$$

$$(a, b_1 + b_2) - (a, b_1) - (a, b_2),$$

$$(a\lambda, b) - (a, \lambda b).$$

The quotient group $F_{A\times B}/\Gamma$ is known as the *tensor product* of A and B and is evidently an abelian group (cf. I.A.4).

There is an operation which may be applied to a p-chain to obtain a $(p - 1)$-chain called the *boundary operation.* It is denoted by ∂ and is defined by the formula

$$\partial C_p = \sum_i g_i \, \partial S_i^p = \sum_j \sum_i g_i [S_i^p : S_j^{p-1}] \, S_j^{p-1},$$

where $C_p = \Sigma_i g_i S_i^p$ and $g_i[S_i^p : S_j^{p-1}]$ is defined by considering G as a Z-module. Moreover, it is linear in $C_p(K, G)$ and hence defines a homomorphism

$$\partial : C_p(K,G) \to C_{p-1}(K,G).$$

The kernel of ∂ is denoted by $Z_p(K, G)$, the elements of which are called *p-cycles*. As a consequence of (iv) in the definition of a complex, $\partial(\partial C_p) = 0$ for any C_p. The image of $C_{p+1}(K, G)$ under ∂ denoted by $B_p(K, G)$ is called the group of *bounding p-cycles* of K over G and its elements are called bounding p-cycles or simply *boundaries*. The quotient group

$$H_p(K,G) = Z_p(K,G)/B_p(K,G)$$

is called the pth *homology group* of K with coefficient group G. The elements of $H_p(K, G)$ are called *homology classes*. Clearly, a *p-cycle* determines a well-defined homology class. Two cycles Γ_1^p and Γ_2^p in the same homology class are said to be *homologous* and we write $\Gamma_1^p \sim \Gamma_2^p$. Obviously, $\Gamma_1^p \sim \Gamma_2^p$, if and only if, $\Gamma_1^p - \Gamma_2^p$ is a boundary.

Assume now that G is the group of integers Z and write $C_p(K) = C_p(K, Z)$, etc. The elements of $C_p(K)$ are called (finite) *integral p-chains* of K. A linear function f^p defined on $C_p(K)$ with values in a commutative topological group G:

$$f^p: C_p(K) \to G$$

is called a *p-dimensional cochain* or simply a *p-cochain*. We define groups dual to the homology groups: The sum of two p-cochains f^p and g^p is defined by the formula

$$(f^p + g^p)(C_p) = f^p(C_p) + g^p(C_p)$$

for any p-chain $C_p \in C_p(K)$. With this definition of addition the p-cochains form a group $C^p(K, G)$. The inverse of the cochain f^p is the cochain $-f^p$ defined by

$$-f^p(C_p) = f^p(-C_p)$$

where $-C_p$ is the p-chain $(-1)C_p$. (This group is actually a topological group with the following topology: For a p-simplex S_i^p and an open set U of G a neighborhood (S_i^p, U) in $C^p(K, G)$ is defined as the set of cochains f^p such that $f^p(S_i^p) \in U$). Since the S_i^p are free generators of the group $C_p(K)$, a p-cochain f^p defines a unique homomorphism of $C_p(K)$ into G.

An operator ∂^* dual to ∂ and called the *coboundary operator* is defined on the p-cochains as follows:

$$(\partial^* f^p)(C_{p+1}) = f^p(\partial C_{p+1}).$$

The image of f^p under ∂^* is a $(p+1)$-cochain called the *coboundary* of f^p. The operator ∂^* has the properties:

$$\text{(i)}\ \partial^*(f^p + g^p) = \partial^* f^p + \partial^* g^p,$$

$$\text{(ii)}\ \partial^*(\partial^* f^p) = 0.$$

This latter property follows from the corresponding property on chains. That ∂^* defines a homomorphism

$$\partial^* : C^p(K,G) \to C^{p+1}(K,G)$$

is clear. The kernel of ∂^* is denoted by $Z^p(K, G)$ and its elements are called *p-cocycles*. The image of $C^{p-1}(K, G)$ under ∂^* denoted by $B^p(K, G)$ is called the group of *cobounding p-cycles* or, simply, *coboundaries*.

The quotient group

$$H^p(K,G) = Z^p(K,G)/B^p(K,G)$$

is called the pth *cohomology group* of K with coefficient group G. (It carries a topology induced by that of $C^p(K, G)$). The elements of $H^p(K, G)$ are called *cohomology classes*. Evidently, a p-cocycle determines a well-defined cohomology class. Two cocycles f^p and g^p in the same cohomology class are said to be *cohomologous* and we write the 'cohomology' $f^p \sim g^p$. Obviously, $f^p \sim g^p$, if and only if, $f^p - g^p$ is a coboundary.

2.2. Singular homology

By a *geometric realization* K_E of an abstract complex K we mean a complex whose simplexes are points, open line segments, open triangles, ... in an Euclidean space E of sufficiently high dimension corresponding, respectively, to the 0, 1, 2, $\cdots$-dimensional objects in K in such a way that distinct simplexes of K correspond to disjoint simplexes of K_E. The point-set union of all the simplexes of the complex K_E written $| K_E |$ is called a *polyhedron* and the complex K is said to be a *covering* of $| K_E |$. Two complexes K and K' are said to be *isomorphic* if there is a 1-1 correspondence between their simplexes $S_i^p \longleftrightarrow S_i'^p$ preserving the incidences (cf. definition of an abstract complex). When K and K' are isomorphic it can be shown that there is an induced homeomorphism $\phi_*: | K_E | \rightarrow | K_E' |$ where K_E and K_E' are geometric realizations of the complexes K and K', respectively such that $\phi S_i^p = S_i'^p$ where $S_i'^p$ is the simplex corresponding to S_i^p under the isomorphism ϕ. It is indeed remarkable that *the corresponding homology groups of any two covering complexes of a polyhedron are isomorphic*. Hence, they are topological invariants of the polyhedron.

If the coefficients G in the definition of the homology groups form a ring F, these groups become modules over F. The rank of $H_p(K, F)$ as a module over F is called the pth *betti number* $b_p(K)$ $(= b_p(K, F))$ of the complex K. If F is a field of characteristic zero, these modules are vector spaces over F. Thus, $b_p(K, F)$ is the dimension of the vector space $H_p(K, F)$, that is the maximum number of p-cycles over F linearly independent of the bounding p-cycles. The expression $\sum_{p=0}^{\dim K} (-1)^p \, b_p(K)$ is called the *Euler-Poincaré characteristic* of K.

Since the homology groups of a covering complex of a polyhedron are topological invariants of the polyhedron so are the betti numbers and hence also the Euler-Poincaré characteristic. This, in turn implies that *if* $|K_E|$ *and* $|K'_E|$ *are homeomorphic, the corresponding homology groups of* K *and* K' *are isomorphic and their betti numbers coincide.*

By a p-simplex $[\varphi : S^p]$, $p = 0, 1, 2, \cdots$ on a differentiable manifold M is understood an *Euclidean* p*-simplex* S^p (point, closed line segment, closed triangle, $\cdots$) together with a differentiable map φ of S^p into M. More precisely, let R^∞ denote the vector space whose points are infinite sequences of real numbers $(x^1, \cdots, x^n, \cdots)$ with only a finite number of coordinates $x^n \neq 0$. The finite-dimensional vector spaces R^p are canonically imbedded in R$^\infty$. Consider the ordered sequence of points $(P_0, \cdots, P_p)$ (necessarily linearly independent) in R^∞ and denote by $\Delta(P_0, \cdots, P_p)$ the smallest convex set containing them, that is $\Delta(P_0, \cdots, P_p) = \{r_0 P_0 + \cdots + r_p P_p \mid r_i \geqq 0, r_0 + \cdots + r_p = 1\}$. Let $\pi(P_0, \cdots, P_p) = \{r'_0 P_0 + \cdots + r'_p P_p \mid r'_0 + \cdots + r'_p = 1\}$, that is, the plane determined by the P_i, $i = 0, \cdots, p$. The numbers $r'_0, \cdots, r'_p$ are called *barycentric coordinates* of a vector in $\pi(P_0, \cdots, P_p)$. By a *singular* p*-simplex* on M we mean a map φ of class 1 of $\Delta(P_0, \cdots, P_p)$ into M. A *singular* p*-chain* is a map of the set of all singular p-simplexes into R usually written as a formal sum $\Sigma g_i s_i^p$ $(g_i \in Z)$ with the singular simplexes s_i^p indexed in some fixed manner.

We denote by $|s^p|$ the *support of* s^p, that is the set of points $\varphi(\Delta(P_0, \cdots, P_p))$. A chain is called *locally finite* if each compact set meets only a finite number of supports with $g_i \neq 0$. We consider only locally finite chains. A singular chain is said to be *finite* if there are only a finite number of non-vanishing g_i. The *support of a p-chain* is the union of all $|s_i^p|$ with $g_i \neq 0$. Singular chains may be added and multiplied by scalars (elements of R) in the obvious manner. Infinite sums are permissible if the result is a locally finite chain.

The *faces* of a p-simplex $s^p \mathrel{\overline{\overline{\cdot}}} [\varphi: P_0, \cdots, P_p]$ $(p > 0)$ are the simplexes $s_i^{p-1} = [\varphi: P_0, \cdots P_{i-1}, P_{i+1}, \cdots P_p]$. A boundary operator ∂ is defined by putting

$$\partial s^p = \sum_{i=0}^{p} (-1)^i s_i^{p-1}.$$

For $p = 0$ we put $\partial s^0 = 0$. The extension to arbitrary singular chains is by linearity. It is easily checked that the condition of local finiteness is fulfilled. Moreover, $\partial\partial = 0$. Note that $[s^p : s_i^{p-1}] = (-1)^i$.

Cycles and boundaries are defined in the usual manner. Let S_p denote the vector space of all finite p-chains, S_p^c the subspace of p-cycles and S_p^b the space of boundaries of finite $(p+1)$-chains. The quotient S_p^c/S_p^b is called the pth *singular homology space or group* of M and is denoted by SH_p.

In this way, it is possible to associate with M a *covering complex* K, that is a complex such that every point of M lies on exactly one simplex of K and every simplex of K lies on M. This important theorem was proved by Cairns [17]. The complex K is, of course, not unique. It follows that M is a polyhedron, that is, M is homeomorphic with $|K|$. Hence, the invariants described above are topological invariants of the manifold. In the sequel, we shall therefore writte $H^p(M, R)$ for $H^p(K, R)$, etc.

2.3. Stokes' theorem

Let φ be a singular p-simplex and α a p-form on the differentiable manifold M. Since φ is continuous, the intersection of the carrier of α and the support of φ is compact. Define the integral of α over $s^p = [\varphi : P_0, \cdots, P_p]$

$$\int_{s^p} \alpha$$

by

$$\int_{s^p} \alpha = \int_{\Delta(P_0,\dots,P_p)} \varphi^*\alpha.$$

For $C_p = \Sigma_i g_i s_i^p$, define the integral of α over C_p

$$\int_{C_p} \alpha$$

by linear extension, that is

$$\int_{C_p} \alpha = \sum_i g_i \int_{s_i^p} \alpha.$$

Now, let α be a $(p-1)$-form over the differentiable manifold M of dimension n and C_p a p-chain of a covering complex K of M. Then, it can be shown in much the same way as the Stokes' formula was established in § 1.6 that

$$\int_{\partial C_p} \alpha = \int_{C_p} d\alpha, \quad 1 \leqq p \leqq n. \tag{2.3.1}$$

Consider the functional L_α defined as follows:

$$L_\alpha(C_p) = \int_{C_p} \alpha. \tag{2.3.2}$$

Clearly, $L_\alpha : C_p(K) \to R$ is a linear functional, that is L_α is a p-cochain with real coefficients. In this way, to a p-form α there corresponds a

p-cochain L_α. It follows from (2.3.1) that if α is a closed form, L_α is a cocycle. Moreover, to an exact form there corresponds a coboundary. This correspondence between differential forms and cochains may be extended by defining a satisfactory product theory for complexes (cf. Appendix B).

2.4. De Rham cohomology

Since any two covering complexes of a differentiable manifold M determine isomorphic homology and cohomology groups we shall call them the homology and cohomology groups, respectively, of M. Now, for a fixed closed differential form α of degree p on M the integral $\int_{\Gamma_p}\alpha$ is a linear functional on SH_p. To see this, put $\Gamma_p' = \Gamma_p + \partial C_{p+1}$; then,

$$\int_{\Gamma_p'} \alpha = \int_{\Gamma_p} \alpha + \int_{\partial C_{p+1}} \alpha = \int_{\Gamma_p} \alpha + \int_{C_{p+1}} d\alpha = \int_{\Gamma_p} \alpha$$

by Stokes' theorem. Hence, there is a unique cohomology class $\{f^p\} \in H^p(M)$ $(= H^p(M, R))$ such that

$$\int_{\Gamma_p} \alpha = f^p(\Gamma_p)$$

for all $\{\Gamma_p\} \in SH_p$ where f^p is a cocycle belonging to the cohomology class $\{f^p\}$. A theorem due to de Rham (cf. Appendix A and [65]) implies that the correspondence $\alpha \rightarrow \{f^p\}$ establishes an isomorphism (provided M is compact), that is

$$D^p(M) \cong H^p(M)$$

(cf. § 2.6). Moreover, the cohomology class associated with the exterior product of two closed differential forms is the cup product of their cohomology classes (cf. Appendix B). Hence, the isomorphism is a ring isomorphism. Since the pth betti number $b_p(M)$ of M is the dimension of the group $H^p(M)$, it follows that *$b_p(M)$ is equal to the number of linearly independent closed differential forms of degree p modulo the exact forms of degree p.* In the remaining sections of this chapter we shall see how this result was extended by Hodge to a more restricted class of forms.

2.5. Periods

General line integrals of the form

$$\int_C p\,dx + q\,dy \tag{2.5.1}$$

are often studied as functionals of the arc (or chain) C under the conditions that the functions $p = p(x, y)$ and $q = q(x, y)$ are of class $k \geqq 1$ in a plane region D and that C is allowed to vary in D. A particularly important type of line integral has the characteristic property that the integral depends only on its end points, that is if C and C' have the same initial and terminal points

$$\int_C p\,dx + q\,dy = \int_{C'} p\,dx + q\,dy. \tag{2.5.2}$$

This is equivalent to the statement that

$$\int_\Gamma p\,dx + q\,dy = 0 \tag{2.5.3}$$

over any closed curve (or cycle) Γ. Now, a necessary and sufficient condition that the line integral (2.5.1) be a function of the end-points of C is that the differential $p\,dx + q\,dy$ be an exact differential, or, in the language of Chapter I that the linear differential form $\alpha = p\,dx + q\,dy$ be an exact differential form. The most important consequence is Cauchy's theorem for simply connected regions. *If α is a holomorphic differential and D a simply connected region, then*

$$\int_{\partial D} \alpha = 0. \tag{2.5.4}$$

If we put

$$f(C) = \int_C \alpha \tag{2.5.5}$$

then f is a linear functional (or cochain) and, in general

$$f(C') = f(C) + f(\Gamma) \tag{2.5.6}$$

where Γ is the cycle $C' - C$. The integral $f(\Gamma)$ is called a *period* of the form α. Hence (2.5.6) may be stated as follows: The values of the line integral (2.5.1) along various chains with the same initial and terminal points are equal to a given value of the integral plus a period.

Conversely, every such sum represents a value of the integral. The study of the cochain f becomes a topological problem by virtue of this result, that is, the problem is to investigate the cycles. As a matter of fact, homology theory has its origin in this fundamental problem. Another important property of the cochain f is the following: If a cycle Γ may be continuously deformed to a point, then $f(\Gamma) = 0$. This is certainly the case if D is simply connected.

Now, if $\Gamma \sim \Gamma'$, $f(\Gamma) = f(\Gamma')$ or, more generally, we may consider the homology

$$\Gamma \sim n_1 \Gamma_1 + \cdots + n_r \Gamma_r, \quad n_i \in Z \tag{2.5.7}$$

and it implies that

$$\begin{aligned} f(\Gamma) &= \sum_{i=1}^{r} n_i f(\Gamma_i) \\ &= \sum_{i=1}^{r} n_i \omega_i \end{aligned} \tag{2.5.8}$$

where the ω_i are the periods of the form α over the cycles Γ_i. The values of the line integral are then all of the form $f(C) + \sum_{i=1}^{b_1(D)} n_i \omega_i$ where $b_1(D)$ is the first betti number of D. This is a well-known expression in analysis. The Cauchy theorem for multiply connected regions may now be stated: *If α is a holomorphic differential and D is a multiply connected region, then*

$$\int_{\Gamma} \alpha = 0 \tag{2.5.9}$$

for every cycle $\Gamma \sim 0$ in D.

2.6. Decomposition theorem for compact Riemann surfaces

The following generalizations can be made here. In the first place, it is possible to consider in place of D a surface with suitably related integrals. The classical example is the study of *abelian integrals*

$$F(z) = \int_{z_0}^{z} R(z,w)\, dz \tag{2.6.1}$$

where $R(z, w)$ is a rational function and $w = w(z)$ is an algebraic function, the integral being evaluated along various paths in the z-plane. A branch of the function $w(z)$ is chosen at z_0 and a path from z_0 to z. The value of $w(z)$ is then determined by analytic continuation along the path of integration. Instead of considering the z-plane we may consider a surface S on which the function $w(z)$ is defined and single-valued. The surface S is called the *Riemann surface* of the algebraic function $w(z)$. It can be shown that the Riemann surface of any algebraic function is homeomorphic to a sphere with g handles. On the other hand, we may consider such a surface and ask for those functions on the surface which correspond to single-valued analytic functions in the z-plane. In this way, we obtain a classification of analytic functions according to their Riemann surfaces. Moreover, the behavior of the integrals of the algebraic functions may be determined from a knowledge of the functions themselves, as well as the topology of the surface. This is Riemann's approach to the study of algebraic functions and their integrals. Since the first betti number of a compact Riemann surface S is $2g$, it can be shown that the periods of an everywhere analytic (henceforth, called holomorphic) integral on S are linear combinations of $2g$ periods. By constructing integrals with prescribed periods on $2g$ independent 1-cycles of a compact Riemann surface S, it can be shown that the de Rham cohomology group $D^1(S)$ is isomorphic to the group $H^1(S)$. This is *de Rham's isomorphism theorem for compact Riemann surfaces.*

Consider now the linear differential form

$$\alpha = p\,dx + q\,dy \tag{2.6.2}$$

over a Riemann surface S and define the operator $*$ by

$$*\alpha = -q\,dx + p\,dy. \tag{2.6.3}$$

That $*\alpha$ has an invariant meaning over S is easily seen by choosing a conformally related coordinate system (x', y'):

$$x = x(x', y'), \quad y = y(x', y')$$

that is

$$\frac{\partial x}{\partial x'} = \frac{\partial y}{\partial y'}, \quad \frac{\partial x}{\partial y'} = -\frac{\partial y}{\partial x'},$$

and checking the transformation law. The operator $*$ has the following properties:

(i) $*(\alpha + \beta) = *\alpha + *\beta, \quad *(f\alpha) = f(*\alpha)$,

(ii) $**\alpha = *(*\alpha) = -\alpha$,

(iii) $\alpha \wedge *\beta = \beta \wedge *\alpha$,

(iv) $\alpha \wedge *\alpha = 0$, if and only if, $\alpha = 0$.

Since $df = (\partial f/\partial x)\,dx + (\partial f/\partial y)\,dy$, we can define the operator $*d$ for functions by

$$(*d)f = *(df) = -\frac{\partial f}{\partial y}\,dx + \frac{\partial f}{\partial x}\,dy. \tag{2.6.4}$$

Define

$$(*d)\alpha = -\,d(*\alpha) \tag{2.6.5}$$

for 1-forms.

If we put

$$\Delta = d{*}d, \tag{2.6.6}$$

then

$$\Delta f = \left(\frac{\partial^2 f}{\partial x^2} + \frac{\partial^2 f}{\partial y^2}\right) dx \wedge dy. \tag{2.6.7}$$

A function f of class 2 is called *harmonic* on S if Δf vanishes on S. Locally, then

$$\frac{\partial^2 f}{\partial x^2} + \frac{\partial^2 f}{\partial y^2} = 0. \tag{2.6.8}$$

A linear differential form α of class 1 on S is called a *harmonic form* if, for each point P of S there is a coordinate neighborhood U of P such that α is the total differential of a harmonic function f in U. This implies that $*\alpha$ is closed. In fact, $\alpha = df$ and $d{*}df = 0$ in U, that is $d{*}\alpha = 0$. Conversely, $d\alpha = 0$ implies that $\alpha = df$, locally (cf. § A. 6). Moreover, $d{*}\alpha = 0$ implies that $d(*df) = 0$. Hence, f is harmonic. We have shown that a linear differential form α of class 1 is harmonic, if and only if, $d\alpha = 0$ and $d{*}\alpha = 0$.

A harmonic differential form $\alpha = p\,dx + q\,dy$ on S, that is a form which satisfies $d\alpha = 0$ and $d{*}\alpha = 0$ defines a holomorphic function $p - iq$ (locally) of $z = x + iy$ $(i = \sqrt{-1})$. Indeed,

$$0 = d\alpha = \left(\frac{\partial q}{\partial x} - \frac{\partial p}{\partial y}\right) dx \wedge dy, \tag{2.6.9}$$

$$0 = d{*}\alpha = \left(\frac{\partial p}{\partial x} + \frac{\partial q}{\partial y}\right) dx \wedge dy, \tag{2.6.10}$$

and so we have locally $(\partial p/\partial y) = (\partial q/\partial x)$ and $(\partial p/\partial x) = -\,(\partial q/\partial y)$, which are the Cauchy-Riemann equations for the functions p and $-\,q$. (A function f of class 1 is *holomorphic* on S if locally $f(x, y) = u(x, y) + iv(x, y)$ and the functions u and v satisfy the Cauchy-Riemann equations). It is an easy matter to show that f is holomorphic on S, if and only if $*df = -\,idf$, that is, if and only if, the differential

df is *pure* (of bidegree (1,0) cf. § 5.2). A linear differential form α on S is said to be a *holomorphic differential* if, in each coordinate neighborhood U it is the differential of a holomorphic function in U. A linear differential form α is locally exact, if and only if, $d\alpha = 0$. Locally, then $\alpha = df$ and in order that f be holomorphic $*df = -idf$ or $*\alpha = -i\alpha$. A differential form satisfying this latter condition is said to be *pure*. Hence, *a linear differential form of class* 1 *is holomorphic on S, if and only if it is closed and pure* (cf. § 5.4). We remark that if α is holomorphic, it is a harmonic form. This is clear from the previous statement.

The formal change of variables $z = x + iy$, $\bar{z} = x - iy$ and the resulting equations $*dz = -idz$, $*d\bar{z} = id\bar{z}$ clarify the nature of pureness: α is pure, if and only if, it is expressible in terms of dz only.

A differential form of class 1 will be called a *regular differential form*. Now, the regular harmonic forms on a compact Riemann surface S form a group $H(S)$ under addition. It can be shown that if α is a closed linear differential form on S, then there is a unique harmonic 1-form homologous to α, that is $H(S)$ is isomorphic to the de Rham cohomology group $D^1(S)$. This is *Hodge's theorem for a compact Riemann surface*. The proof is based on a decomposition of α into a sum of two forms, one of which is exact and the other harmonic. (More generally, a 1-form on a Riemannian manifold may be decomposed into a sum of an exact form, a form which may be expressed as $*df$ for some f and a harmonic form (cf. § 2.7). This is the *decomposition theorem* applied to 1-forms). The de Rham isomorphism theorem together with the Hodge theorem for compact Riemann surfaces implies that *the first betti number of a compact Riemann surface is equal to the number of linearly independent harmonic 1-forms on the surface.*

2.7. The star isomorphism

The geometry of a Riemann surface is conformal geometry. As a possible generalization of the results of the previous section, one might consider more general surfaces, for example, the closed surfaces of § 1.1, the geometry being Riemannian geometry. One might even go further and consider as a replacement for the Riemann surface an n-dimensional Riemannian manifold. To begin with, consider the Euclidean space E^n and let $(u^1, \cdots, u^n)$ be rectangular cartesian coordinates of a point. Let f be a function defined in E^n which is a potential function in some region of the space. In the language of vector analysis,

$$\operatorname{div} \operatorname{grad} f = 0, \qquad (2.7.1)$$

where grad f is the vector field with the components $\partial f/\partial u^i$ relative to the given coordinate system and $-$ div grad f is the scalar

$$\frac{\partial^2 f}{\partial u^{1^2}} + \cdots + \frac{\partial^2 f}{\partial u^{n^2}}\,.$$

Now, in a Riemannian manifold M, the equation

$$\frac{\partial^2 f}{\partial u^{1^2}} + \cdots + \frac{\partial^2 f}{\partial u^{n^2}} = 0 \tag{2.7.2}$$

may hold in a given coordinate neighborhood but it does not have an invariant meaning over M, that is, the left hand side is not a tensor field. A generalization of the concept of a harmonic function is immediately suggested, namely, instead of ordinary (partial) differentiation employ covariant differentiation. Hence, grad f is the covariant vector field $D_i f$ and the divergence of this vector field is the scalar $-\Delta f$ defined by

$$-\Delta f = g^{jk} D_k D_j f \tag{2.7.3}$$

where g_{jk} is the metric tensor field of M and covariant derivatives are taken with respect to the connection canonically defined by the metric. It follows that

$$-\Delta f = \frac{1}{\sqrt{g}}\frac{\partial}{\partial u^i}(\sqrt{g}\, g^{ij} D_j f) \tag{2.7.4}$$

or, alternatively

$$-\Delta f = g^{ij}\left(\frac{\partial^2 f}{\partial u^i\, \partial u^j} - \frac{\partial f}{\partial u^k}\left\{{}^{\,k}_{i\,j}\right\}\right). \tag{2.7.5}$$

Hence, *Laplace's equation* $\Delta f = 0$ is a tensor equation and reduces to (2.7.2) in a Euclidean space in which the u^i $(i = 1, \ldots, n)$ are rectangular cartesian coordinates.

Equation (2.7.4), namely, the condition that the function f be a harmonic function is the condition that the $(n-1)$-form

$$g^{ij} D_j f\, \epsilon_{i(i_1 \ldots i_{n-1})}\, du^{i_1} \wedge \ldots \wedge du^{i_{n-1}} \tag{2.7.6}$$

be closed where $\epsilon_{i_1 i_2 \ldots i_n}$ is the skew-symmetric tensor $\delta^{1\,2 \ldots n}_{i_1 i_2 \ldots i_n}\sqrt{G}$ and $G = \det(g_{ij})$. The discussion of § 2.6 together with the 'interpretation' of a harmonic function as a certain closed $(n-1)$-form suggests the introduction of an operator (defined in terms of the metric) which associates to a p-form α an $(n-p)$-form $*\alpha$ defined as follows: Let

$$\alpha = a_{(i_1 \ldots i_p)}\, du^{i_1} \wedge \ldots \wedge du^{i_p}. \tag{2.7.7}$$

Then

$$*\alpha = a^*_{(j_1 \cdots j_{n-p})} du^{j_1} \wedge \dots \wedge du^{j_{n-p}} \tag{2.7.8}$$

where

$$a^*_{j_1 \cdots j_{n-p}} = \epsilon_{(i_1 \cdots i_p) j_1 \cdots j_{n-p}} a^{(i_1 \cdots i_p)}. \tag{2.7.9}$$

In the last sum, only the terms corresponding to the values of $i_1, \cdots, i_p$ which are different from $j_1, \cdots, j_{n-p}$ can be non-zero. The form $*\alpha$ is called the *adjoint* of the form α. That the form (2.7.6) is the adjoint of the form $df = (\partial f/\partial u^i)\, du^i$ is an easy exercise. The adjoint of the (constant) function 1 (considered as a form of degree 0) is the volume element

$$*1 = \epsilon_{1\dots n} du^1 \wedge \dots \wedge du^n = \sqrt{G}\, du^1 \wedge \dots \wedge du^n. \tag{2.7.10}$$

The adjoint of any function, considered as a 0-form, is its product with the volume element.

If A and B are vectors in E^3 with the natural orientation, and the $*$ operation is defined in terms of the natural Riemannian structure of E^3, then $*\,(A \wedge B)$ is usually called the vector product of A and B. In E^2, the $*$ operator applied to vectors is essentially the operation of a rotation through $\pi/2$ radians.

As in § 2.6 the operator $*$ has the properties:

(i) $*(\alpha + \beta) = *\alpha + *\beta$, $*(f\alpha) = f(*\alpha)$,

(ii) $**\alpha = *(*\alpha) = (-1)^{pn+p}\alpha$,

(iii) $\alpha \wedge *\beta = \beta \wedge *\alpha$,

(iv) $\alpha \wedge *\alpha = 0$, if and only if, $\alpha = 0$ where α and β are forms of degree p and f is a 0-form (function).

Let

$$\alpha = a_{(i_1 \dots i_p)} du^{i_1} \wedge \dots \wedge du^{i_p},$$

and

$$\beta = b_{(i_1 \dots i_p)} du^{i_1} \wedge \dots \wedge du^{i_p};$$

then

$$\alpha \wedge *\beta = a^{(i_1 \dots i_p)} b_{(i_1 \dots i_p)} * 1. \tag{2.7.11}$$

The proof of property (ii) and (2.7.11) follows by choosing an orthonormal coordinate system at a point. Hence, the relation between α and $*\alpha$ is symmetrical, save perhaps for sign.

We define the (global) *scalar product* (α, β) of α and β as the (real) number

$$(\alpha, \beta) = \int_M \alpha \wedge *\beta, \tag{2.7.12}$$

whenever the integral converges as will always be the case in the sequel. (It is assumed that M is orientable and that an orientation of M has been chosen). The scalar product evidently has the properties:

(i) $(\alpha, \alpha) \geqq 0$ and is equal to zero, if and only if $\alpha = 0$,

(ii) $(\alpha, \beta) = (\beta, \alpha)$,

(iii) $(\alpha, \beta_1 + \beta_2) = (\alpha, \beta_1) + (\alpha, \beta_2)$, $(\alpha_1 + \alpha_2, \beta) = (\alpha_1, \beta) + (\alpha_2, \beta)$,

(iv) $(*\alpha, *\beta) = (\alpha, \beta)$

where α, α_1, α_2, β, β_1 and β_2 have the same degree.

If $(\alpha, \beta) = 0$, α and β are said to be *orthogonal.*

It should be remarked that the $*$ operation is an isomorphism between the spaces $\wedge^p (T_P^*)$ and $\wedge^{n-p}(T_P^*)$ at each point P of M.

2.8. Harmonic forms. The operators δ and Δ

There are several well-known examples from classical physics (potential theory) where relations analogous to Laplace's equation hold. The electrical potential due to a system of charges or the vector potential due to a system of currents is not uniquely determined. To the former an arbitrary constant may be added and to the latter an arbitrary vector with vanishing curl. In defining electrical potential we begin with a vector field E representing the electrical intensity which satisfies the equation curl $E = 0$. This is the condition that the electric field be conservative. A function f is then defined as follows:

$$f(P) = \int_{P_0}^{P} E \cdot dr \tag{2.8.1}$$

where r denotes the position vector of a point in E^3 and the $\cdot$ denotes the inner product of vectors in E^3. It follows that $E = \text{grad } f$ and f is determined to within an additive constant.

In defining the vector potential, on the other hand, we begin with the magnetic induction B which satisfies the equation div $B = 0$. As it turns out, this is a sufficient condition for the existence of a vector field A (unique up to a vector field whose curl vanishes) satisfying $B = \text{curl } A$.

We now re-write the above equations as tensor equations in E^3. We may distinguish between covariant and contravariant tensor fields provided the coordinate system is not Euclidean. Let E_i denote the components of the covariant vector field E and B^i the components of the contravariant vector field B. Then,

$$D_j E_i - D_i E_j = 0 \tag{2.8.2}$$

and

$$E_i = D_i f, \tag{2.8.3}$$

locally.

Moreover,

$$D_i B^i = 0 \tag{2.8.4}$$

and

$$B_{ij} = D_j A_i - D_i A_j \tag{2.8.5}$$

where the skew-symmetric tensor field

$$B_{ij} = \epsilon_{ijk} B^k. \tag{2.8.6}$$

In the language of differential forms, if we denote by η and α the 1-forms defined by E and A and by β the 2-form defined by the bivector B_{ij}, then the equations (2.8.2) - (2.8.5) become

$$\begin{aligned} d\eta &= 0, \\ \eta &= df \text{ (locally)}, \\ d\beta &= 0, \\ -\beta &= d\alpha \text{ (locally)}. \end{aligned}$$

We note that $\beta = *\tilde{\beta}$ where $\tilde{\beta}$ is the 1-form corresponding to the covariant vector field $g_{ij}B^j$ where g_{ij} is the metric tensor of E^3.

Now, the theorems of classical potential theory, namely, (a) if η is closed, then η is exact and (b) if β is closed, then β is exact are not necessarily true in an arbitrary 3-dimensional differentiable manifold since the first and second betti numbers may not vanish (cf. § 2.4).

We digress for a moment and consider a Riemannian manifold of dimension n. To a p-form α on M we associate a $(p-1)$-form $\delta\alpha$ defined in terms of the operators d and $*$:

$$\delta\alpha = (-1)^{np+n+1} *d*\alpha. \tag{2.8.7}$$

The form $\delta\alpha$ is called the *co-differential* of α and has the properties:

(i) $\delta(\alpha + \beta) = \delta\alpha + \delta\beta$,

(ii) $\delta\delta\alpha = 0$,

(iii) $*\delta\alpha = (-1)^p\, d*\alpha$, $*d\alpha = (-1)^{p+1}\, \delta *\alpha$.

The form α is said to be *co-closed* if its co-differential is zero. This is equivalent to the statement that its adjoint is closed. If $\alpha = \delta\beta$ we say that α *cobounds* β and that α is *co-exact*.

It should be remarked that in contrast with the differential operator d, the co-differential operator δ involves the metric structure of M in an essential way.

A form α is said to be *harmonic* (or a *harmonic field*) if it is closed and co-closed. This is the definition given by Hodge. K. Kodaira [*46*], on the other hand calls a form α harmonic if $\Delta\alpha = 0$ where Δ is the (Laplace-Beltrami) operator $d\delta + \delta d$. It is evident that the harmonic forms of a given degree form a linear space. However, since the operator Δ is not, in general, a derivation, they do not form an algebra.

If α is the form of degree 1 in E^3 associated with the vector V, then the forms $d\delta\alpha$ and $\delta d\alpha$ are associated with the vectors grad div V and curl curl V and hence the form $\Delta\alpha$ is associated with the vector field $\nabla^2 V \equiv$ grad div V − curl curl V. Now, in the above example, at any point of E^3 where there is no current, the vector potential A satisfies the equation curl curl $A = 0$. Regarding the vector field E, the 1-form associated with it is harmonic, and so from the equation (2.8.1) we conclude that the potential difference between two points in an electrical field is given by the integral of the harmonic form η along 'any' path connecting the points. Moreover, the integral $\int_\Gamma A \cdot dr$ of the vector potential in the magnetic field round a bounding cycle Γ is equal to the integral of the 2-form β over 'any' 2-chain C with $\Gamma = \partial C$, that is,

$$\int_{\partial C} \alpha = \int_C d\alpha = -\int_C \beta. \tag{2.8.8}$$

In § 2.10 we shall sketch a proof of the statement that *there are harmonic p-forms $(0 < p < n)$ on an n-dimensional Riemannian manifold M with the property that the integral*

$$\int \alpha$$

has arbitrarily prescribed periods on $b_p(M)$ independent p-cycles of M. This generalizes the above results for the forms η and β.

2.9. Orthogonality relations

We shall assume in the remaining sections of this chapter that the Riemannian manifold M is compact and orientable. Let α and β be forms of degree p and $p + 1$, respectively. Then, by Stokes' theorem

$$\int_M d(\alpha \wedge *\beta) = 0, \tag{2.9.1}$$

from which

$$\int_M d\alpha \wedge *\beta = (-1)^{p-1} \int_M \alpha \wedge d*\beta. \tag{2.9.2}$$

By (2.8.7), this may also be written as

$$(d\alpha, \beta) = (\alpha, \delta\beta). \tag{2.9.3}$$

Two linear operators A and A' are said to be *dual* if $(A\alpha, \beta) = (\alpha, A'\beta)$ for every pair of forms α and β for which both sides of the relation are defined. Thus, the operators d and δ are dual.

In the same way, we see that, if β is of degree $p - 1$, then

$$(\alpha, d\beta) = (\delta\alpha, \beta). \tag{2.9.4}$$

Hence, *in order that α be closed, it is necessary and sufficient that it be orthogonal to all co-exact forms of degree p.*

The condition is indeed necessary; for, if $d\alpha = 0$, then $(\alpha, \delta\beta) = 0$ for any $(p+1)$-form β. Suppose that α is orthogonal to all co-exact forms of degree p. Then, $(\alpha, \delta d\alpha) = 0$, and so $(d\alpha, d\alpha) = 0$. Hence, from property (i), p. 71, it follows that $d\alpha = 0$.

In order that a form be co-closed, it is necessary and sufficient that it be orthogonal to all exact forms. It follows that if α and β are two p-forms, α being exact and β co-exact, then $(\alpha, \beta) = 0$.

We now show that *in a compact Riemannian manifold the definitions of a harmonic form given by Hodge and Kodaira are equivalent.* Assume that α is a harmonic form in the sense of Kodaira. Then,

$$0 = (\Delta\alpha, \alpha) = (d\delta\alpha, \alpha) + (\delta d\alpha, \alpha) = (d\alpha, d\alpha) + (\delta\alpha, \delta\alpha).$$

Hence, since $(d\alpha, d\alpha) \geqq 0$ and $(\delta\alpha, \delta\alpha) \geqq 0$, it follows that $d\alpha = 0$ and $\delta\alpha = 0$. The converse is trivial.

In particular, *a harmonic function in a compact Riemannian manifold is necessarily a constant.*

We have seen that a harmonic form on a compact manifold is closed. This statement is false if the manifold is not compact. For, a closed form of degree 0 is a constant while in E^n there certainly exist non-constant harmonic functions.

The differential forms of degree p form a linear space $\wedge^p(T^*)$ over R. Denote by $\wedge^p_d(T^*)$, $\wedge^p_\delta(T^*)$ and $\wedge^p_H(T^*)$ the subspaces of $\wedge^p(T^*)$ consisting of those forms which are exact, co-exact and harmonic, respectively. Evidently, these subspaces are orthogonal in pairs, that is forms belonging to distinct subspaces are orthogonal. A p-form orthogonal to the three subspaces is necessarily zero (cf. § 2.10). In other words, the subspaces $\wedge^p_d(T^*)$, $\wedge^p_\delta(T^*)$ and $\wedge^p_H(T^*)$ form a *complete system* in $\wedge^p(T^*)$. (We have previously written $\wedge^p_e(T^*)$ for $\wedge^p_d(T^*)$).

2.10. Decomposition theorem for compact Riemannian manifolds

Let β be a p-form on a compact, orientable Riemannian manifold M. If there is a p-form α such that $\Delta\alpha = \beta$, then, for a harmonic form γ,

$$(\beta, \gamma) = (\Delta\alpha, \gamma) = (\alpha, \Delta\gamma) = 0.$$

Therefore, in order that there exist a form α (of class 2) with the property that $\Delta\alpha = \beta$, it is necessary that β be orthogonal to the subspace $\wedge^p_H(T^*)$. This condition is also sufficient, the proof being given in Appendix C. The original proof given by Hodge in [*39*] depends largely on the Fredholm theory of integral equations.

The dimension of $\wedge^p_H(T^*)$ being finite (cf. Appendix C) we can find an orthonormal basis $\{\varphi_1, \ldots, \varphi_h\}$ for the harmonic forms of degree p:

$$(\varphi_i, \varphi_j) = \delta^i_j.$$

Any other harmonic p-form may then be expressed as a linear combination of these basis forms. Let α be any p-form. The form

$$\alpha_H = \sum_{i=1}^{h} (\alpha, \varphi_i)\varphi_i$$

is harmonic and $\alpha - \alpha_H$ is orthogonal to $\wedge^p_H(T^*)$. In fact,

$$\begin{aligned}(\alpha - \alpha_H, \varphi_j) &= (\alpha, \varphi_j) - (\alpha_H, \varphi_j)\\ &= (\alpha, \varphi_j) - \Big(\sum_{i=1}^{h} (\alpha, \varphi_i)\varphi_i, \varphi_j\Big)\\ &= (\alpha, \varphi_j) - \sum_{i=1}^{h} (\alpha, \varphi_i)\,(\varphi_i, \varphi_j) = 0.\end{aligned}$$

It follows that there exists a form γ such that $\Delta\gamma = \alpha - \alpha_H$. If we set $\alpha_d = d\delta\gamma$ and $\alpha_\delta = \delta d\gamma$, we obtain $\alpha_d + \alpha_\delta = \alpha - \alpha_H$, that is

$$\alpha = \alpha_d + \alpha_\delta + \alpha_H$$

where $\alpha_d \in \wedge^p_d(T^*)$, $\alpha_\delta \in \wedge^p_\delta(T^*)$ and $\alpha_H \in \wedge^p_H(T^*)$. That this decomposition is unique may be seen as follows: Let $\alpha = \alpha'_d + \alpha'_\delta + \alpha'_H$ where $\alpha'_d \in \wedge^p_d(T^*)$, $\alpha'_\delta \in \wedge^p_\delta(T^*)$ and $\alpha'_H \in \wedge^p_H(T^*)$ be another decomposition of α.

Then,

$$(\alpha_d - \alpha'_d) + (\alpha_\delta - \alpha'_\delta) + (\alpha_H - \alpha'_H) = 0$$

and therefore, by the completeness of the system of subspaces $\wedge^p_d(T^*)$, $\wedge^p_\delta(T^*)$ and $\wedge^p_H(T^*)$ in $\wedge^p(T^*)$, $\alpha'_d = \alpha_d$, $\alpha'_\delta = \alpha_\delta$, $\alpha'_H = \alpha_H$. We have proved:

A regular form α of degree p may be uniquely decomposed into the sum

$$\alpha = \alpha_d + \alpha_\delta + \alpha_H$$

where $\alpha_d \in \wedge^p_d(T^*)$, $\alpha_\delta \in \wedge^p_\delta(T^*)$ and $\alpha_H \in \wedge^p_H(T^*)$.

This is the *Hodge-de Rham decomposition theorem* [39].

2.11. Fundamental theorem

At this stage it is appropriate to state the *existence theorems of de Rham* [65]—the proofs of which appear in Appendix A.

(R_1) *Let $\{\Gamma^i_p\}$ $(i = 1, \cdots, b_p(M))$ be a base for the (rational) p-cycles of a compact differentiable manifold M and ω^i_p $(i = 1, \cdots, b_p(M))$ be b_p arbitrary real constants. Then, there exists a regular, closed p-form α on M having the ω^i_p as periods, that is*

$$\int_{\Gamma^i_p} \alpha = \omega^i_p \quad (i = 1, ..., b_p).$$

(R_2) *A closed form having zero periods is an exact form.*

We now establish the existence theorem due to Hodge which is at the very foundation of the subject matter of curvature and homology.

There exists a unique harmonic form α of degree p having arbitrarily assigned periods on b_p independent p-cycles of a compact and orientable Riemannian manifold.

Indeed, let α be a closed p-form having the given periods. The existence of α is assured by the first of de Rham's theorems. By the decomposition theorem $\alpha = \alpha_d + \alpha_H$. (Since α is closed, α_δ is zero and consequently α is orthogonal to $\wedge^p_\delta(T^*)$). Since $\alpha_d \in \wedge^p_d(T^*)$ its periods are zero. Hence the periods of α_H are those of α. The uniqueness follows from (R_2) since a harmonic form whose periods vanish is the zero form.

Let M be a compact and orientable Riemannian manifold. Then, the number of linearly independent real harmonic forms of degree p is equal to the pth betti number of M.

For, let φ_i denote the harmonic p-form whose periods are zero except for the ith which is equal to 1, that is, if $\{\Gamma^i_p\}$ ($i = 1, \cdots, b_p(M)$) is a base for the rational p-cycles of M, then

$$\int_{\Gamma^j_p} \varphi_i = \delta^j_i \quad (i, j = 1, ..., b_p).$$

The existence of the φ_i is assured by the above theorem. The φ_i ($i = 1, ..., b_p$) clearly form a basis for the harmonic forms of degree p and the fundamental theorem is proved.

Although not explicitly mentioned it should be emphasized that the existence theorems of de Rham are valid only for orientable manifolds.

The theorem (R_2) may be deduced from (R_1) and the decomposition theorem of § 2.10.

2.12. Explicit expressions for d, δ, and Δ

In the sequel, unless written otherwise, a p-form α will have the following equivalent representations:

$$\alpha = \frac{1}{p!} a_{i_1 \ldots i_p} du^{i_1} \wedge ... \wedge du^{i_p} = a_{(i_1 \ldots i_p)} du^{i_1} \wedge ... \wedge du^{i_p}$$

in the local coordinates $u^1, \cdots, u^n$. We proceed to obtain formulae for the operators d, δ, and Δ in a Riemannian manifold—the details of the computations being left as an exercise. In the first place,

$$D_i \, a_{i_1 \ldots i_p} = \frac{\partial a_{i_1 \ldots i_p}}{\partial u^i} - \sum_{\rho=1}^{p} a_{i_1 \ldots i_{\rho-1} \, j i_{\rho+1} \ldots i_p} \left\{ {j \atop i_\rho \; i} \right\}. \tag{2.12.1}$$

If we write (cf. (1.4.11))

$$d\alpha = (da)_{(i_1 \ldots i_{p+1})} du^{i_1} \wedge ... \wedge du^{i_{p+1}}$$

then

$$\begin{aligned}(da)_{i_1 \ldots i_{p+1}} &= \delta^{j (j_1 \ldots j_p)}_{i_1 \ldots i_{p+1}} D_j \, a_{(j_1 \ldots j_p)} \\ &= \delta^{j (j_1 \ldots j_p)}_{i_1 \ldots i_{p+1}} \frac{\partial a_{(j_1 \ldots j_p)}}{\partial u^j}\end{aligned} \tag{2.12.2}$$

and

$$\delta\alpha = (\delta a)_{(i_1 \ldots i_{p-1})} du^{i_1} \wedge ... \wedge du^{i_{p-1}}$$

where

$$(\delta a)_{i_1 \ldots i_{p-1}} = (-1)^{np+n+1}(*d*a)_{i_1 \ldots i_{p-1}} \tag{2.12.3}$$

$$= -g^{ij}\,\delta^{(j_1 \ldots j_p)}_{ii_1 \ldots i_{p-1}} D_j a_{(j_1 \ldots j_p)}.$$

Then, the Laplace-Beltrami operator

$$\Delta = d\delta + \delta d$$

is given by

$$(\Delta a)_{i_1 \ldots i_p} = -g^{ij} D_j D_i a_{i_1 \ldots i_p} + \sum_{\rho=1}^{p} a_{i_1 \ldots i_{\rho-1} j i_{\rho+1} \ldots i_p} R^j{}_{i_\rho} \tag{2.12.4}$$

$$+ \tfrac{1}{2} \sum_{\sigma=1}^{p} \sum_{\rho=1}^{p} a_{i_1 \ldots i_{\rho-1} j i_{\rho+1} \ldots i_{\sigma-1} i i_{\sigma+1} \ldots i_p} R^{ij}{}_{i_\rho i_\sigma}$$

where

$$R^{ij}{}_{kl} = g^{jm} R^i{}_{mkl}.$$

In an Euclidean space, the curvature tensor vanishes, and so if the $u^1, \cdots, u^n$ are rectangular coordinates, $g^{ij} = \delta^j_i$ and

$$(\Delta a)_{i_1 \ldots i_p} = -\sum_{i=1}^{n} \frac{\partial^2 a_{i_1 \cdots i_p}}{\partial u^{i^2}}.$$

On the other hand, in a Riemannian manifold M, if we apply Δ to a function f defined over M, we obtain Beltrami's differential operator of the second kind:

$$\Delta f = -g^{ij} D_j D_i f$$

(cf. formula (2.7.3)). The operator Δ is therefore the usual Laplacian.

EXERCISES

A. The star operator

The following seven exercises give rise to an alternate definition of the Hodge star operator.

1. Let V be an n-dimensional vector space over R with an inner product φ: $V \times V \to R$. If $\alpha = v_1 \wedge \ldots \wedge v_p$ and $\beta = w_1 \wedge \ldots \wedge w_p$ are two decomposable p-vectors, let $\langle \alpha, \beta \rangle = \det(\varphi(v_i, w_j))$. Prove that this pairing defines an inner product on $\wedge^p(V)$.

2. Let M be an n-dimensional Riemannian manifold with metric tensor g. In terms of a system of local coordinates (u^i), let $\alpha = a_{(i_1 \ldots i_p)} du^{i_1} \wedge \ldots \wedge du^{i_p}$ and $\beta = b_{(i_1 \ldots i_p)} du^{i_1} \wedge \ldots \wedge du^{i_p}$ be two (anti-symmetrized) p-forms in $\wedge^p(V_P^*)$, P being in the given coordinate neighborhood. Show that

$$\langle \alpha, \beta \rangle = a_{(i_1 \ldots i_p)} b_{(j_1 \ldots j_p)} g^{i_1 j_1}(P) \ldots g^{i_p j_p}(P)$$

where the inner product φ is defined by g.

3. Let $V^p = V \otimes \ldots \otimes V$ (p times) and define $A^p : V^p \to V^p$ by

$$A^p(v_1 \otimes \ldots \otimes v_p) = \frac{1}{p!} \sum_\sigma \operatorname{sgn}(\sigma)\, v_{\sigma(1)} \otimes \ldots \otimes v_{\sigma(p)},$$

the summation being taken over all permutations of the set $(1, \cdots, p)$. Define the map

$$\eta : \wedge^p(V) \to A^p(V^p)$$

by

$$\eta(v_1 \wedge \ldots \wedge v_p) = A^p(v_1 \otimes \ldots \otimes v_p);$$

η is an isomorphism. Furthermore, if we extend φ to an inner product on V^p by

$$\langle v_1 \otimes \ldots \otimes v_p, w_1 \otimes \ldots \otimes w_p \rangle = \langle v_1, w_1 \rangle \ldots \langle v_p, w_p \rangle,$$

then $p! \langle \alpha, \beta \rangle = \langle \eta(\alpha), \eta(\beta) \rangle$.

We have used the notation $\varphi(v,w) = \langle v,w \rangle$, $v,w \in V$. (The correspondence between $w \in V$ and $w^* \in V^*$ given by the condition

$$\langle v, w^* \rangle = \varphi(v,w) \quad \forall\, v \in V$$

defines an isomorphism between V and V^*.)

4. Show that $(\wedge^p(V))^* \cong \wedge^p(V^*)$ under the pairing

$$\langle v_1 \wedge \ldots \wedge v_p, w_1^* \wedge \ldots \wedge w_p^* \rangle = \det(\langle v_i, w_j^* \rangle).$$

5. If the manifold M is oriented, there is a unique n-form e^* in $\wedge^n(V_P^*)$, $P \in M$ such that $\langle e^*, e^* \rangle = 1$ where e^* is positive with respect to the orientation. (Note that the metric tensor g defines an inner product on V_P^*).

6. Define $\lambda : \wedge^p(V_P) \to \wedge^{n-p}(V_P^*)$ by

$$\langle v_1 \wedge \ldots \wedge v_{n-p}, \lambda(w_1 \wedge \ldots \wedge w_p) \rangle = \langle w_1 \wedge \ldots \wedge w_p \wedge v_1 \wedge \ldots \wedge v_{n-p}, e^* \rangle$$

and let $\alpha = a^{(i_1 \ldots i_p)} (\partial/\partial u^{i_1}) \wedge \ldots \wedge (\partial/\partial u^{i_p})$ be an element of $\wedge^p(V_P)$, where the coefficients are anti-symmetrized. Then,

$$\lambda(\alpha) = b_{(j_1 \ldots j_{n-p})} du^{j_1} \wedge \ldots \wedge du^{j_{n-p}},$$

where

$$b_{j_1 \ldots j_{n-p}} = \sqrt{G}\, a^{(i_1 \ldots i_p)} \delta^{1 \ldots n}_{i_1 \ldots i_p\, j_1 \ldots j_{n-p}},$$

and $G = \det(g_{jk})$.

7. Define the map $\mu: \wedge^p(V^*) \to \wedge^{n-p}(V^*)$ by $\mu = \lambda \cdot \gamma$, where $\gamma: \wedge^p(V^*) \to \wedge^p(V)$ is the natural identification map determined by the inner product in $\wedge^p(V^*)$. Then, μ is the star operation of Hodge.

8. Let V be a vector space (over R) with the properties:

(i) V is the direct sum of subspaces V^p where p runs through non-negative integers and

(ii) V has a *coboundary operator* that is an endomorphism d of V such that $d_p V^p \subset V^{p+1}$ with $d_{p+1} d_p = 0$ where d_p denotes the restriction of d to V^p. The vector space

$$H^p(V) = \frac{\text{kernel } d_p}{\text{image } d_{p-1}}$$

is called the pth *cohomology vector space* (*or group*) of V. A theory based on V together with the operator d is usually called a *cohomology theory* or *d-cohomology theory* when emphasis on the coboundary operator is required. We have seen that the Grassman algebra $\wedge(T^*)$ with the exterior differential operator d gives rise to the de Rham cohomology theory. On the other hand, a cohomology theory is defined by the pair $(\wedge(T^*), \delta)$ on a Riemannian manifold by setting $\wedge^{-p} = \wedge^p$, $p = 0,1,2, \cdots$. Prove that the $*$ operator induces an isomorphism between the two cohomology theories.

B. The operators H and G on a compact manifold

1. Show that for any $\alpha \in \wedge^p(T^*)$ there exists a unique p-form $H[\alpha]$ in $\wedge^p_H(T^*)$ with the property $(\alpha,\beta) = (H[\alpha],\beta)$ for all $\beta \in \wedge^p_H(T^*)$.

2. Prove that $H[H[\alpha]] = H[\alpha]$ for any p-form α.

3. For a given p-form α there exists a p-form β satisfying the differential equation $\Delta\beta = \alpha - H[\alpha]$. Show that any two solutions differ by a harmonic p-form and thereby establish the existence of a unique solution orthogonal to $\wedge^p_H(T^*)$. Denote this solution by $G\alpha$ and show that it is characterized by the conditions

$$\alpha = \Delta G\alpha + H[\alpha] \quad \text{and} \quad (G\alpha,\beta) = 0$$

for any $\beta \in \wedge^p_H(T^*)$.

The operator G is called the *Green's operator*.

4. Prove that $H[G\alpha]$ vanishes for any p-form α.

5. Prove:

(a) The operators H and G commute with d, δ, Δ and $*$;

(b) G is *self-dual*, that is

$$(G\alpha,\beta) = (\alpha,G\beta)$$

for any α, β of degree p;

(c) G is *hermitian positive*, that is

$$(G\alpha,\alpha) \geqq 0,$$

equality holding, if and only if, α is harmonic.

C. The second existence theorem of de Rham

1. Establish the theorem (R_2) of § 2.11 from the decomposition theorem of § 2.10.

CHAPTER III

CURVATURE AND HOMOLOGY OF RIEMANNIAN MANIFOLDS

The explicit expression in terms of local coordinates of the Laplace-Beltrami operator Δ (cf. § 2.12) involves the Riemannian curvature tensor in an essential way. It is natural to expect then that the curvature properties of a Riemannian manifold M will affect its homology structure provided we assume that M is compact and orientable. It will be seen that the existence or rather non-existence of harmonic forms of degree p depends largely on the signature of a certain quadratic form defined in terms of the curvature tensor. Hence, by Hodge's theorem (cf. § 2.11), if there are no harmonic p-forms, the pth betti number of the manifold vanishes.

3.1. Some contributions of S. Bochner

If $\tilde{M}$ is a covering manifold of M which is also compact

$$b_p(M) \leqq b_p(\tilde{M}), \quad 0 < p < n \tag{3.1.1}$$

where $n = \dim M$.

This may be seen as follows: If α is a p-form defined on M, then it has a periodic extension $\tilde{\alpha}$ onto $\tilde{M}$, that is $\tilde{\alpha}(\gamma \tilde{P}) = \alpha(P)$ for each element γ in the fundamental group of M and each point $P \in M$ where $\tilde{P} \in \tilde{M}$ lies over P. More simply, if $\pi : \tilde{M} \to M$ is the projection map, then, $\tilde{\alpha} = \pi^*(\alpha)$. Moreover, non-homologous p-forms on M have non-homologous periodic extensions.

Suppose that M is a manifold of positive constant curvature. Then, it can be shown that its universal covering space $\tilde{M}$ is the ordinary sphere. Hence $b_p(\tilde{M})$ vanishes for all p $(0 < p < n)$ and consequently from (3.1.1), $b_p(M) = 0$ $(0 < p < n)$. These spaces are of interest since they provide a source of examples of topological manifolds. They are perhaps the simplest and geometrically the most important Riemannian manifolds. However, constancy of curvature is a very specialized requirement. If, on the contrary, the sectional curvatures are not equal but rather vary within certain definite limits, that is, if the manifold is δ-pinched, the betti numbers of the sphere are retained [*1*]. On the other hand, one of the many applications of the theory of harmonic integrals to global differential geometry made by S. Bochner is to describe families of Riemannian manifolds which from a topological standpoint are homology spheres. For example, a Riemannian manifold of constant curvature is conformally flat (cf. § 3.9). However, the converse is not true. In any case, the betti numbers b_p $(0 < p < n)$ of a conformally flat, compact, orientable Riemannian manifold vanish provided the Ricci curvature is positive definite, that is, the manifold is a homology sphere [*6*, *51*]. In fact, the same conclusion holds even for deviations from conformal flatness provided the deviation is but a fraction of the Ricci (scalar) curvature [*6*, *74*].

In the sequel, by a *homology sphere* we shall mean a homology sphere over the real numbers.

We recall that on a Riemann surface the harmonic differentials are invariant under conformal changes of coordinates. Consider the Riemann surface S of the algebraic function defined by the algebraic equation

$$R(z, w) = 0.$$

The surface is closed and orientable and the (local) geometry is conformal geometry. In fact, in the neighborhood of a 'place' P on S for which $z = a$ let (u, v) be the local coordinates. Then,

$$z - a = (u + iv)^m$$

if the place is the origin of a branch of order m. If z is infinite at the place, $z - a$ is replaced by z^{-1}. Any other local coordinate system $(\bar{u}, \bar{v})$ at P must have the property that $\bar{u} + i\bar{v}$ is a holomorphic function of the complex variable $u + iv$ which is simple in the neighborhood of the place. The local coordinates (u, v) and $(\bar{u}, \bar{v})$ at P are therefore related by analytic functions

$$\bar{u} = \bar{u}(u,v),\ \bar{v} = \bar{v}(u,v),$$

that is as functions of u and v, $\bar{u}$ and $\bar{v}$ satisfy the Cauchy-Riemann equations. We conclude that

$$d\bar{u}^2 + d\bar{v}^2 = \rho^2(du^2 + dv^2)$$

for some (real) analytic function ρ. In this way, a geometry is defined on S in which distance plays no role but angle may be defined, that is angle is invariant under a conformal change of coordinates. After performing a birational transformation of the equation $R(z, w) = 0$ a new algebraic equation is obtained. The Riemann surface S' of the algebraic function thus obtained is homeomorphic to S. Let $f: S \to S'$ denote the homeomorphism and (u, v), (u', v') the local coordinates at $P \in S$ and $P' = f(P) \in S'$, respectively. The functions

$$u' = u'(u,v), \quad v' = v'(u,v)$$

are then analytic, that is f is a holomorphic homeomorphism. It follows that

$$du'^2 + dv'^2 = \sigma^2(du^2 + dv^2)$$

where σ is an analytic function of u and v, that is the homeomorphism is a conformal map of S onto S'.

Conversely, functions whose Riemann surfaces are conformally homeomorphic are birationally equivalent. Their Riemann surfaces are then said to be *equivalent*.

A 2-dimensional Riemannian manifold and a Riemann surface are both topological 2-manifolds. As differentiable manifolds however, they differ in their differentiable structures—the former allowing systems of local parameters related by functions with non-vanishing Jacobian whereas in the latter case only those systems of local parameters which are conformally related are permissible. Clearly then, they differ in their local geometries—the former being Riemannian geometry whereas the latter is conformal geometry. To construct a Riemann surface from a given 2-dimensional Riemannian manifold M we need only restrict the systems of local coordinates so that in the overlap of two coordinate neighborhoods the coordinates are related by analytic functions defining a conformal transformation. That such a covering of M exists follows from the possibility of introducing isothermal parameters on M. The manifold is then said to possess a complex (analytic) structure. We conclude that *conformally homeomorphic 2-dimensional Riemannian manifolds define equivalent Riemann surfaces.* The concept of a complex structure on an $n(= 2m)$-dimensional topological manifold will be discussed in Chapter V.

Two n-dimensional Riemannian manifolds M and M' of class k are said to be *isometric* if there is a differentiable homeomorphism f (of class k) from M onto M' which maps one element of arc into the other. It can be shown that a simply connected, complete Riemannian manifold of constant curvature K is isometric with either Euclidean space ($K = 0$), hyperbolic space ($K < 0$), or spherical space ($K > 0$). Hence, the universal covering manifold of a complete Riemannian manifold of constant curvature K is Euclidean space ($K = 0$), hyperbolic space ($K < 0$), or spherical space ($K > 0$).

Suppose M and M' are not isometric but rather that the map f defines a homeomorphism which reproduces the metric except for a scalar factor. We then say that M and M' are *conformally homeomorphic.*

A Riemannian manifold of constant curvature is called a *space form.* The problem of determining the space forms becomes by virtue of the above remarks a problem in the determination of (discontinuous) groups of motions. A space form may then be regarded as a homogeneous space G/H where G is the group of motions and H the isotropy subgroup leaving a point fixed. It is therefore not surprising that *the curvature properties of a compact Riemannian manifold determine to some extent the structure of its group of motions.* In fact, it is shown that the existence or rather non-existence of 1-parameter groups of motions as well as 1-parameter groups of conformal transformations is dependent upon the Ricci curvature of the manifold [*4*]. On the other hand, the existence of a globally defined 1-parameter group of non-isometric conformal transformations of a compact homogeneous Riemannian manifold is a sufficient condition for it to be a homology sphere. Indeed, it is then isometric with a sphere [*79*].

3.2. Curvature and betti numbers

At this point, it is convenient to employ the symbol denoting a form in the coefficients of the form as well.

Let α be a harmonic 1-form of class 2 defined on a compact, orientable Riemannian manifold M and consider the integral

$$(\Delta\alpha, \alpha) = \int_M \Delta\alpha \wedge *\alpha \tag{3.2.1}$$

over M. Since α is a harmonic form, $\Delta\alpha$ vanishes, and so

$$\int_M \Delta\alpha \wedge *\alpha = 0. \tag{3.2.2}$$

The expression of the integrand in local coordinates is given by

$$\begin{aligned}\Delta\alpha \wedge *\alpha &= (-g^{jk} D_k D_j \alpha_i + R^j_i \alpha_j)\, \epsilon_{sj_1\cdots j_{n-1}} \alpha^s du^i \wedge du^{j_1} \wedge \ldots \wedge du^{j_{n-1}} \\ &= (-\alpha^i g^{jk} D_k D_j \alpha_i + R_{ij} \alpha^i \alpha^j) *1 \end{aligned} \tag{3.2.3}$$

by virtue of the formulae (2.7.8), (2.7.9), and (2.12.4).

Lemma 3.2.1. *For a regular 1-form α on a compact and orientable Riemannian manifold M*

$$\int_M \delta\alpha *1 \equiv \int_M \delta\alpha \wedge *1 = 0. \tag{3.2.4}$$

For,

$$\begin{aligned}\int_M \delta\alpha \wedge *1 &= (\delta\alpha, 1)\\ &= (\alpha, d1)\\ &= (\alpha, 0) = 0.\end{aligned}$$

In the sequel, we employ the notation $\langle t, t'\rangle$ to mean the (*local*) *scalar product* of the tensors t and t' of type $(0, s)$ in case t and t' are simultaneously symmetric or skew-symmetric, that is

$$\langle t,t'\rangle = t_{(i_1\ldots i_s)}\, t'^{(i_1\ldots i_s)}$$

If t and t' are skew-symmetric tensors, $\langle t, t'\rangle = \langle \alpha, \alpha'\rangle$ where α and α' denote the corresponding s-forms (cf. II.A.2). From (2.7.11)

$$(\alpha, \alpha') = \int_M \langle \alpha, \alpha'\rangle *1.$$

Now, consider the integral

$$\int_M g^{jk} D_k D_j(\alpha^i \alpha_i) *1$$

whose value is zero by (3.2.4). Indeed, if we put $\beta = d(\alpha^i\alpha_i)$,

$$\int_M g^{jk} D_j D_k(\alpha^i \alpha_i) *1 = \int_M g^{jk} D_k \beta_j *1 = -\int_M \delta\beta *1 = 0.$$

Then,

$$\begin{aligned}0 = \int_M g^{jk} D_k D_j(\alpha^i \alpha_i) *1 &= \int_M g^{jk} D_k(\alpha_i D_j \alpha^i + \alpha^i D_j \alpha_i) *1\\ &= \int_M g^{jk}(\alpha_i D_k D_j\alpha^i + D_j \alpha^i D_k \alpha_i + D_k \alpha^i D_j \alpha_i + \alpha^i D_k D_j \alpha_i) *1\\ &= 2\int_M (\alpha^i g^{jk} D_k D_j \alpha_i + D_j \alpha_i D^j \alpha^i) *1\end{aligned}$$

where we have put

$$D^j \equiv g^{jk} D_k.$$

Hence,

$$-\int_M \alpha^i g^{jk} D_k D_j \alpha_i *1 = \int_M D_j \alpha_i D^j \alpha^i *1,$$

and so if α is a harmonic 1-form (3.2.2) becomes

$$\int_M (R_{ij} \alpha^i \alpha^j + D_j \alpha_i D^j \alpha^i) *1 = 0. \tag{3.2.5}$$

Denote by Q the operator on 1-forms defined by

$$(Q\alpha)_i = R^j{}_i\alpha_j$$

and assume that the quadratic form

$$\langle Q\alpha,\alpha\rangle \tag{3.2.6}$$

is positive definite. Since the second term in the integrand of (3.2.5) is non-negative we conclude that

$$\langle Q\alpha,\alpha\rangle = 0$$

from which $\alpha = 0$. Since α is an arbitrary harmonic 1-form we have proved

Theorem 3.2.1. *The first betti number of a compact and orientable Riemannian manifold of positive definite Ricci curvature is zero* [4, 62].

If we assume only that $\langle Q\alpha, \alpha\rangle$ is non-negative, then from (3.2.5) $\langle Q\alpha, \alpha\rangle$ as well as $D^j\alpha^i D_j\alpha_i$ must vanish. It follows that $D^j\alpha^i$ vanishes, that is the tangent vectors

$$A(t) = \alpha^i(t)\frac{\partial}{\partial u^i}$$

are parallel along any parametrized curve $u^i = u^i(t)$, $i = 1, \cdots, n$. A vector field with this property is called a *parallel vector field.*

Theorem 3.2.2. *In a compact and orientable Riemannian manifold a harmonic vector field for which the quadratic form* (3.2.6) *is positive semi-definite is necessarily a parallel vector field* [4].

Theorem 3.2.3. *In a coordinate neighborhood of a compact and orientable Riemannian manifold with the local coordinates $u^1, \cdots, u^n$, a necessary and sufficient condition that the 1-form $\alpha = \alpha_i du^i$ be a harmonic form is given by*

$$R^j{}_i\alpha_j - g^{jk} D_k D_j \alpha_i = 0 \tag{3.2.7}$$

[73].

Clearly, if α is harmonic, (3.2.7) holds. Conversely, if the 1-form α is a solution of equation (3.2.7) then, by (3.2.3), $\Delta\alpha \wedge *\alpha = 0$. Hence,

$$0 = (\Delta\alpha, \alpha) = (d\alpha, d\alpha) + (\delta\alpha, \delta\alpha),$$

from which $d\alpha = 0$ and $\delta\alpha = 0$.

We now seek a result analogous to theorem 3.2.1 for b_p $(0 < p < n)$. To this end, let $\alpha = (1/p!)\alpha_{i_1\ldots i_p}\, du^{i_1} \wedge \cdots \wedge du^{i_p}$ be a harmonic form of degree p. Then, again

$$0 = (\Delta\alpha, \alpha) = \int_M \Delta\alpha \wedge *\alpha,$$

and so from (2.12.4) and (2.7.11) we obtain the integral formula

$$\int_M \Big(- g^{jk} D_k D_j \alpha_{i_1\ldots i_p} \alpha^{i_1\ldots i_p} + pR_{ij}\alpha^{ii_2\ldots i_p} \alpha^j{}_{i_2\ldots i_p} + \frac{p(p-1)}{2} R_{ijkl}\, \alpha^{iji_3\ldots i_p} \alpha^{kl}{}_{i_3\ldots i_p}\Big) *1 = 0. \tag{3.2.8}$$

Now,

$$\begin{aligned}
0 &= \int_M g^{jk} D_k D_j (\alpha^{i_1\ldots i_p} \alpha_{i_1\ldots i_p}) *1 \\
&= \int_M g^{jk} D_k(D_j \alpha^{i_1\ldots i_p} \alpha_{i_1\ldots i_p} + \alpha^{i_1\ldots i_p} D_j\alpha_{i_1\ldots i_p}) *1 \\
&= \int_M g^{jk}(D_k D_j \alpha^{i_1\ldots i_p} \alpha_{i_1\ldots i_p} + D_j \alpha^{i_1\ldots i_p} D_k \alpha_{i_1\ldots i_p} \\
&\qquad + D_k D_j \alpha_{i_1\ldots i_p} \alpha^{i_1\ldots i_p} + D_k \alpha^{i_1\ldots i_p} D_j \alpha_{i_1\ldots i_p}) *1 \\
&= 2\int_M (g^{jk}D_kD_j\alpha_{i_1\ldots i_p}\alpha^{i_1\ldots i_p} + D_j\alpha_{i_1\ldots i_p}D^j\alpha^{i_1\ldots i_p}) *1.
\end{aligned}$$

It follows that

$$\int_M (pR_{ij}\alpha^{ii_2\ldots i_p}\alpha^j{}_{i_2\ldots i_p} + \frac{p(p-1)}{2} R_{ijkl}\alpha^{iji_3\ldots i_p}\alpha^{kl}{}_{i_3\ldots i_p} + D_j\alpha_{i_1\ldots i_p}D^j\alpha^{i_1\ldots i_p}) *1 = 0. \tag{3.2.9}$$

Setting

$$F(\alpha) = R_{ij}\alpha^{ii_2\ldots i_p}\alpha^j{}_{i_2\ldots i_p} + \frac{p-1}{2} R_{ijkl}\alpha^{iji_3\ldots i_p}\alpha^{kl}{}_{i_3\ldots i_p} \tag{3.2.10}$$

we obtain

Theorem 3.2.4. *If on a compact and orientable Riemannian manifold M the quadratic form $F(\alpha)$ is positive definite,*

$$b_p(M) = 0, \qquad 0 < p < n$$

[6, 51, 74].

Corollary. *The betti numbers b_p $(0 < p < n)$ of a compact and orientable Riemannian manifold M of positive constant curvature vanish, that is M is a homology sphere.*

Indeed, since the sectional curvatures $R(P, \pi)$ are constant for all two-dimensional sections π at all points P of M the Riemannian curvature tensor is given by

$$R_{ijkl} = K(g_{jk}g_{il} - g_{jl}g_{ik}) \tag{3.2.11}$$

where $K =$ const. is the common sectional curvature. Substituting (3.2.11) into (3.2.10) we obtain

$$F(\alpha) = (n-1)Kg_{ij}\, \alpha^{ii_2\cdots i_p}\, \alpha^j{}_{i_2\cdots i_p} + \frac{p-1}{2} K(g_{jk}\, g_{il} - g_{jl}g_{ik})\, \alpha^{iji_3\cdots i_p}\, \alpha^{kl}{}_{i_3\cdots i_p}$$

$$= p!\,(n-1)\, K\langle \alpha, \alpha\rangle + \frac{p-1}{2} K\alpha^{iji_3\cdots i_p}\, (\alpha_{jii_3\cdots i_p} - \alpha_{iji_3\cdots i_p})$$

$$= p!\,(n-1)\, K\langle \alpha, \alpha\rangle - p!\,(p-1)\, K\langle \alpha, \alpha\rangle = p!\,(n-p)\, K\langle \alpha, \alpha\rangle.$$

Since $K > 0$ the result follows.

If $K = 0$ it follows from (3.2.9) that

$$D_j\, \alpha_{i_1\cdots i_p} = 0. \tag{3.2.12}$$

Since the manifold is locally flat there is a local coordinate system $u^1, \cdots, u^n$ relative to which the coefficients of affine connection $\{{}_j{}^i{}_k\}$ vanish. In these local coordinates (3.2.12) becomes

$$\frac{\partial \alpha_{i_1\cdots i_p}}{\partial u^j} = 0.$$

Thus, there are at most $\binom{n}{p}$ independent harmonic p-forms over M.

Theorem 3.2.5. *The pth betti number of a compact, orientable, locally flat Riemannian manifold is at most the binomial coefficient $\binom{n}{p}$.*

Corollary. *The pth betti number of an n-dimensional torus is $\binom{n}{p}$.*

An n-dimensional manifold M is said to be *completely parallelisable* if there exist n linearly independent differentiable vector fields at each point of M.

Corollary. *The torus is completely parallelisable.*

This follows from the fact that M is locally flat with respect to the metric canonically induced by E^n. For, the torus is the quotient space of E^n by a subgroup of translations and is therefore locally equivalent to ordinary affine space where there is no distinction made between vectors and covectors.

Consider the sectional curvature determined by the plane π defined by the orthonormal tangent vectors $X = \xi^i(\partial/\partial u^i)$ and $Y = \eta^i(\partial/\partial u^i)$ at P. Then,

$$R(P, \pi) = - R_{ijkl}(P)\xi^i(P)\eta^j(P)\xi^k(P)\eta^l(P). \tag{3.2.13}$$

Assume that for all planes π at all points P of M there are constants K_1 and K_2 such that

$$0 < K_1 \leqq R(P, \pi) \leqq K_2. \tag{3.2.14}$$

Let $\{X_1, \cdots, X_n\}$ be an orthonormal frame at P where $X_j = \xi^i_{(j)}(\partial/\partial u^i)$ $(j = 1, \cdots, n)$. Then, since

$$- R_{ijkl}\, \xi^i_{(r)}\, \xi^j_{(r)}\, \xi^k_{(r)}\, \xi^l_{(r)} = 0$$

and

$$K_1 \leqq - R_{ijkl}\, \xi^i_{(r)}\, \xi^j_{(s)}\, \xi^k_{(r)}\, \xi^l_{(s)} \leqq K_2, \quad r \neq s,$$

$r, s = 1, 2, \cdots, n$, it follows that

$$(n-1)\, K_1 \leqq R_{ik}\, \xi^i_{(r)}\, \xi^k_{(r)} \leqq (n-1)\, K_2,$$

the inequalities holding for arbitrary unit tangent vectors X_r. Hence, for any tangent vector $X = \xi^i(\partial/\partial u^i)$

$$(n-1)\, K_1\, \xi^i\, \xi_i \leqq R_{ik}\, \xi^i\, \xi^k \leqq (n-1)\, K_2\, \xi^i\, \xi_i. \tag{3.2.15}$$

It follows from § 1.2 (by taking tensor products) that

$$R_{ik}\, \xi^{ii_2\cdots i_p}\, \xi^k{}_{i_2\cdots i_p} \geqq (n-1)\, K_1\, \xi^{i_1\cdots i_p}\, \xi_{i_1\cdots i_p} \tag{3.2.16}$$

for any tensor whose components $\xi_{i_1\cdots i_p}$ are expressed in the given local coordinates. In terms of the bivector

$$\xi^{ij} = \xi^i\, \eta^j - \xi^j\, \eta^i,$$

where X and Y are orthonormal tangent vectors, the inequalities (3.2.14) become by virtue of (3.2.13)

$$0 < 2K_1 \leqq - R_{ijkl}\, \xi^{ij}\, \xi^{kl} \leqq 2K_2.$$

(The curvature tensor defines a symmetric linear transformation of the space of bivectors (cf. I.I.1). These inequalities say that it is positive definite with eigenvalues between $2K_1$ and $2K_2$). Unfortunately, however, we cannot conclude that for any two independent tangent vectors X and Y

$$0 < 2K_1 \leqq -\frac{R_{ijkl}\,\xi^{ij}\,\xi^{kl}}{\xi^{ij}\,\xi_{ij}} \leqq 2K_2.$$

Assuming that these inequalities are valid for any skew-symmetric tensor field or bivector ξ_{ij} we may conclude that

$$0 < 2K_1 \leqq -\frac{R_{ijkl}\,\xi^{iji_3\ldots i_p}\,\xi^{kl}{}_{i_3\ldots i_p}}{\xi^{i_1\ldots i_p}\,\xi_{i_1\ldots i_p}} \leqq 2K_2 \tag{3.2.17}$$

where $\xi_{i_1\ldots i_p}$ are the components of a tensor, skew-symmetric in its first two indices.

Now, let $\alpha = \alpha_{(i_1\ldots i_p)}du^{i_1} \wedge \ldots \wedge du^{i_p}$ be a harmonic form of degree p. Then, by the inequalities (3.2.16) and (3.2.17)

$$\begin{aligned} F(\alpha) &\geqq (n-1)\,K_1\,\alpha^{i_1\ldots i_p}\,\alpha_{i_1\ldots i_p} - (p-1)\,K_2\,\alpha^{i_1\ldots i_p}\,\alpha_{i_1\ldots i_p} \\ &= p!\,[(n-1)\,K_1 - (p-1)\,K_2]\,\langle\alpha,\alpha\rangle. \end{aligned}$$

The quadratic form $F(\alpha)$ is positive definite if we assume that $(n-1)K_1 > (p-1)K_2$, that is

$$\frac{K_1}{K_2} > \frac{p-1}{n-1}.$$

Since

$$\frac{p-1}{n-1} < \frac{1}{2}, \quad 0 < p \leqq \left[\frac{n}{2}\right],$$

$F(\alpha)$ is a positive definite quadratic form for $0 < p \leqq [\frac{n}{2}]$ provided $K_2 = 2K_1$.

Theorem 3.2.6. *If the curvature tensor of a compact and orientable Riemannian manifold M satisfies the inequalities*

$$0 < K_2 \leqq -\frac{R_{ijkl}\,\xi^{ij}\,\xi^{kl}}{\xi^{ij}\,\xi_{ij}} \leqq 2K_2$$

for any bivector ξ^{ij}, then $b_p(M) = 0$, $0 < p \leqq n-1$ [14].

The conclusion on the betti numbers $b_p(M)$ for $p > [\frac{n}{2}]$ follows by Poincaré duality.

An application of this theorem is given in (III.A.2).

A sharper result in terms of the sectional curvatures is now derived although only partial information on the betti numbers is obtained [1].

A Riemannian manifold with metric g is said to be δ-*pinched* if for any 2-dimensional section π

$$0 < \delta K_1 \leqq R(P, \pi) \leqq K_1.$$

For a suitable normalization of g, the above inequalities may be expressed as

$$0 < \delta \leqq R(P, \pi) \leqq 1.$$

We shall assume this normalization in the sequel.

Theorem 3.2.7. *The second betti number of a δ-pinched, n-dimensional compact and orientable Riemannian manifold vanishes if, either $n = 2m$ and $\delta > \frac{1}{4}$, or $n = 2m + 1$ and $\delta > 2(m - 1)/(8m - 5)$.*

The proof is based on theorem 3.2.4 (with $p = 2$) by obtaining suitable estimates for the various terms in (3.2.10).

Let $\{X_1, \cdots, X_n\}$ be an orthonormal frame in T_P and put

$$K(X_i, X_j) = R(P, \pi)$$

where π is the plane spanned by the vectors X_i and X_j ($i \neq j$). Then, by § 1.10

$$K(X_i, X_j) = - R_{ijij}, \quad i \neq j$$

or

$$K_{ij} = - R_{ijij}, \quad i \neq j.$$

From the inequalities

$$\delta \leqq K(X_i, aX_j + bX_k) \leqq 1$$

where a, b are any two real numbers, we may derive the inequalities

$$a^2(K_{ij} - \delta) - 2abR_{ijik} + b^2(K_{ik} - \delta) \geqq 0$$

and

$$a^2(1 - K_{ij}) + 2abR_{ijik} + b^2(1 - K_{ik}) \geqq 0.$$

Hence,

$$| R_{ijik} | \leqq [(K_{ij} - \delta)\,(K_{ik} - \delta)]^{1/2}$$

and

$$| R_{ijik} | \leqq [(1 - K_{ij})\,(1 - K_{ik})]^{1/2}$$

from which we deduce

$$| R_{ijik} | \leqq \frac{1}{2}(K_{ij} + K_{ik} - 2\delta) \qquad (3.2.18)$$

and

$$|R_{ijik}| \leqq \frac{1}{2}(2 - K_{ij} - K_{ik}). \qquad (3.2.19)$$

Thus,

$$|R_{ijik}| \leqq \frac{1}{2}(1 - \delta), \quad i \neq j, k.$$

In order to obtain estimates for the R_{ijkl} $((i, j) \neq (k, l), i < j, k < l)$ we consider the inequalities

$$\delta \leqq K(aX_i + bX_k, cX_l + dX_j)$$

for any orthonormal set of vectors $\{X_i, X_j, X_k, X_l\}$ and $a, b, c, d \in R$. Put

$$F(a,i; b,k; c,l; d,j) = K(aX_i + bX_k, cX_l + dX_j) - \delta.$$

The function F may be considered as a polynomial in a, b, c and d. As such it is of degree 4 but only of degree 2 in the a, b, c, d taken by themselves. The polynomial

$$G(a,i; b,k; c,l; d,j) = \tfrac{1}{2}[F(a,i; b,k; c,l; d,j) + F(a,i; -b,k; c,l; -d,j)]$$

contains only terms in a^2c^2, a^2d^2, b^2c^2, b^2d^2 and $abcd$. Now, put

$$H(a,i; b,k; c,l; d,j) = G(a,i; b,k; c,l; d,j) + G(-a,i; b,l; c,j; d,k).$$

By employing the identities (1.10.24) and (1.9.20) in the term involving $abcd$ then, by virtue of (3.2.18) and (3.2.19), the polynomial H may be expressed as

$$H = Aa^2c^2 + Ba^2d^2 + Cb^2c^2 + Db^2d^2 + 2Eabcd \geqq 0 \qquad (3.2.20)$$

where

$$A = K_{ij} + K_{il} - 2\delta, \quad B = K_{ij} + K_{ik} - 2\delta,$$

$$C = K_{lk} + K_{lj} - 2\delta, \quad D = K_{kj} + K_{kl} - 2\delta,$$

$$E = 3R_{ijkl}.$$

By a suitable choice for a and b the inequality (3.2.20) gives rise to

$$ACc^4 + (AD + BC - E^2)c^2d^2 + BDd^4 \geqq 0.$$

Since this inequality holds for any c and d

$$|E| \leqq (AD)^{1/2} + (BC)^{1/2},$$

that is,

$$|R_{ijkl}| \leqq \tfrac{1}{3}[(AD)^{1/2} + (BC)^{1/2}]. \tag{3.2.21}$$

Another estimate is obtained from the inequalities

$$K(aX_i + bX_k, cX_l + dX_j) \leqq 1$$

by following a similar procedure. In fact,

$$|R_{ijkl}| \leqq \tfrac{1}{3}[(A'D')^{1/2} + (B'C')^{1/2}] \tag{3.2.22}$$

where

$$A' = 2 - K_{ij} - K_{il}, \quad B' = 2 - K_{ij} - K_{ik},$$
$$C' = 2 - K_{lk} - K_{lj}, \quad D' = 2 - K_{kj} - K_{kl}.$$

From (3.2.21) and (3.2.22) we deduce

$$|R_{ijkl}| \leqq \tfrac{1}{6}(2K_{ij} + 2K_{kl} + K_{ik} + K_{il} + K_{jk} + K_{jl} - 8\delta)$$

and

$$|R_{ijkl}| \leqq \tfrac{1}{6}(8 - 2K_{ij} - 2K_{kl} - K_{ik} - K_{il} - K_{jk} - K_{jl}).$$

Thus,

$$|R_{ijkl}| \leqq \tfrac{2}{3}(1 - \delta), \quad (i,j) \neq (k,l), \quad i < j,\ k < l. \tag{3.2.23}$$

This estimate for the components of the curvature tensor is now applied to (3.2.10). Indeed, for $p = 2$

$$F(\alpha) = R_{ij}\alpha^{ic}\alpha^{j}{}_{c} + \tfrac{1}{2}R_{ijkl}\alpha^{ij}\alpha^{kl}.$$

The right hand side may be evaluated more readily by choosing an orthonormal basis $\{X_s, X_{s*}\}$, $s = 1, \cdots, m$ such that only those components of α of the form α_{ss*} are different from zero. (The existence of such a basis is a standard fact in linear algebra.) Hence,

$$2F(\alpha) = \sum_{i \neq s, s^*} (K_{si} + K_{s*i})(\alpha^{ss^*})^2 + 4\sum_{s<t} R_{ss*tt*}\alpha^{ss^*}\alpha^{tt^*}.$$

Consequently, since $K_{si} \geqq \delta$, $K_{s*i} \geqq \delta$ for all s and i we obtain by virtue of (3.2.23)

$$F(\alpha) \geqq 2(m-1)\delta \sum_{s} (\alpha_{ss*})^2 - \frac{4}{3}(1-\delta) \sum_{s<t} \alpha_{ss*}\alpha_{tt*}$$

for $n = 2m$ and

$$F(\alpha) \geqq (2m-1)\delta \sum_s (\alpha_{ss^*})^2 - \frac{4}{3}(1-\delta) \sum_{s<t} \alpha_{ss^*}\alpha_{tt^*}$$

for $n = 2m + 1$. Finally, from

$$\sum_s (\alpha_{ss^*})^2 = \frac{1}{m-1} \sum_{s<t} [(\alpha_{ss^*})^2 + (\alpha_{tt^*})^2],$$

we obtain

$$F(\alpha) \geqq \sum_{s<t} \left[2\delta\,(\alpha_{ss^*})^2 - \frac{4}{3}(1-\delta)\,\alpha_{ss^*}\alpha_{tt^*} + 2\delta\,(\alpha_{tt^*})^2\right]$$

for $n = 2m$ and

$$F(\alpha) \geqq \sum_{s<t} \left[\frac{2m-1}{m-1}\,\delta\,(\alpha_{ss^*})^2 - \frac{4}{3}(1-\delta)\,\alpha_{ss^*}\alpha_{tt^*} + \frac{2m-1}{m-1}\,\delta\,(\alpha_{tt^*})^2\right]$$

for $n = 2m + 1$ from which for $n = 2m$ and $\delta > \frac{1}{4}$ or $n = 2m + 1$ and $\delta > 2(m-1)/(8m-5)$

$$2F(\alpha) > \sum_{s<t} (\alpha_{ss^*} - \alpha_{tt^*})^2 \geqq 0.$$

This completes the proof.

The following statement is immediately clear from theorem 3.2.1 and Poincaré duality:

Corollary. *A 5-dimensional δ-pinched compact and orientable Riemannian manifold is a homology sphere for $\delta > 2/11$.*

The even dimensional case of the theorem should be compared with theorem 6.4.1.

3.3. Derivations in a graded algebra

The tensor algebra of contravariant (covariant) tensors and the Grassman algebra of differential forms are examples of a type of algebraic structure known as a graded algebra. A *graded algebra* A over a field K is defined by prescribing a set of vector spaces A^p ($p = 0, 1, \cdots$) over K such that the vector space A is the direct sum of the spaces A^p; furthermore, the product of an element of A^p and one of A^q is an element of A^{p+q}, and this product is required to be associative.

The tensor product $A \otimes B$ of the underlying spaces of the graded algebras A and B can be made into a graded algebra by defining a suitable multiplication and graduation in $A \otimes B$.

The exterior differential operator d is an anti-derivation in the ring of exterior differential polynomials, that is for a p-form α and q-form β:

$$d(\alpha \wedge \beta) = d\alpha \wedge \beta + \bar{\alpha} \wedge d\beta \tag{3.3.1}$$

where $\bar{\alpha} = (-1)^p \alpha$. For an element a of A^p the involutive automorphism: $a \to \bar{a} = (-1)^p a$ is called the *bar operation*. An endomorphism θ of the additive structure of A is said to be of *degree* r if for each p, $\theta(A^p) \subset A^{p+r}$. As an endomorphism the operator d is of degree $+1$. An endomorphism θ of A of even degree is called a *derivation* if for any a and b of A

$$\theta(ab) = (\theta a)b + a(\theta b). \tag{3.3.2}$$

It is called an *anti-derivation* if it is of odd degree and

$$\theta(ab) = (\theta a)b + \bar{a}(\theta b). \tag{3.3.3}$$

Evidently, if θ is an anti-derivation, $\theta\theta$ is a derivation. If θ_1 and θ_2 are anti-derivations $\theta_1\theta_2 + \theta_2\theta_1$ is a derivation. The bracket $[\theta_1, \theta_2] = \theta_1\theta_2 - \theta_2\theta_1$ of two derivations is again a derivation. Moreover, for a derivation θ_1 and an anti-derivation θ_2, $[\theta_1, \theta_2]$ is an anti-derivation.

If the algebra A is generated by its elements of degrees 0 and 1, a derivation or anti-derivation is completely determined if it is given in A^0 and A^1.

Let X be an infinitesimal transformation on an n-dimensional Riemannian manifold M. In terms of the natural bases $\{\partial/\partial u^1, \cdots, \partial/\partial u^n\}$ and $\{du^1, \cdots, du^n\}$ relative to the local coordinates $u^1, \cdots, u^n$ write $X = \xi^i(\partial/\partial u^i)$ and $\xi = \xi_i du^i$, ξ being the covariant form of X. Now, for any p-form α, we define the *exterior product operator* $\epsilon(\xi)$:

$$\epsilon(\xi)\alpha = \xi \wedge \alpha, \quad p < n. \tag{3.3.4}$$

Clearly, $\epsilon(\xi)$ is an endomorphism of $\wedge(T^*)$. For any $(p+1)$-form β on M,

$$\begin{aligned}\epsilon(\xi)\alpha \wedge *\beta &= (-1)^p\, \alpha \wedge \epsilon(\xi)*\beta \\ &= (-1)^p\, \alpha \wedge **^{-1}\, \epsilon(\xi)*\beta \\ &= (-1)^{np}\, \alpha \wedge **\, \epsilon(\xi)*\beta\end{aligned}$$

where $*^{-1}$ denotes the inverse of the star operator:

$$*^{-1} = (-1)^{p(n-p)} *$$

for p-forms. We define the operator $i(X)$ on p-forms as follows:

$$i(X) = (-1)^{np+n} * \epsilon(\xi) *. \tag{3.3.5}$$

That $i(X)$ is an endomorphism of $\wedge(T^*)$ is clear. Since

$$\langle \epsilon(\xi)\alpha, \beta \rangle = \langle \alpha, i(X)\beta \rangle \tag{3.3.6}$$

we conclude that $i(X)$ is the dual of the exterior product by ξ operator. Evidently, $i(X)$ lowers the degree by one. The operator $i(X)$ is called the *interior product* by X. From (3.3.5) we obtain

$$\epsilon(\xi) = (-1)^{np+n+1} * i(X) *$$

on forms of degree p.

Lemma 3.3.1. *For every 1-form* α *and infinitesimal transformation* X

$$i(X)\,\alpha = \langle X, \alpha \rangle.$$

From (3.3.5)

$$i(X)\,\alpha = * \,\epsilon(\xi) * \alpha = *(\xi \wedge *\alpha) = * \langle X, \alpha \rangle * 1 = \langle X, \alpha \rangle.$$

Lemma 3.3.2. $i(X)$, $X \in T$ *is an anti-derivation of the algebra* $\wedge(T^*)$.

For, let $\{X_1, \cdots, X_n\}$ and $\{\omega^1, \cdots, \omega^n\}$ be dual bases. Then, by (1.5.1) and (II.A.1)

$$\langle X_1 \wedge \ldots \wedge X_p, \alpha^1 \wedge \ldots \wedge \alpha^p \rangle = \langle \omega^1 \wedge \ldots \wedge \omega^p, \alpha^1 \wedge \ldots \wedge \alpha^p \rangle$$

where $\alpha^1, \ldots, \alpha^p$ are any covectors in T^*. Moreover, from (1.5.1)

$$\langle X_1 \wedge \ldots \wedge X_i \wedge \ldots \wedge X_p, \alpha^1 \wedge \ldots \wedge \alpha^j \wedge \ldots \wedge \alpha^p \rangle = \det(\langle X_i, \alpha^j \rangle).$$

Hence, for any decomposable element $X_2 \wedge \ldots \wedge X_p \in \wedge(T)$, if we apply (3.3.6) and then develop the determinant by the row $i = 1$

$$\begin{aligned}
&\langle X_2 \wedge \ldots \wedge X_i \wedge \ldots \wedge X_p, i(X_1)(\alpha^1 \wedge \ldots \wedge \alpha^j \wedge \ldots \wedge \alpha^p) \rangle \\
&= \langle \omega^2 \wedge \ldots \wedge \omega^i \wedge \ldots \wedge \omega^p, i(X_1)(\alpha^1 \wedge \ldots \wedge \alpha^j \wedge \ldots \wedge \alpha^p) \rangle \\
&= \langle \epsilon(\omega^1)\omega^2 \wedge \ldots \wedge \omega^i \wedge \ldots \wedge \omega^p, \alpha^1 \wedge \ldots \wedge \alpha^j \wedge \ldots \wedge \alpha^p \rangle \\
&= \langle X_1 \wedge \ldots \wedge X_i \wedge \ldots \wedge X_p, \alpha^1 \wedge \ldots \wedge \alpha^j \wedge \ldots \wedge \alpha^p \rangle \\
&= \sum_{j=1}^{p} (-1)^{j+1} \langle X_1, \alpha^j \rangle \langle X_2 \wedge \ldots \wedge X_p, \alpha^1 \wedge \ldots \wedge \hat{\alpha}^j \wedge \ldots \wedge \alpha^p \rangle,
\end{aligned}$$

the circumflex over α^j indicating omission of that symbol. We conclude by linearity that

$$i(X)(\alpha^1 \wedge \dots \wedge \alpha^p) = \sum_{j=1}^{p} (-1)^{j+1} (i(X)\alpha^j)\, \alpha^1 \wedge \dots \wedge \hat{\alpha}^j \wedge \dots \wedge \alpha^p$$

for any $X \in T$, and the lemma now follows easily.

We have shown that a tangent vector field X on M defines an endomorphism $i(X)$ of the exterior algebra $\wedge(T^*)$ of degree -1. It is the unique anti-derivation with the properties:

(i) $i(X)f = 0$ for every function f on M, and

(ii) $i(X)\alpha = \langle X, \alpha \rangle$ for every $X \in T$ and $\alpha \in T^*$.

We remark that $i(X)$ is an anti-derivation whose square vanishes. This is seen as foliows: $i(X)i(X)$ is a derivation annihilating $\wedge^p(T^*)$ for $p = 1,2$. Hence, since $\wedge(T^*)$ is a graded algebra, it is annihilated by $i(X)i(X)$.

3.4. Infinitesimal transformations

Relative to the system of local coordinates $u^1, \cdots, u^n$ at a point P of the differentiable manifold M, the contravariant vectors $(\partial/\partial u^1)_P, \cdots, (\partial/\partial u^n)_P$ form a basis for the tangent space T_P at P. If F denotes the algebra of differentiable functions on M and $f \in F$, the scalar $(\partial f/\partial u^i)_P\, \xi^i(P)$ is the directional derivative of f at P along the tangent vector X_P at P whose components in the local coordinates $(u^i(P))$ are given by $\xi^1(P), \cdots, \xi^n(P)$. We define a linear map which is again denoted by X_P from F into R:

$$X_P f \equiv X_P(f) = \left(\frac{\partial f}{\partial u^i}\right)_P \xi^i(P). \tag{3.4.1}$$

Evidently, it has the property

$$X_P(fg) = X_P f \cdot g(P) + f(P) \cdot X_P g. \tag{3.4.2}$$

In this way, a tangent vector at P may be considered as a linear map of F into R satisfying equation (3.4.2).

Now, an infinitesimal transformation or vector field X is a map assigning to each $P \in M$ a tangent vector $X_P \in T_P$ (cf. § 1.3). If we define the function Xf by $(Xf)(P) = X_P f$ for all $P \in M$, the infinitesimal transformation X may be considered as a linear map of F into the algebra of all real-valued functions on M with the property

$$X(fg) = Xf \cdot g + f \cdot Xg.$$

The infinitesimal transformation X is said to be *differentiable* of class $k - 1$ if Xf is differentiable of class $k - 1$ for every f of class $k \geqq 1$.

We give a geometrical interpretation of vector fields on M in terms of groups of transformations of M which will prove particularly useful when discussing the conformal geometry of a Riemannian manifold as well as the local geometry of a compact semi-simple Lie group (cf. Chapter IV). For a more detailed treatment of the results of this section the reader is referred to [*27*, *63*]. To this end, we define a (*global*) *1-parameter group of differentiable transformations* of M denoted by $\varphi_t \, (-\infty < t < \infty)$ as follows:

(i) φ_t is a differentiable transformation (cf. § 1.5) of $M \, (-\infty < t < \infty)$;

(ii) The map $(t, P) \rightarrow \varphi_t(P)$ is a differentiable map from $R \times M$ into M;

(iii) $\varphi_{s+t} = \varphi_s \, \varphi_t, \; -\infty < s, t < \infty$.

The 1-parameter group φ_t *induces* a (contravariant) vector field X on M defined by the equation

$$(Xf)(P) = \lim_{t \to 0} \frac{f(\varphi_t(P)) - f(P)}{t} \tag{3.4.3}$$

(f: an arbitrary differentiable function) the limit being assured by condition (ii). Under the circumstances, the vector field X is said to be *complete*. On the other hand, a vector field X on M is not necessarily induced by a global 1-parameter group φ_t of M. However, associated with a point P of M there is a neighborhood U of P and a constant $\epsilon > 0$ such that for $|t| < \epsilon$ there is a (local) 1-parameter group of transformations φ_t satisfying the conditions:

(i)′ φ_t is a differentiable transformation of U onto $\varphi_t(U)$, $|t| < \epsilon$;

(ii)′ The map $(t, P) \rightarrow \varphi_t(P)$ is a differentiable map from $(-\epsilon, \epsilon) \times U$ into U;

(iii)′ $\varphi_{s+t}(P) = \varphi_s(\varphi_t(P))$, $P \in U$ provided $|s|$, $|t|$ and $|s + t|$ are each less than ϵ.

Moreover, φ_t induces the vector field X, that is equation (3.4.3) is satisfied for each $P \in U$ and differentiable function f. The vector field X is then said to *generate* φ_t. The proof is omitted. (We shall occasionally write $\varphi_X(P, t)$ for $\varphi_t(P)$ (cf. III.C)). The uniqueness of the local group φ_t is immediate. Hence the existence of a 'flow' in a neighborhood of P is equivalent to that of a 'field of directions' at P.

If M is compact it may be shown that every vector field is complete and in our applications this will usually be the case.

Lemma 3.4.1. *Let ψ be a differentiable map sending M into M' and X a vector field on M. Then, the vector field $\psi_*(X)$ on M' generates the 1-parameter group $\psi\varphi_t\psi^{-1}$ where φ_t is the 1-parameter group generated by X.*

The proof is entirely straightforward.

A vector field X on M is said to be *invariant* by $\psi: M \to M$ if $\psi_*(X) = X$. Therefore, by the lemma, X is invariant by ψ, if and only if ψ commutes with φ_t for every t.

Lemma 3.4.2. *Let f be a differentiable function (of class 2) defined in a neighborhood of $0 \in R$. Assume $f(0) = 0$. Then, there is a differentiable function g defined in the same neighborhood such that $f(t) = tg(t)$ and $g(0) = f'(0)$ where $f' = df/dt$.*

We remark that the lemma is trivial if f is analytic. The proof is given by setting

$$g(t) = \int_0^1 f'(ts)ds.$$

The function g is of class one less than that of f in general. It is important that f be of class 2 at least. For, otherwise g may not be differentiable. To see this, let

$$f(t) = \begin{cases} t^2, & t \geqq 0, \\ -t^2, & t \leqq 0. \end{cases}$$

Then, $g(t) = |t|$.

Corollary. *Let $\tilde{f}$ be a differentiable function on $U \times M$ where U is a neighborhood of $0 \in R$ and M is a differentiable manifold. If $\tilde{f}(0, P) = 0$ for every $P \in M$, then there is a differentiable function g on $U \times M$ with the property that $\tilde{f}(t, P) = tg(t, P)$ and $(\partial\tilde{f}/\partial t)_{(0,P)} = g(0, P)$ for every $P \in M$.*

This is an immediate consequence of lemma 3.4.2.

For any two infinitesimal transformations X and Y of M, YX is not in general an infinitesimal transformation. In fact, if $M = E^n$ and $Xf = \partial f/\partial u^1$, $Yf = \partial f/\partial u^2$, we have $YXf = \partial^2 f/\partial u^2 \partial u^1$. Clearly, the map $f \to (\partial^2 f/\partial u^2 \partial u^1)_P$, $(P \in E^n)$ is not a tangent vector on E^n. However, one may easily check that the map $XY - YX$ is a vector field on M. We shall denote this vector field by $[X, Y]$. The bracket $[X, Y]$ evidently satisfies the Jacobi identity

$$[[X,Y], Z] + [[Y,Z], X] + [[Z,X], Y] = 0,$$

and so *the (differentiable) vector fields on M form a Lie algebra over R.*

Lemma 3.4.3. *For any two infinitesimal transformations X and Y on M,*

$$[X,Y]_P f = \lim_{t\to 0} \frac{(Y - \varphi_{t*} Y)_P f}{t}$$

for any $f \in F$ where φ_t is the 1-parameter group generated by X.

Associated with any $f \in F$, there is a differentiable family of functions g_t on M such that $f\varphi_t = f + tg_t$ where $g_0 = Xf$. This follows from lemma 3.4.2 by putting $\tilde{f}(t, P) = f(\varphi_t(P)) - f(P)$. Hence, if we set $\varphi_{t*} = (\varphi_t)_*$ and $\varphi_{t*}Y = \varphi_{t*}(Y)$

$$\begin{aligned}(\varphi_{t*} Y)_P f &= (Y(f\varphi_t))\,(\varphi_t^{-1}(P)) \\ &= (Yf)\,(\varphi_t^{-1}(P)) + t(Yg_t)\,(\varphi_t^{-1}(P)),\end{aligned}$$

from which

$$\begin{aligned}\lim_{t\to 0} \frac{(Y - \varphi_{t*} Y)_P f}{t} &= \lim_{t\to 0} \frac{(Yf)_P - (Yf)\,(\varphi_t^{-1}(P))}{t} - \lim_{t\to 0} (Yg_t)\,(\varphi_t^{-1}(P)) \\ &= X_P(Yf) - Y_P g_0 \\ &= X_P(Yf) - Y_P(Xf).\end{aligned}$$

Corollary. *If φ_t and ψ_t are the 1-parameter groups generated by X and Y, respectively, then $[X, Y] = 0$, if an only if φ_s and ψ_t commute for every s and t.*

3.5. The derivation $\theta(X)$

We have seen that to each tangent vector field $X \in T$ on a Riemannian manifold M there is associated an anti-derivation $i(X)$ of degree -1 (called the interior product by X) of the exterior algebra $\wedge(T^*)$ of differential forms on M. A derivation $\theta(X)$ of degree 0 of the Grassman algebra $\wedge(T)$ as well as $\wedge(T^*)$ may be defined, and in fact, completely characterized for each $X \in T$ as follows (cf. III.B.3):

(i) $\theta(X)d = d\theta(X)$,

(ii) $\theta(X)f = i(X)df$, $f \in \wedge^0(T^*)$, and

(iii) $\theta(X)Y = [X, Y]$.

Indeed, $\theta(X)f = i(X)df = \langle X, df\rangle = \langle \xi^i(\partial/\partial u^i), (\partial f/\partial u^j)\, du^j\rangle = \xi^i\,(\partial f/\partial u^i) = Xf$ and $\theta(X)df = d\theta(X)f = dXf$; since $\wedge(T^*)$ is generated (locally) by its homogeneous elements of degrees 0 and 1 the derivation $\theta(X)$ may be extended to differential forms of any degree.

On the other hand, by conditions (ii) and (iii), $\theta(X)$ may be extended to all of $\wedge(T)$. In fact, $\theta(X)$ may be extended to the tensor algebras of contravariant and covariant tensors by insisting that (for each X) it be a derivation of these algebras. For example, by lemma 3.4.3

$$\theta(X)Y = \lim_{t\to 0} \frac{Y - \varphi_{t*}Y}{t}$$

where φ_t is the 1-parameter group generated by X. Hence, for any tensor $\mathbf{t}$ of type $(p, 0)$

$$\theta(X)\,\mathbf{t} = \lim_{s\to 0} \frac{\mathbf{t} - \varphi^p_{s*}\,\mathbf{t}}{s}$$

where $\varphi^p_{s*} = \varphi_{s*} \otimes \cdots \otimes \varphi_{s*}$ (p times) is the induced map in T^p_0. (For any $X_1, \cdots, X_p \in T$, $\varphi^p_{s*}(X_1 \otimes \cdots \otimes X_p) = \varphi_{s*}(X_1) \otimes \cdots \otimes \varphi_{s*}(X_p)$).

Since $[\theta(X), \theta(Y)] = \theta(X)\,\theta(Y) - \theta(Y)\,\theta(X)$ is a derivation, it follows from the Jacobi identity that the map $X \to \theta(X)$ is a representation of the Lie algebra of tangent vector fields.

Lemma 3.5.1. *The derivations d, $i(X)$, and $\theta(X)$ are related by the formula*

$$\theta(X) = i(X)d + di(X). \tag{3.5.1}$$

Since both sides are derivations, and since the Grassman algebra of differential forms is generated by its homogeneous elements of degrees 0 and 1, the relation need only be established for differential forms of degrees 0 and 1:

$$(i(X)d + di(X))\,f = i(X)df = \theta(X)f;$$

$$(i(X)d + di(X))\,df = di(X)df = d\theta(X)f = \theta(X)df.$$

Lemma 3.5.2. *For a 1-form α on M and any tangent vector fields X and Y on M:*

$$\langle X \wedge Y, d\alpha\rangle = X\alpha(Y) - Y\alpha(X) - \alpha([X,Y]). \tag{3.5.2}$$

The right hand side is meaningful since at each point P of M, T_P and T^*_P are dual vector spaces. Thus, α is a linear map from T into F. By linearity, it is sufficient to prove the relation for $X = \partial/\partial u^i$, $Y = \partial/\partial u^j$ and $\alpha = gdf$ where f and g are functions expressed in the coordinates (u^i). In fact, if the relation holds for α, $\beta \in \wedge^1(T^*)$, it holds for $\alpha + \beta$ and $f\alpha$ where f is a differentiable function. We may therefore assume $\alpha = du^k$ and in this case, both sides vanish.

Equation (3.5.2) indicates a (local) relationship between the derivations $\theta(X)$ and d. Indeed, if we write

$$d\omega^i = \frac{1}{2} c^i_{jk}\, \omega^j \wedge \omega^k, \quad c^i_{jk} + c^i_{kj} = 0 \tag{3.5.3}$$

and

$$\theta(X_j)\, X_k = -\, b^i_{jk}\, X_i, \quad b^i_{jk} + b^i_{kj} = 0, \tag{3.5.4}$$

where $\{X_i\}$ and $\{\omega^k\}$ are dual bases, then

$$\langle X_j \wedge X_k, d\omega^i\rangle = X_j\, \omega^i(X_k) - X_k\, \omega^i(X_j) - \omega^i([X_j,X_k]),$$

from which by (1.5.1), (3.5.3) and (3.5.4)

$$c^i_{jk} = b^i_{jk}, \quad i, j, k = 1, ..., n.$$

The reader is referred to Chapter IV where this relationship is exploited more fully. We remark that equation (3.5.2) has important implications in the theory of connections as well [*63*].

3.6. Lie transformation groups [27, 63]

A *Lie group* G is a group which is simultaneously a differentiable manifold (the points of the manifold coinciding with the elements of the group) in which the group operation $(a, b) \to ab^{-1}$ $(a, b \in G)$ is a differentiable map of $G \times G$ into G. It is well-known that as a manifold G admits an analytic structure in such a way that the group operations in G are analytic. It follows that the map $x \to ax$ is analytic. We denote this map by L_a and call it the *left translation* in G by a. Hence, every left translation L_a is an analytic homeomorphism of G (as an analytic manifold) with itself. It follows that if x and y are any two elements of G, there exists an element $a = yx^{-1}$ such that the induced map $L_{a*} = (L_a)_*$ maps T_x isomorphically onto T_y.

An infinitesimal transformation X on G is said to be *left invariant* if for every $a \in G$, $L_{a*}X_e = X_a$. Hence, associated with an element $A \in T_e$, where $e \in G$ is the identity, there is a unique left invariant infinitesimal transformation X which takes the value A at e. It can be shown that every left invariant infinitesimal transformation is analytic. Let L denote the set of left invariant infinitesimal transformations of G; L is a vector space over R of dimension equal to that of G. In fact, if to a tangent vector $X_e \in T_e$ we associate the infinitesimal transformation $X \in L$ defined by $X_a = L_{a*}X_e$ $(a \in G)$ it is seen that as vector spaces

T_e and L are isomorphic. Moreover, the conditions $X \in L$, $Y \in L$ imply $[X, Y] \in L$. In fact,

$$L_{a*}[X,Y]_e = [L_{a*}X, L_{a*}Y]_e = [X,Y]_a.$$

It follows that the left invariant infinitesimal transformations of the Lie group G form a Lie algebra L called the *Lie algebra of* G. That the right invariant infinitesimal transformations also form a Lie algebra is clear. However, this Lie algebra is isomorphic with L (cf. Chapter IV).

To an element A of L we associate the local 1-parameter group of transformations φ_t generated by A in a neighborhood of $e \in G$. We show that φ_t is a global 1-parameter group of transformations on G and that it defines a 1-parameter subgroup of G. Since A is invariant by L_{x*} $(x \in G)$, it follows from lemma 3.4.1 that φ_t commutes with L_x for every $x \in G$. Hence, A generates a global 1-parameter group of transformations φ_t on G. The subgroup a_t of G defined by $a_t = \varphi_t(e)$ satisfies $a_{t+s} = a_t\, a_s$; moreover, $\varphi_t(x) = R_{a_t}x (\equiv x\, a_t)$ for every $x \in G$. We call a_t the *1-parameter subgroup of* G *generated by* A.

More generally, we define a *Lie subgroup* G' of G to be a subgroup of G which is simultaneously a submanifold of G. G' is itself a Lie group with respect to the differentiable structure induced by G. Evidently, the subspace L' of left invariant infinitesimal transformations corresponding to the tangent vectors at $e \in G'$ is a subalgebra of L, namely, the Lie algebra of G'.

Let f be an element of the group of automorphisms of a Lie group G. Then, f_* is an automorphism of L: Since $f(e) = e$, if we identify the vector space L with T_e we see that f_* induces an endomorphism of T_e. Since $f^{-1}\, f =$ identity automorphism of G, it follows that f_* is an automorphism. In particular, if f is an *inner automorphism*: $x \rightarrow axa^{-1}$ defined by $a \in G$, the induced automorphism of L is called the *adjoint representation* of G and is denoted by $ad(a)$. For an element $B \in L$, $ad(a)B = R_{a^{-1}*}B$, since $axa^{-1} = R_{a^{-1}}L_a x$. If a_t is the 1-parameter subgroup of G generated by $A \in L$ we conclude from lemma 3.4.3 that

$$[B,A] = \lim_{t \to 0} \frac{ad(a_t^{-1})B - B}{t}$$

for every $B \in L$.

Consider a differentiable manifold M on which a connected Lie group G acts differentiably. G is said to be a *Lie transformation group* on M if the following conditions hold:

(i) To each $a \in G$ there corresponds a homeomorphism R_a of M onto itself such that $R_a R_b = R_{ab}$;

(ii) The point $P \cdot a = R_a P$, $P \in M$ depends differentiably on $a \in G$ and P where $R_a P = R_a(P)$.

Clearly, R_e is the identity transformation of M. Hence, $R_a(R_{a^{-1}}(P)) = P$ for every $a \in G$ and $P \in M$. The group G is said to act *effectively* if $R_a P = P$ for every $P \in M$ implies $a = e$.

Let A be an element of the Lie algebra L of G and a_t the 1-parameter subgroup of G generated by A. A is a left invariant infinitesimal transformation of G. The corresponding 1-parameter group of transformations R_{a_t} on M induces a differentiable vector field A^* on M. Let σ denote the map sending $A \in L$ to $A^* \in L^*$ (the Lie algebra of differentiable vector fields on M).

Lemma 3.6.1. *The map $\sigma : L \to L^*$ is a homomorphism.*

Indeed, for any $P \in M$ denote by σ_P the map from G to M defined by $\sigma_P(x) = P \cdot x$. Then

$$(\sigma_{P*})_e A_e = (\sigma A)_P$$

where $(\sigma_{P*})_e$ is the induced map in T_e (the tangent space at $e \in G$). Clearly, σ is linear. For any two elements A and B of L, set $A^* = \sigma(A)$ and $B^* = \sigma(B)$. Then, from lemma 3.4.3

$$[A^*, B^*] = \lim_{t \to 0} \frac{B^* - R_{a_t}^* B^*}{t}$$

and so, since $R_{a_t}^*(\sigma_{Pa_t^{-1}*})_e B_e = (\sigma_{P*})_e ad(a_t^{-1})\, B_e$ (note that $R_{a_t}\sigma_{Pa_t^{-1}}(x) = P \cdot (a_t^{-1} x a_t)$),

$$[A^*, B^*]_P = \lim_{t \to 0} \frac{\sigma_{P*} B_e - \sigma_{P*} ad(a_t^{-1}) B_e}{t}$$

$$= \sigma_{P*} \lim_{t \to 0} \frac{B_e - ad(a_t^{-1})\, B_e}{t}$$

$$= \sigma_{P*}[A, B]_e = (\sigma[A,B])_P.$$

If G acts effectively on M, σ is an isomorphism. Indeed, if $\sigma(A) = 0$ for some $A \in L$, the associated 1-parameter subgroup R_{a_t} is trivial. Since G is effective we have $a_t = e$, from which $A = 0$.

We remark that the derivations $\theta(A^*)$ correspond to the action of G on M.

3.7. Conformal transformations

Let M be an n-dimensional Riemannian manifold and g the tensor field of type (0,2) defining the Riemannian metric on M. Locally, the metric is given by

$$ds^2 = g_{ij}\, du^i\, du^j$$

where the g_{ij} are the components of g with respect to the natural frames of a local coordinate system (u^i). A metric g^* on M is said to be *conformally related* to g if it is proportional to g, that is, if there is a function $\rho > 0$ on M such that $g^* = \rho^2 g$. By a *conformal transformation* of M is meant a differentiable homeomorphism f of M onto itself with the property that

$$f^*(ds^2) = \rho^2\, ds^2 \tag{3.7.1}$$

where f^* is the induced map in the bundle of frames and ρ is a positive function on M. Clearly, the set of conformal transformations of M forms a group. In fact, it can be shown that it is a Lie transformation group. Let G denote a connected Lie group of conformal transformations of M and L its Lie algebra. To each element $A \in L$ is associated the 1-parameter subgroup a_t of G generated by A. The corresponding 1-parameter group of transformations R_{a_t} on M induces a (right invariant) differentiable vector field A^* on M. A^* in turn defines an infinitesimal transformation $\theta(A^*)$ of the tensor algebra over M corresponding to the action on M of a_t. From the action on the metric tensor g, it follows from (3.7.1) that

$$\theta(A^*)\, g = \lambda g \tag{3.7.2}$$

where λ is a function depending on A^*. On the other hand, a vector field X on M which satisfies (3.7.2) is not necessarily complete (cf. § 3.4). However, X does generate a 1-parameter local group, and for this reason X is called an *infinitesimal conformal transformation* of M. In our applications the manifold M will be compact and therefore the infinitesimal conformal transformations will be complete. In any case, they form a Lie algebra L with the usual bracket $[X, Y] = \theta(X)Y$.

If the scalar λ vanishes, that is, if $\theta(X)g = 0$, the metric tensor g is invariant under the action of $\theta(X)$. The vector field X is then said to define an *infinitesimal motion*. The infinitesimal motions define a subalgebra of the Lie algebra L. For, $\theta([X, Y])g = \theta(X)\theta(Y)g - \theta(Y)\theta(X)g = 0$. Moreover, it can be shown that the group of all the isometries of M onto itself is a Lie group (with respect to the natural topology).

If ξ is the 1-form on M dual to X we shall occasionally write $\theta(\xi)$ for $\theta(X)$.

Proposition 3.7.1. *For any vector field X*

$$(\theta(\xi)g)_{ij} = D_j\, \xi_i + D_i\, \xi_j$$

where ξ is the 1-form on M dual to X.

Let U be a coordinate neighborhood with the local coordinates $u^1, \cdots u^n$. The vector fields $\partial/\partial u^1, \cdots, \partial/\partial u^n$ form a basis of the F-module of vector fields in U where F is the algebra of differentiable functions on U. Denoting the components of the metric tensor g by g_{ij} we have $g = g_{ij}\, du^i \otimes du^j$. Applying the derivation $\theta(X)$ to g we obtain

$$\theta(X)g = (Xg_{ij})\, du^i \otimes du^j + g_{ij}(X\, du^i) \otimes du^j + g_{ij}\, du^i \otimes (X du^j)$$

$$= \xi^k \frac{\partial g_{ij}}{\partial u^k} du^i \otimes du^j + g_{ij}\, d(Xu^i) \otimes du^j + g_{ij}\, du^i \otimes d(Xu^j)$$

$$= \xi^k \frac{\partial g_{ij}}{\partial u^k} du^i \otimes du^j + g_{ij} \frac{\partial \xi^i}{\partial u^l} du^l \otimes du^j + g_{ij} \frac{\partial \xi^j}{\partial u^l} du^i \otimes du^l$$

$$= \left(\xi^k \frac{\partial g_{ij}}{\partial u^k} + g_{kj} \frac{\partial \xi^k}{\partial u^i} + g_{ik} \frac{\partial \xi^k}{\partial u^j}\right) du^i \otimes du^j.$$

It follows that

$$(\theta(\xi)g)_{ij} = \xi^k \frac{\partial g_{ij}}{\partial u^k} + g_{kj} \frac{\partial \xi^k}{\partial u^i} + g_{ik} \frac{\partial \xi^k}{\partial u^j},$$

and, since the right hand side is equal to $D_j\xi_i + D_i\xi_j$ we may write

$$(\theta(\xi)g)_{ij} = D_j\, \xi_i + D_i\, \xi_j. \tag{3.7.3}$$

Corollary. *An infinitesimal conformal transformation X on an n-dimensional Riemannian manifold satisfies the equation*

$$\theta(\xi)g + \frac{2}{n}(\delta\xi)g = 0. \tag{3.7.4}$$

Indeed,

$$\lambda g_{ij} = (\vartheta(\xi)g)_{ij} = D_j\, \xi_i + D_i\, \xi_j.$$

Transvecting this equation with g^{ij}

$$\lambda = \frac{2}{n} D_i\, \xi^i = -\frac{2}{n}\, \delta\xi.$$

Corollary. *A necessary and sufficient condition that an infinitesimal conformal transformation X be a motion is given by $\delta\xi = 0$.*

If the vector field X has constant divergence, that is, if $\delta\xi =$ const., the transformation is said to be *homothetic*.

Assume that the vector field X defines an infinitesimal motion on M. Then, $\theta(X)g$ vanishes, that is

$$D_j \xi_i + D_i \xi_j = 0. \tag{3.7.5}$$

It follows that

$$D_k D_j \xi_i + D_i D_k \xi_j + D_j D_i \xi_k + D_k D_i \xi_j + D_j D_k \xi_i + D_i D_j \xi_k = 0.$$

Hence, applying the Bianchi identity (1.10.24) and the interchange formula (1.7.19) for covariant derivatives

$$\begin{aligned} 0 &= D_k D_j \xi_i + D_i D_k \xi_j + D_j D_i \xi_k \\ &= D_k D_j \xi_i + D_i D_k \xi_j + D_i D_j \xi_k - \xi_l R^l{}_{kij}. \end{aligned}$$

We conclude that

$$D_k D_j \xi^i + \xi^l R^i{}_{jkl} = 0. \tag{3.7.6}$$

(This means that the Lie derivative of the affine connection vanishes or, what is the same, $\theta(X)$ commutes with the operator of covariant differentiation (cf. § 3.10)). On the other hand, if X is a solution of these equations it need not be an infinitesimal motion (cf. § 3.10).

In the case where M is E^n, if we choose a cartesian coordinate system $(x^1, ..., x^n)$ equations (3.7.5) and (3.7.6) reduce to

$$\frac{\partial \xi_i}{\partial x^j} + \frac{\partial \xi_j}{\partial x^i} = 0 \quad \text{and} \quad \frac{\partial^2 \xi_i}{\partial x^j \, \partial x^k} = 0.$$

Integrating, we obtain

$$\xi_i = \Sigma a_{ij} x_j + a_i, \quad a_{ij} = - a_{ji}.$$

The vector whose components are the a_i is the translation part of the motion whereas the tensor with components a_{ij} defines a rotation about the origin.

The infinitesimal motion X is usually called a *Killing vector field.*

Let L be a subalgebra of the Lie algebra T of tangent vector fields on M. A p-form on M is said to be *L-invariant* if it is a zero of all the derivations $\theta(X)$ for $X \in L$. Clearly, the L-invariant differential forms constitute a subalgebra of the Grassman algebra of differential forms on M. Moreover, this subalgebra is stable under the operator d. This follows from property (i) of § 3.5.

Let α and β be any two p-forms on the compact and orientable

Riemannian manifold M. Then, by Stokes' theorem and formula (3.5.1), if X is an infinitesimal transformation

$$\int_M \theta(X)\,(\alpha \wedge *\beta) = \int_M di(X)\,(\alpha \wedge *\beta) = 0.$$

Since $\theta(X)$ is a derivation,

$$(\theta(X)\alpha, \beta) = -\int_M \alpha \wedge \theta(X)*\beta.$$

If, therefore, we put

$$*\bar{\theta}(X) = -\,\theta(X)*, \tag{3.7.7}$$

that is

$$\bar{\theta}(X) = (-\,1)^{np+p+1} *\theta(X)*, \tag{3.7.8}$$

we have

$$(\theta(X)\alpha, \beta) = (\alpha, \bar{\theta}(X)\,\beta). \tag{3.7.9}$$

It follows that the operator $\bar{\theta}(X)$ is the dual of $\theta(X)$. One thus obtains

$$\bar{\theta}(X) = \delta\epsilon(\xi) + \epsilon(\xi)\delta \tag{3.7.10}$$

where ξ is the covariant form for X. Since the operators $\theta(X)$ and d commute, so do their duals as one may easily see from (3.7.10):

$$\delta\bar{\theta}(X) = \bar{\theta}(X)\delta.$$

Moreover, if g denotes the metric tensor of M

$$(\theta(X) + \bar{\theta}(X))\,\alpha \tag{3.7.11}$$

$$= \delta\xi \cdot \alpha + \sum_{r=1}^{p} g^{jk}(\theta(X)g)_{ki_r}\, \alpha_{i_1 \ldots i_{r-1} j i_{r+1} \ldots i_p}\, du^{i_1} \wedge \ldots \wedge du^{i_p}$$

where the $\alpha_{i_1 \ldots i_p}$ are the coefficients of α in the local coordinates (u^i).

The proof of (3.7.11) is a lengthy but entirely straightforward computation and is therefore left as an exercise for the reader.

Theorem 3.7.1. *The harmonic forms on a compact and orientable Riemannian manifold M are K-invariant differential forms where K is the Lie algebra of infinitesimal motions on M[73, 35].*

The proof depends on the fact that $\theta(X) + \bar{\theta}(X)$, $X \in K$ annihilates differential forms. Indeed, since X is an infinitesimal motion, $\theta(X)g = 0$ and, therefore, $\delta\xi = 0$. Let α be a harmonic form. Then, $d\theta(X)\alpha = \theta(X)d\alpha = 0$ and $\delta\theta(X)\alpha = -\,\delta\bar{\theta}(X)\alpha = -\,\bar{\theta}(X)\delta\alpha = 0$. Hence, $\theta(X)\alpha$ is a harmonic form; but $\theta(X)\alpha = di(X)\alpha$, from which by the Hodge-de Rham decomposition of a differential form (cf. § 2.10), $\theta(X)\alpha = 0$.

Corollary. *In a compact and orientable Riemannian manifold the inner product of a harmonic vector field and a Killing vector field is a constant.*

In fact, if α is a harmonic 1-form and X an element of K, $0 = \theta(X)\alpha = di(X)\alpha$.

The corollary may be generalized as follows:

Theorem 3.7.2. *The inner product of a K-invariant closed 1-form and an element X of K is a constant equal to* $< X, H[\alpha] >$.

For, $0 = \theta(X)\alpha = di(X)\alpha$. By the Hodge-de Rham decomposition of a 1-form, $\alpha = df + H[\alpha]$ for some function f, from which $0 = \theta(X)\alpha = \theta(X)df = di(X)df$. Hence, $\langle X, df\rangle = k =$ const. We conclude that $(\xi, df) = \int *k = 0$ since $(\xi, df) = (\delta\xi, f) = 0$.

Let X be an element of the Lie algebra L of infinitesimal conformal transformations of M. Then, equation (3.7.11) reduces to

$$(\theta(X) + \bar{\theta}(X))\,\alpha = \left(1 - \frac{2p}{n}\right)\delta\xi \cdot \alpha \tag{3.7.12}$$

in view of formula (3.7.4), and we have the following generalization of theorem 3.7.1:

Theorem 3.7.3. *Let M be a compact and orientable Riemannian manifold of dimension n. Then, a harmonic k-form* α *is L-invariant, if and only if,* $n = 2k$ *or,* $\delta\xi \cdot \alpha$ *is co-closed [35].*

Corollary. *On a compact and orientable 2-dimensional Riemannian manifold the inner product of a harmonic vector field and an infinitesimal transformation defining a 1-parameter group of conformal transformations is a constant.*

This is clearly the case if M is a Riemann surface (cf. Chap. V).

Since formula (3.7.12) is required in the proof of theorem 3.7.5 and again in Chapter VII a proof of it is given below:

Applying $\theta(X)$ to $\langle\alpha, \beta\rangle = g^{i_1j_1}\cdots g^{i_pj_p}\,\alpha_{(i_1\ldots i_p)}\,\beta_{(j_1\ldots j_p)}$ we obtain

$$\theta(X)\,\langle\alpha, \beta\rangle = \langle\theta(X)\,\alpha, \beta\rangle + \langle\alpha, \theta(X)\,\beta\rangle + \frac{2p}{n}\delta\xi\,\langle\alpha, \beta\rangle. \tag{3.7.13}$$

We also have

$$\theta(X)*1 = -\,\delta\xi *1. \tag{3.7.14}$$

From (3.7.13) and (3.7.14), we obtain

$$\theta(X)\,(\langle\alpha, \beta\rangle *1) = \langle\theta(X)\,\alpha, \beta\rangle *1 + \langle\alpha, \theta(X)\,\beta\rangle *1 + \left(\frac{2p}{n} - 1\right)\delta\xi\,\langle\alpha, \beta\rangle *1. \tag{3.7.15}$$

The integral of the left side of (3.7.15) over M vanishes by Stokes' theorem. Hence, integrating (3.7.15) gives

$$0 = (\theta(X)\,\alpha, \beta) + (\alpha, \theta(X)\,\beta) + \left(\left(\frac{2p}{n} - 1\right)\delta\xi \cdot \alpha, \beta\right).$$

Thus,

$$(\theta(X)\,\alpha + \bar{\theta}(X)\,\alpha, \beta) = \left(\left(1 - \frac{2p}{n}\right)\delta\xi \cdot \alpha, \beta\right);$$

and so, since α and β are arbitrary

$$\theta(X)\,\alpha + \bar{\theta}(X)\,\alpha = \left(1 - \frac{2p}{n}\right)\delta\xi \cdot \alpha.$$

Let M be a Riemannian manifold, $C_0(M)$ the largest connected group of conformal transformations of M and $I_0(M)$ the largest connected group of isometries of M. (Note that L and K are the Lie algebras of $C_0(M)$ and $I_0(M)$, respectively.) We shall prove the following:

Theorem 3.7.4. *Let M be a compact Riemannian manifold. If $C_0(M) \neq I_0(M)$, then, there is no harmonic form of degree p, $0 < p < n$ ($n = \dim M$) whose length is a non-zero constant* [78].

Since a harmonic form on a compact Riemannian manifold is invariant by $I_0(M)$, a harmonic form on a compact homogeneous Riemannian manifold (cf. VI. E) is of constant length. (A *Riemannian homogeneous manifold* is a Riemannian manifold whose group of isometries is transitive.) Hence, as an immediate consequence of theorem 3.7.4 we have

Theorem 3.7.5. *Let M be a compact homogeneous Riemannian manifold. If $C_0(M) \neq I_0(M)$, then M is a homology sphere* [78].

Since we are interested in connected groups, the hypothesis of theorem 3.7.4 may be replaced by the following: *Let M be a compact Riemannian manifold admitting an infinitesimal non-isometric conformal transformation.* We may also assume that M is orientable; for, if M is not orientable, we need only take an orientable two-fold covering space of M.

Proof of Theorem 3.7.4. Let α be a harmonic form of degree p. We shall first prove

$$(\bar{\theta}(X)\,\alpha, \theta(X)\,\alpha) = 0. \tag{3.7.16}$$

Since α is closed, $\theta(X)\,\alpha = di(X)\,\alpha$. On the other hand, since α is co-closed, $\delta\bar{\theta}(X)\,\alpha = \bar{\theta}(X)\,\delta\alpha = 0$. Thus,

$$(\bar{\theta}(X)\,\alpha, \theta(X)\,\alpha) = (\delta\bar{\theta}(X)\,\alpha, i(X)\,\alpha) = 0.$$

Applying (3.7.12) and (3.7.16) we obtain

$$\begin{aligned}(\theta(X)\,\alpha, \theta(X)\,\alpha) &= (\theta(X)\,\alpha + \bar{\theta}(X)\,\alpha, \theta(X)\,\alpha)\\ &= \left(1 - \frac{2p}{n}\right)(\delta\xi \cdot \alpha, \theta(X)\,\alpha) \qquad (3.7.17)\\ &= \left(1 - \frac{2p}{n}\right)\int_M \delta\xi\,\langle\alpha, \theta(X)\,\alpha\rangle *1.\end{aligned}$$

From now on, we assume that α is not only harmonic but is also of constant length, that is, $\langle\alpha, \alpha\rangle$ is constant. Hence, $\theta(X)\langle\alpha, \alpha\rangle = 0$, and so, from (3.7.13)

$$\langle\theta(X)\,\alpha, \alpha\rangle = -\frac{p}{n}\delta\xi\,\langle\alpha, \alpha\rangle. \qquad (3.7.18)$$

Substituting (3.7.18) into (3.7.17) we obtain

$$(\theta(X)\,\alpha, \theta(X)\,\alpha) = -\left(1 - \frac{2p}{n}\right)\frac{p}{n}\int_M (\delta\xi)^2\,\langle\alpha, \alpha\rangle *1. \qquad (3.7.19)$$

If $2p \leqq n$, the right hand side of (3.7.19) is non-positive; but the left hand side is non-negative. Consequently, $\theta(X)\,\alpha = 0$ and by (3.7.18) either $\delta\xi = 0$ or $\alpha = 0$. If X is not an infinitesimal isometry, $\delta\xi \neq 0$. We have therefore proved that if M admits an infinitesimal non-isometric conformal transformation, then there is no harmonic form of constant length and degree p, $0 < p \leqq n/2$. If α is a harmonic form of constant length and degree $p > n/2$, then its adjoint $*\alpha$ is a harmonic form of constant length and of degree $n - p < n/2$. This completes the proof.

By employing theorem 3.7.5, it can be shown that M is, in fact, isometric with a sphere (cf. III. F).

3.8. Conformal transformations (continued)

In this section we characterize the infinitesimal conformal transformations and motions of a compact and orientable Riemannian manifold M as solutions of a system of differential equations on M. Moreover, we investigate the existence of (global) 1-parameter groups of conformal transformations of M and find that when the Ricci curvature tensor is positive definite no such groups except $\{e\}$ exist.

For a 1-form α on M we define the symmetric tensor field

$$t(\alpha) = \theta(\alpha)g + \frac{2}{n}(\delta\alpha)\,g \qquad (3.8.1)$$

of type (0,2) where we have written $\theta(\alpha)$ for $\theta(A)$—the vector field A being defined by duality. Clearly, the elements A of L satisfy the equation $t(\alpha)=0$. In a coordinate neighborhood U with local coordinates $u^1, \cdots, u^n$ the tensor $t(\alpha)$ has the components

$$(t(\alpha))_{ij} = D_j\,\alpha_i + D_i\,\alpha_j + \frac{2}{n}(\delta\alpha)\,g_{ij},$$

the divergence of which is given by

$$\begin{aligned}(\delta' t(\alpha))_j &= g^{ik}\,D_k(D_j\,\alpha_i + D_i\,\alpha_j) + \frac{2}{n}(d\delta\alpha)_j\\ &= 2D_i\,D_j\,\alpha^i - (\delta d\alpha)_j + \frac{2}{n}(d\delta\alpha)_j\end{aligned}$$

since

$$(\delta d\alpha)_j = g^{ik}\,D_k(D_j\,\alpha_i - D_i\,\alpha_j).$$

The operator δ' is used in place of $-\,\delta$ since $t(\alpha)$ is symmetric. From the Ricci identity (1.7.19) we obtain

$$D_i\,D_j\,\alpha^i = (Q\alpha)_j - (d\delta\alpha)_j,$$

and so

$$\begin{aligned}\delta' t(\alpha) &= 2Q\alpha - \delta d\alpha - \left(2 - \frac{2}{n}\right) d\delta\alpha\\ &= 2Q\alpha - \Delta\alpha - \left(1 - \frac{2}{n}\right) d\delta\alpha.\end{aligned} \tag{3.8.2}$$

Now, since the tensor $t(\alpha)$ is symmetric and is annihilated by g, that is, since $\langle g, t(\alpha)\rangle = 0$,

$$\begin{aligned}\langle t(\alpha), t(\alpha)\rangle &= (t(\alpha))_{ij}\,D^j\,\alpha^i = g^{jk}\,(t(\alpha))_{ij}\,D_k\,\alpha^i\\ &= g^{jk}\,D_k[\alpha^i(t(\alpha))_{ij}] - \langle \delta' t(\alpha), \alpha\rangle\\ &= -\,\delta(\alpha^i(t(\alpha))_{ij}\,du^j) - \langle \delta' t(\alpha), \alpha\rangle.\end{aligned}$$

Integrating both sides of this relation and applying Stokes' formula we obtain the integral formula

$$(\delta' t(\alpha), \alpha) + (t(\alpha), t(\alpha)) = 0 \tag{3.8.3}$$

where we have put

$$(t(\alpha), t(\alpha)) = \int_M \langle t(\alpha), t(\alpha)\rangle *1.$$

An application of (3.8.2) together with (3.8.3) yields:

Theorem 3.8.1. *There are no non-trivial (global) 1-parameter groups of conformal transformations on a compact and orientable Riemannian manifold M of dimension $n \geqq 2$ with negative definite Ricci curvature* [*4, 73*].

For, let X be the infinitesimal conformal transformation induced by a given 1-parameter group of conformal transformations of M and ξ the 1-form defined by X by duality. Then $t(\xi)$ vanishes, and so by (3.8.2) and (3.8.3)

$$(\Delta\xi + \left(1 - \frac{2}{n}\right)d\delta\xi - 2Q\xi, \xi) = 0.$$

A computation gives

$$(d\xi, d\xi) + 2\left(1 - \frac{1}{n}\right)(\delta\xi, \delta\xi) = 2\,(Q\xi, \xi),$$

and consequently, if $\langle Q\xi, \xi\rangle \leqq 0$ then, for $n \geqq 2$, we must have

$$\langle Q\xi, \xi\rangle = 0, \ \delta\xi = 0, \ DX = 0.$$

Moreover, if the Ricci curvature is negative definite we conclude that $\xi = 0$, that is X vanishes.

We have proved in addition that *if the Ricci quadratic form is negative semi-definite, then a vector field X on M which generates a 1-parameter group of conformal transformations of M is necessarily a parallel field.*

Corollary. *There are no (global) 1-parameter groups of motions on a compact and orientable Riemannian manifold of negative definite Ricci curvature.*

We have seen that an infinitesimal conformal transformation on a Riemannian manifold M must satisfy the differential equation

$$\Delta\alpha + \left(1 - \frac{2}{n}\right)d\delta\alpha = 2Q\alpha. \tag{3.8.4}$$

Conversely, if M is compact and orientable, and ξ is a 1-form on M which is a solution of equation (3.8.4), then by (3.8.2) and (3.8.3) $(t(\xi), t(\xi)) = 0$ from which $t(\xi) = 0$, that is $\theta(\xi)g + (2/n)(\delta\xi)g = 0$. It follows that the vector field X dual to ξ is an infinitesimal conformal transformation. We have proved [*73*]

Theorem 3.8.2. *On a compact and orientable Riemannian manifold a necessary and sufficient condition that the vector field X be an infinitesimal conformal transformation is given by*

$$\Delta\xi + \left(1 - \frac{2}{n}\right)d\delta\xi = 2Q\xi.$$

Corollary. *On a compact and orientable Riemannian manifold, a necessary and sufficient condition that the infinitesimal transformation X generate a 1-parameter group of motions is given by the equations*

$$\Delta\xi = 2Q\xi \quad \text{and} \quad \delta\xi = 0.$$

3.9. Conformally flat manifolds

Let M be a Riemannian manifold with metric tensor g. Consider the Riemannian manifold M^* constructed from M as follows: (i) $M^* = M$ as a differentiable manifold, that is, as differentiable manifolds M and M^* have equivalent differentiable structures which we identify; (ii) the metric tensor g^* of M^* is conformally related to g, that is, $g^* = \rho^2 g$ $(\rho > 0)$. Since the quadratic form ds^2 for $n = 2$ is reducible to the form $\lambda[(du^1)^2 + (du^2)^2]$ (in infinitely many ways) the metric tensors of any two 2-dimensional Riemannian manifolds are conformally related. In the sequel, we shall therefore assume $n > 2$.

For convenience we write $\rho = e^\sigma$. It follows that the components g_{ij} and g^*_{ij} of the tensors g and g^* are related by the equations

$$g^*_{ij} = e^{2\sigma} g_{ij}. \tag{3.9.1}$$

The components of the Levi-Civita connections associated with the metric tensors g and g^* are then related as follows:

$$\Gamma^{*i}_{jk} = \Gamma^i_{jk} + \delta^i_j D_k\sigma + \delta^i_k D_j\sigma - g_{jk} g^{il} D_l\sigma.$$

A computation gives

$$\begin{aligned} e^{-2\sigma} R^*_{ijkl} = R_{ijkl} &- g_{il}\,\sigma_{jk} - g_{jk}\,\sigma_{il} + g_{ik}\,\sigma_{jl} + g_{jl}\,\sigma_{ik} \\ &- (g_{jk} g_{il} - g_{jl} g_{ik}) \langle d\sigma, d\sigma\rangle \end{aligned} \tag{3.9.2}$$

where we have put

$$\sigma_{ij} = D_j D_i\sigma - D_i\sigma D_j\sigma.$$

Transvecting (3.9.2) with g^{il} we see that the components of the corresponding Ricci tensors are related by

$$R^*_{jk} = R_{jk} - (n-2)\,\sigma_{jk} + [\Delta\sigma - (n-2)\langle d\sigma, d\sigma\rangle]\, g_{jk}. \tag{3.9.3}$$

Again, transvecting (3.9.3) with g^{jk} we obtain the following relation between the scalar curvatures R and R^*:

$$R^* = e^{-2\sigma}[R + 2(n-1)\Delta\sigma - (n-1)(n-2)\langle d\sigma, d\sigma\rangle]. \tag{3.9.4}$$

Eliminating $\Delta\sigma$ from (3.9.3) and (3.9.4) we obtain

$$\begin{aligned}\sigma_{ij} = &-\frac{1}{n-2}(R^*_{ij} - R_{ij}) + \frac{1}{2(n-1)(n-2)}(g^*_{ij}R^* - g_{ij}R) \\ &-\frac{1}{2}\langle d\sigma, d\sigma\rangle g_{ij}.\end{aligned} \tag{3.9.5}$$

Transvecting (3.9.2) with g^{*ir} and substituting (3.9.5) in the resulting equation we obtain $C^{*i}{}_{jkl} = C^i{}_{jkl}$ where

$$\begin{aligned}C^i{}_{jkl} = R^i{}_{jkl} &- \frac{1}{n-2}(R_{jk}\,\delta^i_l - R_{jl}\,\delta^i_k + g_{jk}R^i{}_l - g_{jl}R^i{}_k) \\ &+ \frac{R}{(n-1)(n-2)}(g_{jk}\,\delta^i_l - g_{jl}\,\delta^i_k).\end{aligned} \tag{3.9.6}$$

Evidently, the $C^i{}_{jkl}$ are the components of a tensor called the *Weyl conformal curvature tensor*. Moreover, this tensor remains invariant under a conformal change of metric. The case $n = 3$ is interesting. Indeed, by choosing an orthogonal coordinate system ($g_{ij} = 0$, $i \neq j$) at a point (cf. § 1.11), it is readily shown that the Weyl conformal curvature tensor vanishes.

Consider a Riemannian manifold M with metric g and let g^* be a conformally related locally flat metric. Under the circumstances M is said to be (locally) *conformally flat*. Clearly then, the Weyl conformal curvature tensor of M vanishes. Conversely, if the tensor $C^i{}_{jkl}$ is a zero tensor on M, there exists a function σ such that $g^* = e^{2\sigma}g$ is a locally flat metric on M. For, from (3.9.6)

$$\begin{aligned}D_i C^i{}_{jkl} = D_i R^i{}_{jkl} &- \frac{1}{n-2}(D_l R_{jk} - D_k R_{jl} + g_{jk} D_i R^i{}_l - g_{jl} D_i R^i{}_k) \\ &+ \frac{1}{(n-1)(n-2)}(g_{jk} D_l R - g_{jl} D_k R).\end{aligned} \tag{3.9.7}$$

Applying (1.10.21) and (1.10.22) we deduce

$$D_i C^i{}_{jkl} = (n-3)C_{jkl}$$

where we have put

$$\begin{aligned}C_{jkl} = &\frac{1}{n-2}(D_l R_{jk} - D_k R_{jl}) \\ &- \frac{1}{2(n-1)(n-2)}(g_{jk} D_l R - g_{jl} D_k R).\end{aligned} \tag{3.9.8}$$

Hence, for $n > 3$, $C_{ijk} = 0$.

If $g^* = e^{2\sigma}g$ is a locally flat metric, both R^*_{ij} and R^* vanish, and so from (3.9.5)

$$D_j\, D_i\, \sigma = D_i\, \sigma\, D_j\, \sigma - \frac{1}{n-2}\left(\frac{Rg_{ij}}{2(n-1)} - R_{ij}\right) - \frac{1}{2}\, g_{ij}\, \langle d\sigma, d\sigma\rangle. \quad (3.9.9)$$

The integrability conditions of the system (3.9.9) are evidently given by

$$D_k\, D_j\, D_i\, \sigma - D_j\, D_k\, D_i\, \sigma = -\, R^r{}_{ijk}\, D_r\, \sigma. \quad (3.9.10)$$

It follows after substitution from (3.9.9) into (3.9.10) that $C_{ijk} = 0$. Thus, the equations (3.9.9) are integrable.

Proposition 3.9.1. *A necessary and sufficient condition that a Riemannian manifold of dimension $n > 3$ be conformally flat is that its Weyl conformal curvature tensor vanish. For $n = 3$, it is necessary and sufficient that the tensor $C_{ijk} = 0$.*

The conformal curvature tensor of a Riemannian manifold of constant curvature is readily seen to vanish. Thus,

Corollary. *A Riemannian manifold of constant curvature is conformally flat provided $n \geqq 3$.*

We now show that a compact and orientable conformally flat Riemannian manifold M whose Ricci curvature is positive definite is a homology sphere. This is certainly the case if M is a manifold of positive constant curvature.

Indeed, since M is conformally flat, its Weyl conformal curvature tensor vanishes. Hence, from formula (3.2.10), for a harmonic p-form α

$$F(\alpha) = \frac{n-2p}{n-2}\, R_{ij}\, \alpha^{ii_2\cdots i_p}\, \alpha^j{}_{i_2\cdots i_p} + p!\, \frac{p-1}{(n-1)\,(n-2)}\, R\langle\alpha, \alpha\rangle. \quad (3.9.11)$$

Since the operator Q is positive definite let λ_0 denote the greatest lower bound of the smallest eigenvalues of Q on M. Then, for any 1-form β, $\langle Q\beta, \beta\rangle \geqq \lambda_0\, \langle\beta, \beta\rangle$ and the scalar curvature $R = g^{ij}\, R_{ij} \geqq n\, \lambda_0 > 0$. This latter statement follows from the fact that at the pole of a geodesic coordinate system the scalar curvature R is the trace of the matrix (R_{ij}), $(g_{ij}(P) = \delta_{ij})$.

Again, at a point $P \in M$ if a geodesic coordinate system is chosen it follows from (3.9.11) that

$$\begin{aligned} F(\alpha) &\geqq p!\,\frac{n-2p}{n-2}\, \lambda_0\, \langle\alpha, \alpha\rangle + p!\, \frac{n(p-1)}{(n-1)\,(n-2)}\, \lambda_0\, \langle\alpha, \alpha\rangle \\ &= p!\, \frac{n-p}{n-1}\, \lambda_0\, \langle\alpha, \alpha\rangle \end{aligned} \quad (3.9.12)$$

at P from which we conclude that $F(\alpha)$ is a positive definite quadratic form. We thus obtain the following generalization of cor., theorem 3.2.4:

Theorem 3.9.1. *The betti numbers $b_p (0 < p < n)$ of a compact and orientable conformally flat Riemannian manifold of positive definite Ricci curvature vanish* [6, 51].

For $n = 2, 3$ this is, of course, evident from theorem 3.2.1 and Poincaré duality.

If M is a Riemannian manifold which is not conformally flat, that is, if for $n > 3$ its conformal curvature tensor does not vanish, we may introduce a quantity which measures its deviation from conformal flatness and ask under what conditions M remains a homology sphere. To this end, let

$$2C = \sup_{\xi \in \wedge^2(T)} \frac{| C_{ijkl}\, \xi^{ij}\, \xi^{kl} |}{\langle \xi, \xi \rangle} \tag{3.9.13}$$

for all skew-symmetric tensors of type (2,0) at all points P of M. C is a measure of the deviation of M from conformal flatness. Substituting for the Riemannian curvature tensor from (3.9.6) into equation (3.2.10) we find

$$F(\alpha) = \frac{n-2p}{n-2} R_{ij}\, \alpha^{ii_2\cdots i_p}\, \alpha^{j}{}_{i_2\cdots i_p}$$

$$+ p! \frac{(p-1)R}{(n-1)(n-2)} \langle \alpha, \alpha \rangle + \frac{p-1}{2} C_{ijkl}\, \alpha^{iji_3\cdots i_p}\, \alpha^{kl}{}_{i_3\cdots i_p}$$

where α is a harmonic p-form. Applying (3.9.12) and (3.9.13) we have at the pole P of a geodesic coordinate system

$$F(\alpha) \geqq p! \frac{n-p}{n-1} \lambda_0 \langle \alpha, \alpha \rangle + \frac{p-1}{2} C_{ijkl}\, \alpha^{iji_3\cdots i_p}\, \alpha^{kl}{}_{i_3\cdots i_p}$$

$$\geqq p! \frac{n-p}{n-1} \lambda_0 \langle \alpha, \alpha \rangle - p! \frac{p-1}{2} C \langle \alpha, \alpha \rangle$$

$$\geqq p! \left(\frac{n-p}{n-1} \lambda_0 - \frac{p-1}{2} C \right) \langle \alpha, \alpha \rangle.$$

Hence, $F(\alpha)$ is a positive definite quadratic form provided $((n-p)/(n-1))\lambda_0 > ((p-1)/2)C$ and, in this case, if M is compact and orientable, $b_p(M) = 0$.

Theorem 3.9.2. *Let M be a compact and orientable Riemannian manifold of positive Ricci curvature.* If

$$\frac{n-p}{n-1}\lambda_0 > \frac{p-1}{2}C, \tag{3.9.14}$$

then, $b_p(M)$ vanishes [*6*, *74*].

Corollary. *M is a homology sphere if* (3.9.14) *holds for all* p, $0 < p < n$.

This generalizes theorem 3.9.1.

3.10. Affine collineations

Let M be a Riemannian manifold with metric tensor g and $C = C(t)$ a geodesic on M defined by the parametric equations $u^i = u^i(t)$, $i = 1, \cdots, n$. Denoting the arc length by s, that is $ds^2 = g_{ij}du^i du^j$, the equations of C are given by

$$\frac{d^2u^i}{dt^2} + \Gamma^i_{jk}\frac{du^j}{dt}\frac{du^k}{dt} = \lambda(t)\frac{du^i}{dt} \tag{3.10.1}$$

where $\lambda(t) = (d^2s/dt^2)/(ds/dt)$ and the Γ^i_{jk} are the coefficients of the Levi Civita connection (associated with the metric). By an *affine collineation* of M we mean a differentiable homeomorphism f of M onto itself which maps geodesics into geodesics, the arc length receiving an affine transformation:

$$s \rightarrow as + b$$

for some constants $a \neq 0$ and b. Clearly, if f is a motion it is an affine collineation. The converse, however, is not true in general, but, if we assume that M is compact and orientable, an affine collineation is necessarily a motion (theorem 3.10.1).

It can be shown that the affine collineations of M form a Lie group. Let G denote a connected Lie group of affine collineations of M and L its Lie algebra. To each element A of L we associate the 1-parameter subgroup a_τ of G generated by A. The corresponding 1-parameter group of transformations R_{a_τ} on M induces a (right invariant) vector field A^* on M. The vector field A^* in turn defines an infinitesimal transformation $\theta(A^*)$ of M corresponding to the action on M of a_τ. Since the elements of G map geodesics into geodesics the Lie derivative of the left hand side of (3.10.1) with s as parameter must vanish.

We evaluate the Lie derivative of the Levi Civita connection forms ω_j^k with respect to a vector field X defining an infinitesimal affine collineation:

$$\theta(X)\omega_j^k = di(X)\omega_j^k + i(X)d\omega_j^k$$

$$= di(X)\omega_j^k + i(X)\,[\omega_j^l \wedge \omega_l^k - \tfrac{1}{2}R^k{}_{jlm}\, du^l \wedge du^m]$$

$$= d\langle \xi^r \frac{\partial}{\partial u^r}, \Gamma^k_{jt}\, du^t\rangle + i(X)\omega_j^l\, \omega_l^k - \omega_j^l\, i(X)\omega_l^k$$

$$- \tfrac{1}{2}R^k{}_{jlm}\,[i(X)du^l\, du^m - du^l\, i(X)du^m] \tag{3.10.2}$$

$$= d(\xi^r\, \Gamma^k_{jr}) + \xi^r[\Gamma^l_{jr}\, \Gamma^k_{lm} - \Gamma^l_{jm}\, \Gamma^k_{lr}]\, du^m + \xi^l\, R^k{}_{jml}\, du^m$$

$$= \left[\frac{\partial \xi^r}{\partial u^m}\, \Gamma^k_{jr} + \xi^r\, \frac{\partial \Gamma^k_{jr}}{\partial u^m} + \xi^r(\Gamma^l_{jr}\, \Gamma^k_{lm} - \Gamma^l_{jm}\, \Gamma^k_{lr} + R^k{}_{jmr})\right] du^m.$$

Consequently,

$$0 = \theta(X)\,\frac{d^2u^k}{ds^2} + \frac{\theta(X)\omega_j^k}{ds}\frac{du^j}{ds} + \frac{\omega_j^k}{ds}\,\frac{\theta(X)\, du^j}{ds}$$

$$= \left(\frac{\partial^2\, \xi^k}{\partial u^l\, \partial u^m} - \frac{\partial \xi^k}{\partial u^r}\, \Gamma^r_{lm} + \Gamma^k_{jm}\, \frac{\partial \xi^j}{\partial u^l}\right)\frac{du^l}{ds}\frac{du^m}{ds} + \frac{\theta(X)\omega_l^k}{ds}\,\frac{du^l}{ds}$$

$$= \left(\frac{\partial^2 \xi^k}{\partial u^l \partial u^m} - \frac{\partial \xi^k}{\partial u^r}\, \Gamma^r_{lm} + \Gamma^k_{jm}\, \frac{\partial \xi^j}{\partial u^l}\right)\frac{du^l}{ds}\,\frac{du^m}{ds} \tag{3.10.3}$$

$$+ \left[\frac{\partial \xi^r}{\partial u^m}\, \Gamma^k_{lr} + \xi^r\, \frac{\partial \Gamma^k_{lr}}{\partial u^m} + \xi^r\left(\Gamma^s_{lr}\, \Gamma^k_{sm} - \Gamma^s_{lm}\, \Gamma^k_{sr} + R^k{}_{lmr}\right)\right]\frac{du^l}{ds}\,\frac{du^m}{ds}\,.$$

Hence, by (3.10.3) for an infinitesimal affine collineation $X = \xi^i(\partial/\partial u^i)$

$$D_k\, D_j\, \xi^i + \xi^r\, R^i{}_{jkr} = 0. \tag{3.10.4}$$

Transvecting (3.10.4) with g^{jk} we see that

$$g^{jk}\, D_k\, D_j\, \xi^i + R^i_r\, \xi^r = 0$$

or

$$\Delta \xi = 2Q\xi.$$

Again, if we transvect (3.10.4) with δ^j_i we obtain $D_k D_i \xi^i = 0$, that is

$$d\delta\xi = 0.$$

Hence, if M is compact and orientable

$$0 = (d\delta\xi, \xi) = (\delta\xi, \delta\xi)$$

from which $\delta\xi = 0$. We conclude (by theorem 3.8.2, cor.)

Theorem 3.10.1. *In a compact and orientable Riemannian manifold an infinitesimal affine collineation is a motion* [73].

Corollary. *There exist no (non-trivial) 1-parameter groups of affine collineations on a compact and orientable Riemannian manifold of negative definite Ricci curvature.*

This follows from theorem 3.8.1.

More generally, it can be shown that an infinitesimal affine collineation defined by a vector field of bounded length on a complete but not compact Riemannian manifold is an infinitesimal motion. We remark that compactness implies completeness (cf. § 7.7).

3.11. Projective transformations

We have defined an affine collineation of a Riemannian manifold M as a differentiable homeomorphism f of M onto M preserving the geodesics and the affine character of the parameter s denoting arc length along a geodesic. If, more generally, f leaves the geodesics invariant, the affine character of the parameter s not necessarily being preserved, f is called a *projective transformation.*

A transformation f of M is *affine*, if and only if

$$f^*\omega = \omega$$

where ω is the matrix of forms defining the affine connection of M, or, equivalently in terms of a system of local coordinates

$$\Gamma^{*i}_{\ jk} = \Gamma^i_{jk},$$

where the $\Gamma^{*i}_{\ jk}$ are given by $f^*\omega^i_j = \Gamma^{*i}_{\ jk}\, du^k$, f^* denoting the induced dual map on forms. A transformation f of M is projective, if and only if there exists a covector $p(f)$ depending on f such that

$$\Gamma^{*i}_{\ jk} = \Gamma^i_{jk} + p_j(f)\delta^i_k + p_k(f)\delta^i_j \qquad (3.11.1)$$

where the $p_i(f)$ are the components of $p(f)$ with respect to the given local coordinates. Under the circumstances, ω and $f^*\omega$ are called *projectively related* affine connections. On the other hand, two affine connections ω and ω^* are said to be *projectively related* if there exists a covariant vector field p_i such that in the given local coordinates

$$\Gamma^{*i}_{\ jk} = \Gamma^i_{jk} + p_j \delta^i_k + p_k \delta^i_j.$$

Let M be a Riemannian manifold with metric g. If there exists a metric g^* on M such that the connections ω and ω^* canonically defined by g and g^* are projectively related, then, by means of a straightforward computation, the tensor w whose components are

$$W^i_{\ jkl} = R^i_{\ jkl} - \frac{1}{n-1}(R_{jk}\,\delta^i_l - R_{jl}\,\delta^i_k), \quad n > 1 \tag{3.11.2}$$

is an invariant of the projectively related affine connections, that is, the tensor w^* corresponding to the connection ω^* projectively related to ω coincides with w. This tensor is known as the *Weyl projective curvature tensor*. Its vanishing is of particular interest. Indeed, if $w = 0$, the curvature of M (relative to g or g^*) has the representation

$$R^i_{\ jkl} = \frac{1}{n-1}(R_{jk}\,\delta^i_l - R_{jl}\,\delta^i_k).$$

Hence,

$$R_{ijkl} = \frac{1}{n-1}(R_{jk}\,g_{il} - R_{jl}\,g_{ik}) \tag{3.11.3}$$

from which, by the symmetry properties of the Riemannian curvature tensor

$$R_{jk}\,g_{il} - R_{jl}\,g_{ik} + R_{ik}\,g_{jl} - R_{il}\,g_{jk} = 0.$$

Transvecting with g^{il} we deduce that

$$R_{jk} = \frac{R}{n}\,g_{jk}. \tag{3.11.4}$$

Substituting the expression (3.11.4) for the Ricci curvature in (3.11.3) gives

$$R_{ijkl} = \frac{R}{n(n-1)}(g_{jk}\,g_{il} - g_{jl}\,g_{ik}). \tag{3.11.5}$$

Thus, M is a manifold of constant curvature.

Conversely, assume that M (with metric g or g^*) has constant curvature. Then, its curvature has the representation (3.11.5) and its Ricci curvature is given by (3.11.4). Substituting from (3.11.4) and (3.11.5) into (3.11.2), we conclude that the tensor w vanishes.

Let M be a Riemannian manifold with metric g. If M may be given a locally flat metric g^* such that the Levi Civita connections ω and ω^* defined by g and g^*, respectively, are projectively related, then M is said to be *locally projectively flat*. Under the circumstances, the geodesics of the manifold M with metric g correspond to 'straight lines' of the manifold M with metric g^*. For $n > 3$, it can be shown that a necessary and sufficient condition for M to be locally projectively flat is that its Weyl projective curvature tensor vanishes. Thus, *a necessary and sufficient condition for a Riemannian manifold to be locally projectively flat is that it have constant curvature.*

We have shown that a compact and orientable Riemannian manifold M of positive constant curvature is a homology sphere. Moreover, (from a local standpoint) M is locally projectively flat, that is its Weyl projective curvature tensor vanishes. It is natural, therefore, to inquire into the effect on homology in the case where this tensor does not vanish. With this purpose in mind, a measure W of the deviation from projective flatness is introduced. Indeed, we define

$$2W = \sup_{\xi \in \wedge^2(T)} \frac{|W_{ijkl}\, \xi^{ij}\, \xi^{kl}|}{\langle \xi, \xi \rangle}$$

the least upper bound being taken over all skew-symmetric tensors of order 2.

Theorem 3.11.1. *In a compact and orientable Riemannian manifold of dimension n with positive Ricci curvature, if*

$$\frac{n-p}{n-1}\lambda_0 > \frac{p-1}{2} W \tag{3.11.6}$$

(where λ_0 has the meaning previously given) for all $p = 1, \cdots, n-1$, then M is a homology sphere [6, 74].

Indeed, substituting for the Riemannian curvature tensor from (3.11.2) into equation (3.2.10) we obtain

$$F(\alpha) \geqq p! \left(\frac{n-p}{n-1}\lambda_0 - \frac{p-1}{2} W\right) \langle \alpha, \alpha \rangle$$

by virtue of the fact that at the pole of a geodesic coordinate system

$$R_{ij}\, \alpha^{ii_2 \cdots i_p}\, \alpha^{j}{}_{i_2 \cdots i_p} \geqq p!\, \lambda_0 \langle \alpha, \alpha \rangle$$

and

$$W_{ijkl}\, \alpha^{iji_3 \cdots i_p}\, \alpha^{kl}{}_{i_3 \cdots i_p} \geqq -p!\, W \langle \alpha, \alpha \rangle.$$

Hence, $F(\alpha)$ is non-negative provided

$$\frac{n-p}{n-1}\lambda_0 \geqq \frac{p-1}{2} W.$$

If strict inequality holds, M is a homology sphere.

Corollary. *Under the conditions of the theorem, if*

$$W < \frac{2\lambda_0}{(n-1)(n-2)},$$

M is a homology sphere.

We have proved that the betti numbers of the sphere are retained even for deviations from projective flatness, that is from constant curvature. This, however, is not surprising as we need only compare with theorem 3.2.6. In a certain sense, however, theorem 3.11.1 is a stronger result. Indeed, the function W need only be bounded above but need not be uniformly bounded below.

Theorem 3.11.1 implies that the homology structure of a compact and orientable Riemannian manifold with metric of positive constant curvature is preserved under a variation of the metric preserving the signature of the Ricci curvature as well as the inequality (3.11.6), that is, a manifold carrying the varied metric is a homology sphere.

EXERCISES

A. Locally convex hypersurfaces [58, 14]. Minimal varieties [4]

1. Let M be a Riemannian manifold of dimension n locally isometrically imbedded (without singularities) in E^{n+1} with the canonical (Euclidean) metric. The manifold M is then said to be a *local hypersurface* of E^{n+1}. Let a_{ij} denote the coefficients of the second fundamental form of M in terms of the cartesian coordinates of E^{n+1}. Then, the curvature of M is given by the (Gauss) equations

$$R_{ijkl} = a_{jk}\, a_{il} - a_{jl}\, a_{ik}.$$

M is said to be *locally convex* if the second fundamental form is definite, that is, if the principal curvatures $\kappa_{(r)}$ are of the same sign everywhere. Under the circumstances, every point of M admits a neighborhood in which the vectors tangent to the lines of curvature are the vectors of an orthonormal frame.

Consequently,

$$a_{ij} = \kappa_{(i)}\,\delta_{ij}, \quad (i\text{: not summed})$$

from which we derive

$$R_{ijkl} = \kappa_{(i)}\,\kappa_{(j)}\,(\delta_{jk}\,\delta_{il} - \delta_{jl}\,\delta_{ik}).$$

Hence,

$$-R_{jk} = [\kappa^2_{(j)} - \Big(\sum_{r=1}^{n} \kappa_{(r)}\Big)\kappa_{(j)}]\,\delta_{jk}.$$

By employing theorem 3.2.4 show that if M is compact and orientable, then $b_1(M) = b_2(M) = 0$.

2. If at each point of M, the ratio of the largest to the smallest principal curvature is at most $\sqrt{2}$, M is a homology sphere.

Hint: Apply theorem 3.2.6.

3. If M is locally isometrically imbedded in an $(n + 1)$-dimensional space of positive constant curvature K, the Gauss equations are given by

$$R_{ijkl} = (a_{jk}\,a_{il} - a_{jl}\,a_{ik}) + K(g_{jk}\,g_{il} - g_{jl}\,g_{ik}).$$

Show that the assertions in A.1 and A.2 are also valid in this case.

4. If the *mean curvature* of the hypersurface vanishes, that is, if, in terms of the metric g of M,

$$g^{ij}\,a_{ij} = 0$$

then, from the representation of the curvature tensor given in A.1

$$R_{jk}\,\xi^j\,\xi^k = -\,g^{jk}\,\eta_j\,\eta_k$$

where

$$\eta_i = a_{il}\,\xi^l.$$

In this case, M is called a *minimal hypersurface* or a *minimal variety* of E^{n+1}.

Show that the only groups of motions of a compact and orientable minimal variety are groups of translations.

5. Show that the only groups of motions of a compact and orientable minimal variety (hypersurface of zero mean curvature) imbedded in a manifold of constant negative curvature are translation groups.

6. If all the geodesics of a hypersurface M are also geodesics of the space in which it is imbedded, M is called a *totally geodesic hypersurface.* It is known that a totally geodesic hypersurface is a minimal variety. Hence, if it is compact and orientable and, if the imbedding space is a manifold of constant non-positive curvature its only groups of motions are translation groups.

B. 1-parameter local groups of local transformations

1. Let P be a point on the differentiable manifold M and U a coordinate neighborhood of P on which a vector field $X \neq 0$ is given. Denote the components of X at P with respect to the natural basis in U by ξ^i. There exists at P a local coordinate system $v^1, \cdots, v^n$ such that the corresponding parametrized curves with v^1 as parameter have at each point Q the vector X_Q as tangent vector. If we put $v^1 = t$, the equations

$$u^i = u^i(v^2, \cdots, v^n, t), \quad i = 1, \cdots, n$$

defining the coordinate transformations at P are the equations of the 'integral curves' (cf. I. D.8) when the v^i, $i = 2, \cdots, n$ are regarded as constants and t as the parameter, that is, the coordinate functions u^i, $i = 1, \cdots, n$ are solutions of the system of differential equations

$$\frac{du^i}{dt} = \xi^i(t) \tag{*}$$

with $\xi^i(0) = \xi^i$, the point P corresponding to $t = 0$. More precisely, it is possible to find a neighborhood $U(Q)$ of Q and a positive number $\epsilon(Q)$ for every $Q \in U$ such that the system (*) has a solution for $|t| < \epsilon(Q)$. Denoting this solution by

$$u^i(v^2, \cdots, v^n, t) = \exp(tX)u^i(v^2, \cdots, v^n, 0)$$

show that

$$\exp(sX)\exp(tX)\, u^i(v^2, \cdots, v^n, 0) = \exp((s+t)\, X)\, u^i(v^2, \cdots, v^n, 0)$$

provided both sides are defined. In this way, we see that the 'exp' map defines a local 1-parameter group $\exp(tX)$ of (local) transformations.

2. Conversely, every 1-parameter local group of local transformations φ_t may be so defined. Indeed, for every $P \in M$ put

$$P(t) = \varphi_t(P)$$

and consider the vector field X defined by the initial conditions

$$\xi^i = \left(\frac{du^i}{dt}\right)_{t=0}$$

(or, $X_P = (dP(t)/dt)_{t=0}$). It follows that

$$\varphi_t = \exp(tX).$$

3. The map $\exp(tX)$ is defined on a neighborhood $U(Q)$ for $|t| < \epsilon(Q)$ and induces a map $\exp(tX)_*$ which is an isomorphism of T_P onto $T_{P(t)}$—the tangent space at $P(t) = \exp(tX)P$. The induced dual map $\exp(tX)^*$ sends $\wedge(T^*_{P(t)})$ into $\wedge(T^*_P)$. For an element $\alpha_{P(t)} \in \wedge^p(T^*_{P(t)})$

$$\frac{\exp(tX)^* \, \alpha_{P(t)} - \alpha_P}{t}$$

is an element of $\wedge^p(T^*_P)$. Show that

$$[\theta(X)\alpha]_P = \lim_{t \to 0} \frac{\exp(tX)^* \, \alpha_{P(t)} - \alpha_P}{t}$$

and that consequently

$$\theta(X)\,(\alpha \wedge \beta) = \theta(X)\,\alpha \wedge \beta + \alpha \wedge \theta(X)\,\beta$$

for any elements α, $\beta \in \wedge^*(T)$.

Hint: Show that

$$\exp(tX)^* \, (\alpha \wedge \beta)_{P(t)} = \exp(tX)^* \, \alpha_{P(t)} \wedge \exp(tX)^* \, \beta_{P(t)}.$$

C. Frobenius' theorem and infinitesimal transformations

1. Show that the conditions in Frobenius theorem (I. D.4) may be expressed in the following form: If the basis of the tangent space T_P at $P \in M$ is chosen so that the subspace $F(P)$ of T_P of dimension r is spanned by the vectors $X_A (A = q + 1, \cdots, n)$ then, if we take $\Theta = \theta^1 \wedge \ldots \wedge \theta^q$ the conditions of complete integrability are given by

$$c^i_{AB} = 0, \; A, B = q + 1, \cdots, n, \; i = 1, \cdots, q.$$

This is equivalent to the condition that $[X_A, X_B]$ is a linear combination of $X_{q+1}, \cdots, X_n$ only. In other words, F is completely integrable, if and only if, for any two infinitesimal transformations X, Y such that X_P, $Y_P \in F(P)$ for all $P \in U$ the bracket $[X,Y]_P \in F(P)$.

2. Associated with the vector fields X and Y are the local one parameter groups $\varphi_X(P,t)$ and $\varphi_Y(P,t)$. Then, $[X,Y]_P$ is the tangent at $t = 0$ to the curve

$$\varphi_{[X,Y]}(P, t) = \varphi_Y\left(\varphi_X\left(\varphi_Y\left(\varphi_X\left(P, \frac{t}{\sqrt{t}}\right), \frac{t}{\sqrt{t}}\right), -\frac{t}{\sqrt{t}}\right), -\frac{t}{\sqrt{t}}\right).$$

This formula shows, geometrically, the necessity of the integrability conditions for F. For, if X_P and Y_P are contained in $F(P)$ for all $P \in U$ and F is integrable, the integral curves of X and Y must be contained in the integral manifold. Hence, the formula shows that the above curves must also be contained in the integral manifold from which it follows that $[X,Y]_P \in F(P)$.

D. The third fundamental theorem of Lie

By differentiating the equations (3.5.3), the relations

$$c^i_{jk}c^j_{rs} + c^i_{jr}c^j_{sk} + c^i_{js}c^j_{kr} = 0 \tag{3.D.1}$$

are obtained. Conversely, assuming n^3 constants c^i_{jk} are given with the property that

$$c^i_{jk} + c^i_{kj} = 0,$$

show that the conditions (3.D.1) are sufficient for the existence of n linear differential forms, linearly independent at each point of a region in R^n, and which satisfy the relations (3.5.3).

This may be shown in the following way:

Consider the system

$$\frac{\partial h^i_j}{\partial t} = \delta^i_j + a^k c^i_{kr} h^r_j \quad (i, j = 1, \cdots, n) \tag{3.D.2}$$

of n^2 linear partial differential equations in n^2 variables h^i_j in the space R^{n+1} of independent variables $t, a^1, \cdots, a^n$—the $a^1, \cdots, a^n$ being treated as parameters. Given the initial conditions

$$h^i_j(0, a^1, \cdots, a^n) = 0,$$

the equations (3.D.2) have unique (analytic) solutions $h^i_j(t, a^1, \cdots, a^n)$ valid throughout R^{n+1}.

Observe that

$$\left(\frac{\partial h^i_j}{\partial t}\right)_{(t,0,\ldots,0)} = \delta^i_j.$$

Hence,

$$h^i_j(t, 0, \cdots, 0) = \delta^i_j t.$$

In particular,

$$h^i_j(1, 0, \cdots, 0) = \delta^i_j.$$

Now, define n linear differential forms ω^i by

$$\omega^i = h^i_j da^j.$$

In terms of the ω^i, 2-forms λ^i and 1-forms α^i (both sets independent of dt) are defined by the equations

$$d\omega^i = \lambda^i + dt \wedge \alpha^i. \tag{3.D.3}$$

Indeed,

$$\alpha^j = i\left(\frac{\partial}{\partial t}\right) d\omega^j = \frac{\partial h^j_k}{\partial t}\, da^k = da^j + a^k c^j_{kr}\omega^r.$$

Differentiating the equations (3.D.3) we obtain

$$d\lambda^i = dt \wedge d\alpha^i.$$

On the other hand,

$$d\alpha^i = c^i_{jr}\alpha^j \wedge \omega^r - a^k c^j_{ks} c^i_{jr}\omega^s \wedge \omega^r + a^k c^i_{kr}\lambda^r + a^k c^i_{kr}\, dt \wedge \alpha^r.$$

It follows that

$$i\left(\frac{\partial}{\partial t}\right) d\alpha^j = a^k c^j_{kr}\alpha^r$$

and

$$i\left(\frac{\partial}{\partial t}\right) d\lambda^j = c^j_{kr}\alpha^k \wedge \omega^r + a^k c^j_{kr}\lambda^r - a^k c^j_{rs} c^r_{kt}\omega^t \wedge \omega^s.$$

Put

$$\theta^i = \lambda^i - \tfrac{1}{2}\, c^i_{jk}\omega^j \wedge \alpha^k.$$

Thus,

$$i\left(\frac{\partial}{\partial t}\right) d\theta^j = a^k c^j_{kr}\theta^r.$$

On the other hand, by setting

$$\theta^i = \tfrac{1}{2} f^i_{jk} da^j \wedge da^k,$$

$$\frac{\partial f^i_{jk}}{\partial t} = a^r c^i_{rs} f^s_{jk}. \tag{3.D.4}$$

Since $h^i_j(0, a^1, \cdots, a^n) = 0$, it follows that $f^i_{jk}(0, a^1, \cdots, a^n) = 0$. Consequently, by (3.D.4) the f^i_{jk} vanish for all t, and so the θ^i vanish identically. Hence,

$$\lambda^i = \tfrac{1}{2}\, c^i_{jk}\omega^j \wedge \omega^k.$$

Now, consider the map

$$\phi : R^n \to R^{n+1}$$

defined by

$$\phi(x^1, \cdots, x^n) = (1, x^1, \cdots, x^n)$$

and set

$$\sigma^i = \phi^*\omega^i.$$

Then, the σ^i are 1-forms in R^n and

$$d\sigma^i = \frac{1}{2} c^i_{jk}\sigma^j \wedge \sigma^k.$$

The linear independence of the σ^i is shown by making use of the fact that when $a^i = 0$, $i = 1, \cdots, n$,

$$\sigma^i = h^i_j(1, 0, \cdots, 0)\, dx^j.$$

E. The homogeneous space $SU(3)/SO(3)$

1. Show that a compact symmetric space admitting a vector field generating globally a 1-parameter group of non-isometric conformal transformations is isometric with a sphere.

Hint: Apply the following theorem: *If a compact simply connected symmetric space is a rational homology sphere, it is isometric with a sphere except for* $SU(3)/SO(3)$ [82]. The exceptional case may be disposed of as follows: Let G be a compact simple Lie group, $\sigma \neq$ identity an involutary automorphism of G (cf. VI.E.1) and H the subgroup of G consisting of all elements fixed by σ. Then, there exists a unique (up to a constant factor) Riemannian metric on G/H invariant under G. With respect to this metric, G/H is an irreducible symmetric space (that is, the linear isotropy group is irreducible). Hence, G/H is an Einstein space. *But a compact Einstein space admitting a non-isometric conformal transformation is isometric with a sphere* [77].

Let $\mathbf{G}$ be the Lie algebra of $SU(3)$ consisting of all skew-hermitian matrices of trace 0 and $\mathbf{H}$ the Lie algebra of $SO(3)$ consisting of all real skew-hermitian matrices of trace 0. Let σ denote the map sending an element of $SU(3)$ into its complex conjugate. Since $SU(3)/SO(3)$ is symmetric and simply connected, its homogeneous holonomy group is identical with $\mathbf{G}/\mathbf{H}$. It follows that the action of $SO(3)$ on $\mathbf{G}/\mathbf{H}$ is irreducible. Hence $SU(3)/SO(3)$ is irreducible.

That $SU(3)/SO(3)$ does not admit a non-isometric conformal transformation is a consequence of the fact that it is not isometric with a sphere in the given metric.

F. The conformal transformation group [79]

1. Show that a compact homogeneous Riemannian manifold M of dimension $n > 3$ which admits a non-isometric conformal transformation, that is, for which $C_0(M) \neq I_0(M)$ (cf. §3.7) is isometric with a sphere.

To see this, let $G = I_0(M)$ and $M = G/K$. The subgroup K need not be connected. Since G is compact, it can be shown that the fundamental group of M is finite. Indeed, the first betti number of M is zero by theorem 3.7.5. Secondly, M is conformally flat provided $n > 3$. For, if X is an infinitesimal conformal transformation

$$\begin{aligned}\theta(X)\langle C, C\rangle &= \langle\theta(X)\,C, C\rangle + \frac{4}{n}\,\delta\xi\,\langle C, C\rangle \\ &= \frac{4}{n}\,\delta\xi\,\langle C, C\rangle\end{aligned}$$

where C is the conformal curvature tensor. This formula is an immediate consequence of (3.7.4) and the fact that $\theta(X)\,C = 0$. The manifold M being homogeneous, and the tensor C being invariant by $I_0(M)$, $\langle C, C\rangle$ is a constant. Therefore, if X is not an infinitesimal isometry, $\delta\xi \neq 0$, from which $\langle C, C\rangle = 0$, that is, C must vanish. Hence, if $n > 3$, M is conformally flat.

Let $\tilde{M}$ be the universal covering space of M. Since, the fundamental group of M is finite, $\tilde{M}$ is compact. Since M is conformally flat, so is $\tilde{M}$. Thus, $\tilde{M}$ is isometric with a sphere. We have invoked the theorem that *a compact, simply connected, conformally flat Riemannian manifold is conformal with a sphere* [*83*]. The manifold M is consequently an Einstein space. It is therefore isometric with a sphere (cf. III.E.1).

CHAPTER IV

COMPACT LIE GROUPS

The results of the previous chapter are now applied to the problem of determining the betti numbers of a compact semi-simple Lie group G. On the one hand, we employ the facts on curvature and betti numbers already established, and on the other hand, the theory of invariant differential forms. It turns out that the harmonic forms on G are precisely those differential forms invariant under both the left and right translations of G. The conditions of invariance when expressed analytically reduce the problem of the determination of betti numbers to a purely algebraic one. No effort is made to compute the betti numbers of the four main classes of simple Lie groups since this discussion is beyond the scope of this book. However, for the sake of completeness, we give the Poincaré polynomials in these cases omitting those for the five exceptional simple Lie groups.

Locally, G has the structure of an Einstein space of positive curvature and this fact is used to prove that the first and second betti numbers vanish. These results are also obtained from the theory of invariant differential forms. The existence of a harmonic 3-form is established from differential geometric considerations and this fact allows us to conclude that the third betti number is greater than or equal to one. It is also shown that the Euler-Poincaré characteristic is zero.

4.1. The Grassman algebra of a Lie group

Consider a compact (connected) Lie group G. Its Lie algebra L has as underlying vector space the tangent space T_e at the identity $e \in G$. We have seen (§ 3.6) that an element $A \in T_e$ determines a unique left invariant infinitesimal transformation which takes the value A at e; moreover, these infinitesimal transformations are the elements of L. Let $X_\alpha(\alpha = 1, \cdots, n)$ be a base of the Lie algebra L and $\omega^\alpha(\alpha = 1, \cdots, n)$

the dual base for the *forms of Maurer-Cartan*, that is the base such that $\omega^\alpha(X_\beta) = \delta^\alpha_\beta\,(\alpha, \beta = 1, \cdots, n)$. (In the sequel, Greek indices refer to vectors, tensors, and forms on T_e and its dual.) A differential form α is said to be *left invariant* if it is invariant by every $L_a (a \in G)$, that is, if $L_a^* \alpha = \alpha$ for every $a \in G$ where L_a^* is the induced map in $\wedge(T^*)$. The forms of Maurer-Cartan are left invariant pfaffian forms. For an element $X \in L$ and an element α in the dual space, $\alpha(X)$ is constant on G. Hence, by lemma 3.5.2

$$\langle X \wedge Y, d\alpha\rangle = -\alpha([X,Y]) \tag{4.1.1}$$

where X, Y are any elements of L and α any element of the dual space. If we write

$$[X_\beta, X_\gamma] = C_{\beta\gamma}{}^\alpha X_\alpha, \tag{4.1.2}$$

then, from (4.1.1)

$$d\omega^\alpha = -\frac{1}{2} C_{\beta\gamma}{}^\alpha \omega^\beta \wedge \omega^\gamma. \tag{4.1.3}$$

The constants $C_{\beta\gamma}{}^\alpha$ are called the *constants of structure* of L with respect to the base $\{X_1, \cdots, X_n\}$. These constants are not arbitrary since they must satisfy the relations

$$[X_\alpha, X_\beta] + [X_\beta, X_\alpha] = 0 \tag{4.1.4}$$

and

$$[X_\alpha, [X_\beta, X_\gamma]] + [X_\beta, [X_\gamma, X_\alpha]] + [X_\gamma, [X_\alpha, X_\beta]] = 0, \tag{4.1.5}$$

$\alpha, \beta, \gamma = 1, \cdots, n$, that is

$$C_{\beta\gamma}{}^\alpha + C_{\gamma\beta}{}^\alpha = 0 \tag{4.1.6}$$

and

$$C_{\alpha\beta}{}^\rho C_{\gamma\rho}{}^\delta + C_{\beta\gamma}{}^\rho C_{\alpha\rho}{}^\delta + C_{\gamma\alpha}{}^\rho C_{\beta\rho}{}^\delta = 0. \tag{4.1.7}$$

The equations (4.1.3) are called the *equations of Maurer-Cartan.*

Since the induced dual maps L_a^* $(a \in G)$ commute with d, we have

$$L_a^* \cdot d\alpha = dL_a^* \alpha = d\alpha$$

for any Maurer-Cartan form α, that is, if α is a left invariant 1-form, $d\alpha$ is a left invariant 2-form. This also follows from (4.1.3). More generally, if $A_{\alpha_1 \ldots \alpha_p}$ are any constants, the p-form $A_{\alpha_1 \ldots \alpha_p} \omega^{\alpha_1} \wedge \ldots \wedge \omega^{\alpha_p}$ is a left invariant differential form on G. That any left invariant differential form of degree $p > 0$ may be expressed in this manner is clear. A left invariant form may be considered as an alternating multilinear form on the Lie algebra L of G. We may therefore identify the left invariant forms with the homogeneous elements of the Grassman algebra associated with L. The number of linearly independent left invariant p-forms is therefore equal to $\binom{n}{p}$.

Lemma 4.1.1. *The underlying manifold of the Lie group G is orientable.*

Indeed, the n-form $\omega^1 \wedge \ldots \wedge \omega^n$ on G is continuous and different from zero everywhere. G may then be oriented by the requirement that this form is positive everywhere (cf. § 1.6).

The Lie group G is thus a compact, connected, orientable analytic manifold.

4.2. Invariant differential forms

For any $X \in L$, let $ad(X)$ be the map $Y \to [X, Y]$ of L into itself. It is clear that $X \to ad(X)$ is a linear map, and so, since

$$ad([X_1,X_2])Y = [[X_1,X_2],Y] = -[[X_2,Y],X_1] - [[Y,X_1],X_2]$$
$$= (ad(X_1)ad(X_2) - ad(X_2)ad(X_1))Y$$

we conclude that $X \to ad(X)$ is a representation. It is called the *adjoint representation* of L (cf. § 3.6).

Let $\bar{\theta}(X)$ be the (unique) derivation of $\wedge(T_e)$ which coincides with $ad(X)$ on $T_e = \wedge^1(T_e)$ defined by

$$\bar{\theta}(X)(X_1 \wedge \ldots \wedge X_p) = \sum_{\alpha=1}^{p} X_1 \wedge \ldots \wedge [X,X_\alpha] \wedge \ldots \wedge X_p.$$

Define the endomorphism $\theta(X)$ $(X \in L)$ of $\wedge(T_e^*)$ by

$$\langle X_1 \wedge \ldots \wedge X_p, -\theta(X)(\alpha^1 \wedge \ldots \wedge \alpha^p)\rangle$$
$$= \langle \bar{\theta}(X)(X_1 \wedge \ldots \wedge X_p), \alpha^1 \wedge \ldots \wedge \alpha^p\rangle$$

where $\alpha^1, \cdots, \alpha^p$ are any elements of $\wedge^1(T_e^*)$ (cf. II.A.4).

Lemma 4.2.1. *$\theta(X)$ is a derivation.*

If Δ^α_β denotes the minor obtained by deleting the row α and column β of the matrix $(\langle X_\alpha, \alpha^\beta\rangle)$,

$$\langle X_1 \wedge \ldots \wedge X_p, -\theta(X)(\alpha^1 \wedge \ldots \wedge \alpha^p)\rangle$$
$$= \langle \bar{\theta}(X)(X_1 \wedge \ldots \wedge X_p), \alpha^1 \wedge \ldots \wedge \alpha^p\rangle$$
$$= \sum_{\gamma=1}^{p} \langle X_1 \wedge \ldots \wedge [X,X_\gamma] \wedge \ldots \wedge X_p, \alpha^1 \wedge \ldots \wedge \alpha^p\rangle$$

$$= \langle ad(X)\, X_\rho, \alpha^\sigma \rangle\, \Delta^\rho_\sigma$$
$$= \langle X_\rho, -\theta(X)\, \alpha^\sigma \rangle\, \Delta^\rho_\sigma$$
$$= \sum_{\gamma=1}^{p} \langle X_1 \wedge \dots \wedge X_p, \alpha^1 \wedge \dots \wedge -\theta(X)\, \alpha^\gamma \wedge \dots \wedge \alpha^p \rangle.$$

It follows that

$$\theta(X)\,(\alpha^1 \wedge \dots \wedge \alpha^p) = \sum_{\gamma=1}^{p} \alpha^1 \wedge \dots \wedge \theta(X)\, \alpha^\gamma \wedge \dots \wedge \alpha^p,$$

that is, $\theta(X)$ is a derivation.

Lemma 4.2.2. $\theta(X_\beta)\omega^\alpha = C_{\gamma\beta}{}^\alpha \omega^\gamma$.

Indeed,

$$\begin{aligned}\langle X_\gamma, -\theta(X_\beta)\omega^\alpha \rangle &= \langle \bar{\theta}(X_\beta)\, X_\gamma, \omega^\alpha \rangle \\ &= \langle [X_\beta, X_\gamma], \omega^\alpha \rangle \\ &= C_{\beta\gamma}{}^\rho \langle X_\rho, \omega^\alpha \rangle \\ &= C_{\beta\gamma}{}^\alpha.\end{aligned}$$

Lemma 4.2.3. $\theta(X) = di(X) + i(X)d$.

It suffices to verify this formula for forms of degree 0 and 1 in $\wedge(T_e^*)$ —the Grassman algebra associated with L. The identity is trivial for forms of degree 0 since they are constant functions. In degree 1 we need only consider the forms ω^α. Then, $\theta(X_\beta)\omega^\alpha = C_{\gamma\beta}{}^\alpha \omega^\gamma$. But,

$$\begin{aligned}(di(X_\beta) + i(X_\beta)d)\, \omega^\alpha &= i(X_\beta)\, d\omega^\alpha \\ &= -\tfrac{1}{2} i(X_\beta)\, C_{\gamma\rho}{}^\alpha\, \omega^\gamma \wedge \omega^\rho \\ &= -\tfrac{1}{2} C_{\gamma\rho}{}^\alpha\, (i(X_\beta)\, \omega^\gamma \wedge \omega^\rho - \omega^\gamma\, i(X_\beta)\, \omega^\rho) \\ &= -\tfrac{1}{2} (C_{\beta\rho}{}^\alpha\, \omega^\rho - C_{\gamma\beta}{}^\alpha\, \omega^\gamma) \\ &= C_{\gamma\beta}{}^\alpha\, \omega^\gamma.\end{aligned}$$

Corollary 4.2.3. $\theta(X)d = d\theta(X)$.

Lemma 4.2.4. $d = \frac{1}{2}\epsilon(\omega^\alpha)\theta(X_\alpha)$.

It is only necessary to verify this formula for the forms of degrees 0 and 1 in $\wedge(T_e^*)$. Again, since the forms of degree 0 are the constant functions on G both sides vanish. For a form of Maurer-Cartan ω^β

$$\begin{aligned}\tfrac{1}{2}\epsilon(\omega^\alpha)\theta(X_\alpha)\, \omega^\beta &= \tfrac{1}{2}\epsilon(\omega^\alpha)\, C_{\gamma\alpha}{}^\beta\, \omega^\gamma \\ &= -\tfrac{1}{2} C_{\alpha\gamma}{}^\beta\, \omega^\alpha \wedge \omega^\gamma \\ &= d\omega^\beta.\end{aligned}$$

Let β be an element of $\wedge^p(T_e^*)$. Then, β is a left invariant p-form on G, and so may be expressed in the form $\beta = B_{\alpha_1 \ldots \alpha_p}\, \omega^{\alpha_1} \wedge \ldots \wedge \omega^{\alpha_p}$ where the coefficients are constants. Applying lemma 4.2.2 we obtain the formula

$$\theta(X_\alpha)\beta = \sum_{r=1}^{p} B_{\alpha_1 \cdots \alpha_{r-1}\rho\alpha_{r+1} \cdots \alpha_p}\, C_{\alpha_r \alpha}{}^{\rho}\, \omega^{\alpha_1} \wedge \cdots \wedge \omega^{\alpha_p}.$$

It follows from lemma 4.2.4 that

$$d\beta = \frac{1}{2}\sum_{r=1}^{p} B_{\alpha_1 \cdots \alpha_{r-1}\rho\alpha_{r+1} \cdots \alpha_p}\, C_{\alpha_r \alpha}{}^{\rho}\, \omega^{\alpha} \wedge \omega^{\alpha_1} \wedge \ldots \wedge \omega^{\alpha_p}.$$

An element β of the *Grassman* algebra of G is said to be *L-invariant* or, simply, *invariant* if it is a zero of every derivation $\theta(X)$, $X \in L$, that is, if $\theta(X)\beta = 0$ for every left invariant vector field X. Hence, an invariant differential form is bi-invariant.

Proposition 4.2.1. *An invariant form is a closed form.*

This is an immediate consequence of lemma 4.2.4.

Remark: Note that the operator $\theta(X)$ of § 3.5 coincides with the operator $\theta(X)$ defined here on forms only.

4.3. Local geometry of a compact semi-simple Lie group

From (4.1.2) it is seen that the structure constants are the components of a tensor on T_e of type (1,2). A new tensor on T_e is defined by the components

$$g_{\alpha\beta} = C_{\alpha\sigma}{}^{\rho}\, C_{\rho\beta}{}^{\sigma} \tag{4.3.1}$$

relative to the base $X_\alpha(\alpha = 1, \cdots, n)$. It follows from (4.1.6) and (4.1.7) that this tensor is symmetric. It can be shown that a necessary and sufficient condition for G to be semi-simple is that the rank of the matrix $(g_{\alpha\beta})$ is n. (A Lie group is said to be *semi-simple* if the fundamental bilinear symmetric form—trace *ad X ad Y* is non-degenerate). Moreover, since G is compact it can be shown that $(g_{\alpha\beta})$ is positive definite.

The tensor defined by the equations (4.3.1) may now be used to raise and lower indices and for this purpose we consider the inverse matrix $(g^{\alpha\beta})$. The structure constants have yet another symmetry property. Indeed, if we multiply the identities (4.1.7) by $C_{\sigma\delta}{}^{\alpha}$ and contract we find that the tensor

$$C_{\alpha\beta\gamma} = g_{\gamma\sigma}\, C_{\alpha\beta}{}^{\sigma} \tag{4.3.2}$$

is skew-symmetric.

In terms of a system of local coordinates $u^1, \cdots, u^n$ the vector fields $X_\alpha(\alpha = 1, \cdots, n)$ may be expressed as $X_\alpha = \xi^i_\alpha(\partial/\partial u^i)$. Since G is completely parallelisable, the $n \times n$ matrix (ξ^i_α) has rank n, and so, if we put

$$g^{ij} = \xi^i_\alpha \, \xi^j_\beta \, g^{\alpha\beta} \tag{4.3.3}$$

the matrix (g^{ij}) is positive definite and symmetric. We may therefore define a metric g on G by means of the quadratic form

$$ds^2 = g_{jk} \, du^j \, du^k \tag{4.3.4}$$

where the g_{jk} are elements of the matrix inverse to (g^{jk}). Again, the metric tensor g may be used to raise and lower indices in the usual manner. It should be remarked that the metric is completely determined by the group G.

We now define n covariant vector fields $v^\alpha(\alpha = 1, \cdots, n)$ on G with components $\xi^\alpha_i(i = 1, \cdots, n)$ (relative to the given system of local coordinates) by the formulae

$$\xi^\alpha_i = g^{\alpha\beta} \, \xi^j_\beta \, g_{ij}. \tag{4.3.5}$$

It follows easily that

$$\xi^i_\alpha \, \xi^\alpha_j = \delta^i_j \quad \text{and} \quad \xi^i_\alpha \, \xi^\beta_i = \delta^\beta_\alpha. \tag{4.3.6}$$

However, it does not follow that, in the metric g the $X_\alpha(\alpha = 1, \cdots, n)$ are orthonormal vectors at each point of G.

A set of n^2 linear differential forms $\omega^i_j = \Gamma^i_{jk} du^k$ is introduced in each coordinate neighborhood by putting

$$\Gamma^i_{jk} = \xi^i_\alpha \frac{\partial \xi^\alpha_j}{\partial u^k}. \tag{4.3.7}$$

By virtue of the equations (4.3.6) the Γ^i_{jk} may be written as

$$\Gamma^i_{jk} = -\frac{\partial \xi^i_\alpha}{\partial u^k} \xi^\alpha_j.$$

It is easily verified that equations (1.7.3) are satisfied in the overlap of two coordinate neighborhoods. The n^2 forms ω^i_j in each coordinate neighborhood define therefore an affine connection on G. The torsion tensor $T_{jk}{}^i$ of this connection may be written as

$$T_{jk}{}^i = \frac{1}{2} \xi^i_\alpha \left(\frac{\partial \xi^\alpha_j}{\partial u^k} - \frac{\partial \xi^\alpha_k}{\partial u^j}\right). \tag{4.3.8}$$

(The factor $\frac{1}{2}$ is introduced for reasons of convenience (cf. 1.7.18)).

Since the equations (4.1.2) may be expressed in terms of the local coordinates (u^i) in the form

$$\xi^r_\beta \frac{\partial \xi^i_\gamma}{\partial u^r} - \xi^r_\gamma \frac{\partial \xi^i_\beta}{\partial u^r} = C_{\beta\gamma}{}^\alpha\, \xi^i_\alpha, \tag{4.3.9}$$

it is easy to check that

$$T_{jk}{}^i = \tfrac{1}{2} C_{\beta\gamma}{}^\alpha\, \xi^i_\alpha\, \xi^\beta_j\, \xi^\gamma_k \tag{4.3.10}$$

from which we conclude that the covariant torsion tensor

$$T_{jkl} = g_{il}\, T_{jk}{}^i \tag{4.3.11}$$

is skew-symmetric. It follows from (1.9.12) that

$$\Gamma^i_{jk} = \left\{{}^{\,i}_{j\,k}\right\} + T_{jk}{}^i$$

where the $\left\{{}^{\,i}_{j\,k}\right\}$ are the coefficients of the Levi Civita connection. Hence, from (4.3.7)

$$\left\{{}^{\,i}_{j\,k}\right\} = \frac{1}{2}\, \xi^i_\alpha \left(\frac{\partial \xi^\alpha_j}{\partial u^k} + \frac{\partial \xi^\alpha_k}{\partial u^j}\right). \tag{4.3.12}$$

Lemma 4.3.1. *The elements of the Lie algebra L of G define translations in G.*

Indeed, from (4.3.12) and (4.3.8)

$$D_k\, \xi^\alpha_j = T_{jk}{}^i\, \xi^\alpha_i \tag{4.3.13}$$

where D_k is the operator of covariant differentiation with respect to the Levi Civita connection. Multiplying these equations by ξ^l_α and contracting we obtain

$$-\, \xi^\alpha_j\, D_k\, \xi^l_\alpha = T_{jk}{}^l.$$

Again, if we multiply by ξ^j_β and contract, the result is

$$D_k\, \xi^l_\beta = T_{kj}{}^l\, \xi^j_\beta. \tag{4.3.14}$$

These equations may be rewritten in the form

$$D_k\, \xi^\beta_l = T_{lk}{}^j\, \xi^\beta_j$$

from which we conclude that $\theta(X_\beta)g = 0$.

4.4. Harmonic forms on a compact semi-simple Lie group

In terms of the metric (4.3.4) on G the star operator may be defined and we are then able to prove the following

Proposition 4.4.1. *Let α be an invariant p-form on G. Then,*

(i) *$d\alpha$ is invariant;*
(ii) *$*\alpha$ is invariant, and*
(iii) *if $\alpha = d\beta$, β is invariant.*

Let X be an element of the Lie algebra L of G. Then, $\theta(X)d\alpha = d\theta(X)\alpha = 0$; $\theta(X)*\alpha = *\theta(X)\alpha = 0$ by formulae (3.7.7) and (3.7.11). Hence, (i) and (ii) are established. By the decomposition theorem of § 2.9 we may write $\alpha = d\delta G\alpha$ where G is the Green's operator (cf. II.B.4). Since $\delta = (-1)^{np+n+1}*d*$ on p-forms we may put $\alpha = d*d\gamma$ where γ is some $(n-p)$-form. Then, $0 = \theta(X)\alpha = \theta(X)d*d\gamma = d\theta(X)*d\gamma = d*\theta(X)d\gamma = d*d\theta(X)\gamma$, from which $\delta d\theta(X)\gamma = (-1)^{np+1}*d*d\theta(X)\gamma = 0$. Since $(\delta d\theta(X)\gamma, \theta(X)\gamma) = (d\theta(X)\gamma, d\theta(X)\gamma)$ and $\theta(X)d\gamma = d\theta(X)\gamma$, $d\gamma$ is invariant. Thus, from (ii), $*d\gamma$ is invariant. This completes the proof of (iii).

Proposition 4.4.2. *The harmonic forms on G are invariant.*

This follows from lemma 4.3.1 and theorem 3.7.1.

Proposition 4.4.3. *The invariant forms on G are harmonic.*

Indeed, if β is an invariant p-form it is co-closed. For, by lemma 4.2.4,

$$\begin{aligned}\delta\beta = (-1)^{np+n+1}*d*\beta &= \tfrac{1}{2}(-1)^{np+n+1}*\epsilon(\omega^\alpha)\,\theta(X_\alpha)*\beta \\ &= \tfrac{1}{2}(-1)^{np+n+1}*\epsilon(\omega^\alpha)**^{-1}\,\theta(X_\alpha)*\beta \\ &= -\tfrac{1}{2}\sum_{\alpha=1}^{n} i(X_\alpha)\,\theta(X_\alpha)\,\beta = 0. \end{aligned} \tag{4.4.1}$$

Hence, by prop. 4.2.1, β is harmonic.

Note that prop. 4.4.1 is a trivial consequence of prop. 4.4.3.

Therefore, in order to find the harmonic forms β on a compact Lie group G we need only solve the equations

$$\sum_{r=1}^{p} B_{\alpha_1\cdots\alpha_{r-1}\rho\alpha_{r+1}\cdots\alpha_p}\, C_{\alpha_r\alpha}{}^{\rho} = 0 \tag{4.4.2}$$

where $\beta = B_{\alpha_1\cdots\alpha_p}\,\omega^{\alpha_1}\wedge\ldots\wedge\omega^{\alpha_p}$. The problem of determining the betti numbers of G has as a result been reduced to purely algebraic considerations.

Remarks : In proving prop. 4.4.3 we obtained the formula

$$\delta = -\frac{1}{2}\sum_{\alpha=1}^{n} i(X_\alpha)\theta(X_\alpha)$$

thereby showing that δ is an anti-derivation in $\wedge(T_e^*)$. (The proposition could have been obtained by an application of the Hodge-de Rham decomposition of a form). *It follows that the exterior product of harmonic forms on a compact semi-simple Lie group is also harmonic.*

Theorem 4.4.1. *The first and second betti numbers of a compact semi-simple Lie group G vanish.*

Let $\beta = B_\alpha\omega^\alpha$ be a harmonic 1-form. Then, from (4.4.2), $B_\rho\, C_{\alpha_1\alpha}{}^\rho = 0$. Multiplying these equations by $C_\gamma{}^{\alpha\alpha_1} = g^{\alpha\rho}\, C_{\gamma\rho}{}^{\alpha_1}$ and contracting results in $B_\gamma = 0$, $\gamma = 1, \cdots, n$.

If $\alpha = A_{\alpha\beta}\,\omega^\alpha \wedge \omega^\beta$ is a harmonic 2-form, then by (4.4.2)

$$A_{\rho\beta}\, C_{\alpha\gamma}{}^\rho + A_{\alpha\rho}\, C_{\beta\gamma}{}^\rho = 0, \quad \gamma = 1, \cdots, n. \tag{4.4.3}$$

Permuting α, β and γ cyclically and adding the three equations obtained gives

$$A_{\rho\beta}\, C_{\alpha\gamma}{}^\rho + A_{\rho\alpha}\, C_{\gamma\beta}{}^\rho + A_{\rho\gamma}\, C_{\beta\alpha}{}^\rho = 0,$$

and so from (4.4.3)

$$A_{\rho\gamma}\, C_{\beta\alpha}{}^\rho = 0, \quad \gamma = 1, \cdots, n.$$

Multiplying these equations by $C_\delta{}^{\alpha\beta}$ results in $A_{\gamma\delta} = 0$ ($\gamma, \delta = 1, \cdots, n$).

Suppose G is a compact but not necessarily semi-simple Lie group. We have seen that the number of linearly independent left invariant differential forms of degree p on G is $\binom{n}{p}$. If we assume that $b_p(G) = \binom{n}{p}$, then the Euler characteristic $\chi(G)$ of G is zero. For,

$$\chi(G) = \sum_{p=0}^{n}(-1)^p \binom{n}{p} = 0.$$

(This is not, however, a special implication of $b_p(G) = \binom{n}{p}$ (cf. theorem 4.4.3)).

A compact (connected) abelian Lie group G has these properties. For, since G is abelian so is its Lie algebra L. Therefore, by (4.1.2) its structure constants vanish. A metric g is defined on G as follows:

$$g^{ij} = \sum_{\alpha=1}^{n} \xi_\alpha^i\, \xi_\alpha^j.$$

Now, by lemma 4.2.2, $\theta(X_\beta)\omega^\alpha = 0$, $\alpha, \beta = 1, \cdots, n$, that is the ω^α are invariant. Hence, by the proof of prop. 4.4.3 they are harmonic with respect to g. Since $\theta(X)$, $X \in L$ is a derivation, $\theta(X)\dot{\alpha} = 0$ for any left invariant p-form α. We conclude therefore that $b_p(G) = \binom{n}{p}$.

Theorem 4.4.2. *A compact connected abelian Lie group G is a multi-torus.*

To prove this we need only show that the vector fields $X_\alpha(\alpha = 1, \cdots, n)$ are parallel in the constructed metric. (This is left as an exercise for the reader.) For, by applying the interchange formulae (1.7.19) to the $X_\alpha(\alpha = 1, \cdots, n)$ and using the fact that the X_α are linearly independent vector fields we conclude that G is locally flat. However, a compact connected group which is locally isomorphic with E^n (as a topological group) is isomorphic with the n-dimensional torus.

We have seen that the Euler characteristic of a torus vanishes. It is now shown that for a compact connected semi-simple Lie group G, $\chi(G) = 0$. Indeed, the proof given is valid for any compact Lie group. Let ν_p denote the number of linearly independent left invariant p-forms no linear combination of which is closed; ν_{p-1} is then the number of linearly independent exact p-forms. Since the dimension of $\wedge^p(T_e^*)$ is $\binom{n}{p}$ we have by the decomposition of a p-form

$$\binom{n}{p} = b_p(G) + \nu_p + \nu_{p-1}.$$

Hence,

$$\begin{aligned}\chi(G) &= \sum_{p=0}^{n} (-1)^p\, b_p(G)\\ &= \sum_{p=0}^{n} (-1)^p \binom{n}{p} + \sum_{p=0}^{n} (-1)^{p+1}\, \nu_p + \sum_{p=1}^{n} (-1)^{p+1}\, \nu_{p-1}\\ &= (-1)^{n+1}\, \nu_n - \nu_0,\end{aligned}$$

and so, since $\nu_0 = \nu_n = 0$, $\chi(G) = 0$.

Theorem 4.4.3. *The Euler characteristic of a compact connected Lie group vanishes.*

4.5. Curvature and betti numbers of a compact semi-simple Lie group G

In this section we make use of the curvature properties of G in order to prove theorem 4.4.1. We begin by forming the curvature tensor defined by the connection (4.3.7). Denoting the components of this tensor by $E^i{}_{jkl}$ with respect to a given system of local coordinates $u^1, \cdots, u^n$ we obtain

$$E^i{}_{jkl} = R^i{}_{jkl} + D_l\, T_{jk}{}^i - D_k\, T_{jl}{}^i + T_{jk}{}^r\, T_{rl}{}^i - T_{jl}{}^r\, T_{rk}{}^i$$

where the $R^i{}_{jkl}$ are the components of the Riemannian curvature tensor. Since the $E^i{}_{jkl}$ all vanish and since $D_l T_{jk}{}^i = 0$, it follows from the Jacobi identity that

$$R^i{}_{jkl} = T_{kl}{}^r\, T_{rj}{}^i. \tag{4.5.1}$$

By virtue of the equations (4.3.1) and (4.3.10)

$$T_{rk}{}^s\, T_{js}{}^r = \tfrac{1}{4} g_{jk}.$$

Hence, forming the Ricci tensor by contracting on i and l in (4.5.1) we conclude that

$$R_{jk} = \tfrac{1}{4} g_{jk}. \tag{4.5.2}$$

It follows that G is locally an Einstein space with positive scalar curvature, and so by theorem 3.2.1, the first betti number of G is zero.

In order to prove that $b_2(G)$ is also zero we establish the following

Lemma 4.5.1. *In a coordinate neighborhood U of G with the local coordinates (u^i) $(i = 1, \cdots, n)$, we have the inequalities*

$$0 \geqq R_{ijkl}\, \xi^{ij}\, \xi^{kl} \geqq -\tfrac{1}{2} \langle \xi, \xi \rangle$$

where the $\xi_{ij} = -\xi_{ji}$ are functions in U defining a skew-symmetric tensor field ξ of type (0,2) and $\xi = \xi_{(ij)}\, du^i \wedge du^j$ [74].

In general, the curvature tensor defines a symmetric linear transformation of the space of bivectors (cf. I.I.). The above inequality says it is negative definite with eigenvalues between 0 and $-\frac{1}{4}$.

Since the various sides of the inequalities are scalar functions on G the lemma may be proved by choosing a special system of local coordinates. In fact, we fix a point O of G and choose (geodesic) coordinates so that at O, $g_{ij} = \delta_{ij}$. Then, since

$$\sum_{r,s=1}^{n} T_{jrs}\, T_{krs} = \tfrac{1}{4}\, \delta_{jk},$$

$$\sum_{r<s} (2\sqrt{2}\, T_{jrs})\, (2\sqrt{2}\, T_{krs}) = \delta_{jk},$$

and so the $2\sqrt{2}\, T_{jrs}$ $(r < s,\, j = 1, \cdots, n)$ represent n orthonormal vector fields in $E^{n(n-1)/2}$. We denote by T_{Ars} $(r < s,\ A = n+1, \cdots, n(n-1)/2)$, $(n(n-1)/2) - n$ orthonormal vectors in $E^{n(n-1)/2}$ orthogonal to the vectors T_{jrs}. Hence,

$$\sum_{s=1}^{n} (2\sqrt{2}\, T_{ijs})\,(2\sqrt{2}\, T_{kls}) + \sum_{A=n+1}^{n(n-1)/2} T_{ijA}\, T_{klA} = \delta_{(ij)\,(kl)}$$

for $i < j,\ k < l$ ($\delta_{(ij)(kl)} = 1$ if $i = k,\ j = l$ and vanishes otherwise), and so

$$8\sum_{s=1}^{n} \Big(\sum_{i<j} T_{ijs}\, \xi_{ij}\Big)^2 + \sum_{A=n+1}^{n(n-1)/2} \Big(\sum_{i<j} T_{ijA}\, \xi_{ij}\Big)^2 = \sum_{i<j} (\xi_{ij})^2.$$

We may therefore conclude that

$$\sum_{s=1}^{n}\sum_{i,j}\sum_{k,l} T_{ijs}\, T_{kls}\, \xi_{ij}\, \xi_{kl} \leqq \tfrac{1}{2}\, \langle \xi, \xi \rangle.$$

This completes the proof.

A straightforward application of theorem 3.2.4 shows that $b_2(G) = 0$ by virtue of the lemma and formula (4.5.2).

Theorem 4.5.1. $b_3(G) \geqq 1$.

For, the torsion tensor (4.3.11) defines a harmonic 3-form on G.

For more precise information on $b_3(G)$ the reader is referred to (IV.B).

4.6. Determination of the betti numbers of the simple Lie groups

We have seen that a p-form on a compact semi-simple Lie group G is harmonic, if and only if, it is invariant and therefore, in order to find the harmonic forms β on G, it is sufficient to solve the equations (4.4.2) for the coefficients $B_{\alpha_1 \cdots \alpha_p}$ of β.

A semi-simple group is the direct product of a finite number of simple non-commutative groups. (A Lie group is said to be simple if there are no non-trivial normal subgroups). Hence, in order to give a complete classification of compact semi-simple Lie groups it is sufficient to classify the compact simple Lie groups. There are four main classes of simple Lie groups:

1) The group A_l of unitary transformations in $(l+1)$-space of determinant $+1$;

2) The group B_l: this is the orthogonal group in $(2l + 1)$-space the elements of which have determinant $+ 1$;

3) The group C_l: this is the symplectic group in $2l$-space, that is C_l is the group of unitary transformations leaving invariant the skew-symmetric bilinear form $\sum_{i,j=1}^{2l} a_{ij}x_iy_j$ where the coefficients are given by $a_{2r-1,2r} = - a_{2r,2r-1} = 1$ with all other $a_{ij} = 0$;

4) The group D_l of orthogonal transformations in $2l$-space ($l = 3, 4, \cdots$), the elements of which have determinant $+ 1$.

There are also five exceptional compact simple Lie groups whose dimensions are 14, 52, 78, 133, and 248 commonly denoted by G_2, F_4, E_6, E_7, and E_8, respectively.

The polynomial $p_G(t) = b_0 + b_1t + \cdots + b_nt^n$ where the b_i ($i = 0, \cdots, n$) are the betti numbers of G is known as the *Poincaré polynomial* of G. Let $G = G_1 \times \cdots \times G_k$ where the G_i ($i = 1, \cdots, k$) are simple. Then, it can be shown that

$$p_G(t) = p_{G_1}(t) \dots p_{G_k}(t) \tag{4.6.1}$$

where $p_{G_i}(t)$ is the Poincaré polynomial of G_i. Therefore, in order to find the betti numbers of a compact semi-simple Lie group we first express it as the direct product of simple Lie groups, and then compute the Poincaré polynomials of these groups, after which we employ the formula (4.6.1).

Regarding the topology of a compact simple Lie group we already know that (a) it is orientable; (b) $b_1 = b_2 = 0$, $b_3 \geqq 1$ and, therefore, since the star operator is an isomorphism (or, by Poincaré duality) $b_{n-2} = b_{n-1} = 0$, $b_{n-3} \geqq 1$; (c) the Euler characteristic vanishes.

We conclude this chapter by giving (without proof) the Poincaré polynomials of the four main classes of simple Lie groups:

$$p_{A_l}(t) = (1 + t^3)(1 + t^5) \dots (1 + t^{2l+1}),$$

$$p_{B_l}(t) = (1 + t^3)(1 + t^7) \dots (1 + t^{4l-1}),$$

$$p_{C_l}(t) = (1 + t^3)(1 + t^7) \dots (1 + t^{4l-1}),$$

$$p_{D_l}(t) = (1 + t^3)(1 + t^7) \dots (1 + t)^{2l-1}(1 + t^{4l-5}),\ l > 2.$$

Remark: $A_1 = B_1 = C_1$, $B_2 = C_2$ and $A_3 = D_3$.

EXERCISES

A. The second betti number of a compact semi-simple Lie group

1. Prove that $b_2(G) = 0$ by showing that if α is an harmonic 2-form, then $i(X)\alpha$ vanishes for any $X \in L$. Make use of the fact that $b_1(G) = 0$.

B. The third betti number of a compact simple Lie group [48]

1. Let $Q(L)$ denote the vector space of invariant bilinear symmetric forms on L, that is, the space of those forms q such that

$$q(X,Y) = q(Y,X) \quad \text{and} \quad q(\theta(Z)X,Y) = q(X,\theta(Z)Y)$$

for any $X,Y,Z \in L$. To each $q \in Q(L)$ we associate a 3-form $\alpha(q)$ by the condition

$$\langle X \wedge Y \wedge Z, \alpha(q)\rangle = q(\theta(X)Y,Z).$$

Evidently, the map

$$q \to \alpha(q)$$

is linear.

2. For each $q \in Q(L)$ show that $\alpha(q)$ is invariant, and hence harmonic.

3. Since the derived algebra $[L,L] = \{[X,Y] \mid X,Y \in L\}$ coincides with L, the map $q \to \alpha(q)$ of

$$Q(L) \to \wedge^3_H(T^*)$$

is an isomorphism into. Show that it is an isomorphism onto. Hence, $b_3(G) = \dim Q(L)$.

Hint: For any element $\alpha \in \wedge^3_H(T^*)$ and $X \in L$, $i(X)\alpha$ is closed. Since $b_2(G) = 0$, there is a 1-form $\beta = \beta_X$ such that $i(X)\alpha = d\beta_X$. Now, show that

$$d\theta(Y)\beta_X = d\beta_{[Y,X]}$$

that is

$$\theta(Y)\beta_X = \beta_{[Y,X]}.$$

Finally, show that the bilinear function

$$q(X,Y) = -\langle X, \beta_Y\rangle$$

is invariant.

4. Prove that if G is a simple Lie group, then $b_3(G) = 1$.

CHAPTER V

COMPLEX MANIFOLDS

In a well-known manner one can associate with an irreducible curve V_1 a real analytic manifold M^2 of two dimensions called the Riemann surface of V_1. Since the geometry of a Riemann surface is conformal geometry, M^2 is not a Riemannian manifold. However, it is possible to define a Riemannian metric on M^2 in such a way that the harmonic forms constructed with this metric serve to establish topological invariants of M^2. In his book on harmonic integrals [*39*], Hodge does precisely this, and in fact, in seeking to associate with any irreducible algebraic variety V_n a Riemannian manifold M^{2n} of $2n$ dimensions he is able to obtain the sought after generalization of a Riemann surface referred to in the introduction to Chapter I. The metric of an M^{2n} has certain special properties that play an important part in the sequel insofar as the harmonic forms constructed with it lead to topological invariants of the manifold. The approach we take is more general and in keeping with the modern developments due principally to A. Weil [*70*, *72*]. We introduce first the concept of a complex structure on a separable Hausdorff space M in analogy with § 1.1. In terms of a given complex structure a Riemannian metric may be defined on M. If this metric is torsion free, that is, if a certain 2-form associated with the metric and complex structure is closed, the manifold is called a Kaehler manifold. As examples, we have complex projective n-space P_n and, in fact, any *projective variety*, that is irreducible algebraic variety holomorphically imbedded without singularities in P_n.

The local geometry of a Kaehler manifold is studied, and in Chapter VI, from these properties, its global structure is determined to some extent. In Chapter VII we further the discussions of Chapter III by considering groups of transformations of Kaehler manifolds—in particular, Kaehler-Einstein manifolds. It may be said that of the diverse applications of the theory of harmonic integrals, those made to Kaehler manifolds are amongst the most interesting.

5.1. Complex manifolds

A *complex analytic* or, simply, a *complex manifold* of complex dimension n is a $2n$-dimensional topological manifold endowed with a complex analytic structure. This concept may be defined in the same way as the concept of a differentiable structure (cf. § 1.1)—the notion of a holomorphic function replacing that of a differentiable function. Indeed, a separable Hausdorff space M is said to have a *complex analytic structure*, or, simply, a *complex structure* if it possesses the properties:

(i) Each point of M has an open neighborhood homeomorphic with an open subset in C_n, the (number) space of n complex variables; that is, there is a finite or countable open covering $\{U_\alpha\}$, and for each α, a homeomorphism $u_\alpha : U_\alpha \to C_n$;

(ii) For any two open sets U_α and U_β with non-empty intersection the map $u_\beta u_\alpha^{-1} : u_\alpha(U_\alpha \cap U_\beta) \to C_n$ is defined by holomorphic functions of the complex coordinates with non-vanishing Jacobian.

The n complex functions defining u_α are called *local complex coordinates* in U_α. The concept of a *holomorphic function* on M or on an open subset of M is defined in the obvious way (cf. V.A.). Every open subset of M has a complex structure, namely, the complex structure induced by that of M (cf. § 5.8).

A complex manifold possesses an underlying real analytic structure. Indeed, corresponding to local complex coordinates $z^1, \cdots, z^n$ we have real coordinates $x^1, \cdots, x^n, y^1, \cdots, y^n$ where

$$z^k = x^k + \sqrt{-1}\, y^k;$$

moreover, in the overlap of two coordinate neighborhoods the real coordinates are related by *analytic* functions with non-vanishing Jacobian (cf. V.A.).

Any real analytic function may be expressed as a formal power series in $z^1, \cdots, z^n, \bar{z}^1, \cdots, \bar{z}^n$ by putting

$$x^k = \frac{1}{2}(z^k + \bar{z}^k), \quad y^k = \frac{1}{2\sqrt{-1}}(z^k - \bar{z}^k),$$

where $\bar{z}^k$ denotes the complex conjugate of z^k. Consequently, whenever real analytic coordinates are required we may employ the coordinates $z^1, \cdots, z^n, \bar{z}^1, \cdots, \bar{z}^n$.

For reasons of motivation we sacrifice details in the remainder of this section, clarifying any misconceptions beginning with § 5.2.

We consider differential forms of class ∞ with complex values on a complex manifold. Let U be a coordinate neighborhood with (complex) coordinates $z^1, \cdots, z^n$. Then, the differentials $dz^1, \cdots, dz^n$ constitute a (complex) base for the differential forms of degree 1. It follows that a differential form of degree p may be expressed in U as a linear combination (with complex-valued coefficients of class ∞) of exterior products of p-forms belonging to the sets $\{dz^i\}$ and $\{d\bar{z}^i\}$. A term consisting of q of the $\{dz^i\}$ and r of the $\{d\bar{z}^i\}$ with $q + r = p$ is said to be of *bidegree* (q, r). A differential form of bidegree (q, r) is a sum of terms of bidegree (q, r). The notion of a form of bidegree (q, r) is independent of the choice of local coordinates since in the overlap of two coordinate neighborhoods the coordinates are related by holomorphic functions. A differential form on M is said to be of bidegree (q, r) if it is of bidegree (q, r) in a neighborhood of each point.

It is now shown that *a complex manifold is orientable*. For, let $z^1, \cdots, z^n$ be a system of local complex coordinates and set $z^k = x^k + \sqrt{-1}y^k$. Then, the x^i and y^j together form a real system of local coordinates. Since $dz^k \wedge d\bar{z}^k = -2\sqrt{-1}dx^k \wedge dy^k$,

$$dx^1 \wedge \ldots \wedge dx^n \wedge dy^1 \wedge \ldots \wedge dy^n = \frac{\sqrt{-1}^n}{2^n} dz^1 \wedge \ldots \wedge dz^n \wedge d\bar{z}^1 \wedge \ldots \wedge d\bar{z}^n.$$

It follows that the form

$$\Theta(z) = \frac{\sqrt{-1}^n}{2^n} dz^1 \wedge \ldots \wedge dz^n \wedge d\bar{z}^1 \wedge \ldots \wedge d\bar{z}^n$$

is real. That M is orientable is a consequence of the fact that Θ is defined globally up to a positive factor. For, let $w^1, \cdots, w^n$ be another system of local complex coordinates. Then,

$$dw^1 \wedge \ldots \wedge dw^n = J\, dz^1 \wedge \ldots \wedge dz^n$$

where

$$J = \det \frac{\partial(w^1, \ldots, w^n)}{\partial(z^1, \ldots, z^n)} .$$

Hence, $d\bar{w}^1 \wedge \ldots \wedge d\bar{w}^n = \bar{J}\, d\bar{z}^1 \wedge \ldots \wedge d\bar{z}^n$ from which

$$\Theta(w) = J\bar{J}\Theta(z).$$

To define Θ globally we choose a locally finite covering and a partition of unity subordinated to the covering.

We have seen that a complex manifold is by definition even dimensional and have proved that it is orientable. These topological properties however, are *not* sufficient to ensure that a separable Hausdorff space has a complex structure as may be shown by the example of the 4-sphere due to Hopf and Ehresmann [*30*]. It is beyond the scope of this book to display this example as it involves some familiarity with the theory of characteristic classes.

Examples of complex manifolds :

1) The space of n complex variables C_n: It has one coordinate neighborhood, namely, the space of the variables $z^1, \cdots, z^n$.

2) An oriented surface S admits a complex structure. For, consider a Riemannian metric ds^2 on S. Locally, the metric is 'conformal', that is, there exist *isothermal parameters* u, v such that $ds^2 = \lambda(du^2 + dv^2)$ with $\lambda > 0$. We define complex (isothermal) coordinates z, $\bar{z}$ by putting $z = u + iv$ where the orientation of S is determined by the order (u, v). In these local coordinates $ds^2 = \lambda\, dz\, d\bar{z}$. In terms of another system of isothermal coordinates $(w, \bar{w})$, $ds^2 = \mu\, dw\, d\bar{w}$. Since $dw = a\, dz + b\, d\bar{z}$ it follows that $a\bar{b} = \bar{a}b = 0$, from which, by the given orientation $b = 0$ and $dw = a\, dz$. We conclude that w is a holomorphic function of z. Hence, condition (ii) for a complex structure is satisfied.

3) The Riemann sphere S^2: Consider S^2 as $C \cup \infty$, that is as the one point compactification of the complex plane. A complex structure is defined on S^2 by means of the atlas:

$(U_1, u_1) = (C, \iota)$ where ι is the identity map of C,

$(U_2, u_2) = (C - 0 \cup \infty, \zeta)$ where

$$\zeta(z) = \frac{1}{z}, \; z \neq \infty$$

$$\zeta(\infty) = 0.$$

In the overlap $U_1 \cap U_2 = C - 0$, $u_2 u_1^{-1}$ is given by the holomorphic function $\zeta = 1/z$.

4) Complex projective space P_n: P_n may be considered as the space of complex lines through the origin of C_{n+1} (cf. § 5.9 for details). It is the proper generalization to n dimensions of the Riemann sphere P_1.

5) Let Γ be a discrete subgroup of maximal rank of the group of translations of C_n and consider the manifold which is the quotient of C_n by Γ; this is a complex multi-torus—the coordinates of a point of C_n serving as local coordinates of C_n/Γ (cf. § 5.9).

6) $S^{2n-1} \times S^1$: Let G denote the group generated by the transformation of $C_n - 0$ given by $(z^1, \cdots, z^n) \to (2z^1, \cdots, 2z^n)$. Evidently,

$(C_n - 0)/G$ is homeomorphic with $S^{2n-1} \times S^1$ and has a complex structure induced by that of $C_n - 0$. The group G is properly discontinuous and acts without fixed points (cf. § 5.8). The quotient manifold $(C_n - 0)/G$ is called a *Hopf manifold* (see p. 167 and VII D).

7) Every covering of a complex manifold has a naturally induced complex structure (cf. § 5.8).

5.2. Almost complex manifolds

The concept of a complex structure is but an instance of a more general type of structure which we now consider. This concept may be defined from several points of view—the choice made here being geometrical, that is, in terms of fields of subspaces of the complexified tangent space. Indeed, a 'choice' of subspace of the 'complexification' of the tangent space at each point is made so that the union of the subspace and its 'conjugate' is the whole space. The given subspace is then said to define a complex structure in the tangent space at the given point. More precisely, if at each point P of a differentiable manifold, a complex structure is given in the tangent space at that point, which varies differentiably with P, the manifold is said to have an almost complex structure and is itself called an almost complex manifold.

With a vector space V over R of dimension n we associate a vector space V^c over C of complex dimension n called its *complexification* as follows: Let V^c be the space of all linear maps of V^* into C where, as usual, V^* denotes the dual space of V. Then, V^c is a vector space over C, and since $(V^*)^*$ can be identified with V, $V^c \supset V$. An element $v \in V^c$ belongs to V, if and only if, for all $\alpha \in V^*$, $\alpha(v) \in R$. Briefly, V^c is obtained from V by extending the field R to the field C.

Let ϕ be an isomorphism of C_n onto V^c. The vector $\bar{v} = \phi(\overline{\phi^{-1}(v)})$, $v \in V^c$ is called the conjugate of v. The vector v is said to be *real* if $\bar{v} = v$. Clearly, the real vectors of V^c form a vector space of dimension n over R. To a linear form α on V^c we associate a form $\bar{\alpha}$ on V^c defined by

$$\bar{\alpha}(v) = \overline{\alpha(\bar{v})}, \quad v \in V^c.$$

The map $\alpha \to \bar{\alpha}$ is evidently an involutory automorphism of $(V^c)^*$.

On the space V^c, tensors may be defined in the obvious way. The involutory automorphism $v \to \bar{v}$, $v \in V^c$ defines an involutory automorphism $t \to \bar{t}$, $t \in (V^c)^r_0$ (the linear space of tensors of type $(r, 0)$ on V^c). Every tensor on V (relative to $GL(n, R)$) defines a tensor on V^c, namely, the tensor coinciding with its conjugate, with which it may be identified. Such a tensor on V^c is said to be *real.*

Now, let V be a real vector space of even dimension $2n$. A subspace W^c of the complexification V^c of V of complex dimension n is said to define a *complex structure* on V if

$$V^c = W^c \oplus \bar{W}^c$$

where $\bar{W}^c$ is the space consisting of all conjugates of vectors in W^c. In this case, an element $v \in V^c$ has the unique representation

$$v = w_1 + \bar{w}_2, \quad w_1, w_2 \in W^c.$$

Since

$$\bar{v} = \bar{w}_1 + w_2,$$

the (real) vectors v of V are those elements of V^c which may be written in the form

$$v = w + \bar{w}, \quad w \in W^c. \tag{5.2.1}$$

We proceed to show that a complex structure on V may be defined equivalently by means of a certain tensor on V. Indeed, to every vector $v \in V$ there corresponds a real vector $Jv \in V$ defined by

$$Jv = \sqrt{-1}w + \overline{\sqrt{-1}w}$$

where $v = w + \bar{w}$, $w \in W^c$.

The operator J has the properties:

(i) J is linear

and

(ii) $J^2 v \equiv J(Jv) = -v.$

Moreover, J may be extended to V^c by linearity. The operator J is a 'quadrantal versor', that is, it has the effect of multiplying w by $\sqrt{-1}$ and $\bar{w}$ by $-\sqrt{-1}$. Thus W^c is the eigenspace of J for the eigenvalue $\sqrt{-1}$ and $\bar{W}^c$ that for the eigenvalue $-\sqrt{-1}$. Hence, a complex structure on V defines a linear endomorphism J of V, that is, by § 1.2, a tensor on V, with the property

$$J^2 = -I, \tag{5.2.2}$$

where I is the identity operator on V.

Conversely, let V be a real vector space of dimension m and J a linear endomorphism of V satisfying (5.2.2). Since a tensor on V defines a real tensor on the complexification V^c of V, J may be extended to V^c. We seek the eigenvectors and eigenvalues in V^c of the operator J. For this purpose put

$$Jv = zv, \quad v \in V^c.$$

Applying J to both sides of this relation gives

$$-v = z^2 v.$$

Hence, the eigenvalues are $\sqrt{-1}$ and $-\sqrt{-1}$, and so since J is a real operator, that is $\overline{Jv} = J\bar{v}$, the eigenvectors of $-\sqrt{-1}$ are the conjugates of those of $\sqrt{-1}$. The vector space V must therefore be even dimensional, that is $m = 2n$. The eigenvectors of $\sqrt{-1}$ form a vector space of complex dimension n which we denote by $V^{1,0}$ and those corresponding to $-\sqrt{-1}$ form the vector space $V^{0,1} = \bar{V}^{1,0}$; moreover,

$$V^{1,0} \cap V^{0,1} = \{0\},$$

that is $V^c = V^{1,0} \oplus V^{0,1}$ (direct sum). Thus, the tensor J defines a complex structure on V.

An element of the eigenspace $V^{1,0}$ will be called *a vector of bidegree* (or *type*) (1,0) and an element of $V^{0,1}$ *a vector of bidegree* (*or type*) (0,1).

A complex structure may be defined on the dual space of V in the obvious manner. The tensor product

$$\underbrace{V^c \otimes \dots \otimes V^c}_{s} \otimes \underbrace{(V^c)^* \otimes \dots \otimes (V^c)^*}_{t}$$

may then be decomposed into a direct sum of tensor products of vector spaces each identical with one of the spaces $V^{1,0}$, $V^{0,1}$, $V^{*1,0}$ and $V^{*0,1}$. A term in this decomposition is said to be a *pure tensor space* and an element of such a space is called a *tensor of type* $\binom{q_1 r_1}{q_2 r_2}$ if $V^{1,0}$ occurs q_1 times, $V^{0,1}$ – r_1 times, $V^{*1,0}$ – q_2 times and $V^{*0,1}$ – r_2 times. A skew-symmetric tensor or, equivalently, an element of the Grassman algebra over V^c (or $(V^c)^*$) is a sum of *pure forms* each of which is said to be of bidegree (q_1, r_1) (or (q_2, r_2)). For example,

$$V^c \otimes V^c = \underset{(2,0)}{V^{1,0} \otimes V^{1,0}} \oplus \underset{(1,1)}{V^{1,0} \otimes V^{0,1}} \oplus \underset{(1,1)}{V^{0,1} \otimes V^{1,0}} \oplus \underset{(0,2)}{V^{0,1} \otimes V^{0,1}},$$

that is, an element of the tensor space $V^c \otimes V^c$ is a sum of tensors of types $\binom{2\,0}{0\,0}$, $\binom{1\,1}{0\,0}$ and $\binom{0\,2}{0\,0}$. We denote by $\wedge^{q,r}$ the space of forms of bidegree (q, r).

In the sequel, we shall employ the following systems of indices unless otherwise indicated: upper case Latin letters $A, B, \cdots$ run from $1, \cdots, 2n$ and lower case Latin letters $i, j, \cdots$ run from $1, \cdots, n$; moreover, $i^* = i + n$ and $(i + n)^* = i$.

Let $\{e_1, \cdots, e_n\}$ be a basis of $V^{1,0}$. Denote the conjugate vectors $\bar{e}_i$ by e_{i^*}, $i = 1, \cdots, n$. Apparently, they form a basis of $V^{0,1}$, and since $V^c = V^{1,0} \oplus V^{0,1}$, the $2n$ vectors $\{e_i, e_{i^*}\}$ form a basis of V^c. Such a basis will be called a *J-basis* where J is the linear endomorphism defining the complex structure of V. Any two J-bases $\{e_i, e_{i^*}\}$, $\{e'_i, e'_{i^*}\}$ are related by equations of the form

$$e'_i = a^j_i\, e_j, \quad e'_{i^*} = a^{j^*}_{i^*}\, e_{j^*} \tag{5.2.3}$$

where (a^i_j) is a non-singular $n \times n$ matrix with complex coefficients, that is (a^i_j) is an element of the general linear group $GL(n, C)$ satisfying $a^{i^*}_{j^*} = \overline{a^i_j}$. With respect to a J-basis the tensor J has components $F_B{}^A$ where

$$F_j{}^i = \sqrt{-1}\,\delta^i_j, \quad F^{i^*}_{j^*} = -\sqrt{-1}\,\delta^i_j, \quad F^i_{j^*} = F_j{}^{i^*} = 0. \tag{5.2.4}$$

Hence, an element $v \in V$ (as a subset of V^c) has the components (v^i, v^{i^*}) where $v^{i^*} = \bar{v}^i$ and its image by J the components $(Jv)^i = \sqrt{-1}v^i$, $(Jv)^{i^*} = -\sqrt{-1}v^{i^*}$.

Consider the real basis $\{f_i, f_{i^*}\}$ defined in terms of the J-basis $\{e_i, e_{i^*}\}$ of V^c:

$$f_i = \frac{1}{\sqrt{2}}(e_i + e_{i^*}), \quad f_{i^*} \equiv Jf_i = \frac{\sqrt{-1}}{\sqrt{2}}(e_i - e_{i^*}). \tag{5.2.5}$$

Since

$$e_i = \frac{1}{\sqrt{2}}(f_i - \sqrt{-1}\, f_{i^*}), \quad e_{i^*} = \frac{1}{\sqrt{2}}(f_i + \sqrt{-1}\, f_{i^*}), \tag{5.2.6}$$

the vectors $\{f_i, f_{i^*}\}$, $i = 1, \cdots, n$ define a basis of V^c as well as V. Conversely, from a basis of V of the type $\{f_i, f_{i^*}\}$, where $f_{i^*} = Jf_i$ we obtain from (5.2.6) a basis of V^c, since $\bar{e}_i = e_{i^*}$.

If the matrix (a^i_j) in (5.2.3) is written as $(a^i_j) = (b^i_j) + \sqrt{-1}(c^i_j)$ where (b^i_j), (c^i_j) are $n \times n$ matrices, any two real bases of the type defined by (5.2.5) are related by a matrix of the form

$$\begin{pmatrix} (b^i_j) & (c^i_j) \\ -(c^i_j) & (b^i_j) \end{pmatrix} \in GL(2n, R)$$

called the *real representation* of the matrix (a^i_j). We remark that the determinant of the real representation of (a^i_j) is $|\det(a^i_j)|^2 > 0$.

With respect to the real basis $\{f_i, f_{i^*}\}$ the tensor J is given by the matrix

$$J_n = \begin{pmatrix} 0 & I_n \\ -I_n & 0 \end{pmatrix}.$$

It is easy to see that an element of $GL(2n, R)$ belongs to the real representation of $GL(n, C)$, if and only if, it commutes with J_n.

A metric may be defined on V by prescribing a positive definite symmetric tensor g on V (cf. § 1.9). In terms of a given basis of V we denote the components of g by g_{AB}. Suppose V is given the complex structure J. Then, an *hermitian structure* is given to V by insisting that J be an isometry, that is, for any $v \in V$

$$g(Jv, Jv) = g(v, v). \tag{5.2.7}$$

An equivalent way of expressing this in terms of the prescribed base is given by

$$F_A{}^C F_B{}^D g_{CD} = g_{AB} \quad \text{or} \quad F_B{}^A g_{CA} = -g_{AB} F_C{}^A. \tag{5.2.8}$$

The tensors g and J are then said to *commute*. The space V endowed with the hermitian structure defined by J and the *hermitian metric* g is called an *hermitian vector space*. It is immediate from (5.2.7) and $J^2 = -I$ that for any vector v, the vectors v and Jv are orthogonal.

Let $F_{AB} = F_A{}^C g_{BC}$ and consider the 2-form Ω on V defined in terms of a given basis of V by

$$\Omega = \tfrac{1}{2} F_{AB}\, \omega^A \wedge \omega^B \tag{5.2.9}$$

where the $\omega^A (A = 1, \cdots, 2n)$ are elements of the dual base. We define an operator which is again denoted by J on the space of real tensors t of type (0, 2) by

$$(Jt)_{AB} = t_{AC} F_B{}^C. \tag{5.2.10}$$

Denoting by J once again the induced map on 2-forms and taking account of (5.2.8) we may write $J\Omega = g$.

The metric of any Euclidean vector space with a complex structure can be modified in such a way that the space is given an hermitian structure. To see this, let V be an Euclidean vector space with a complex structure defined by the linear transformation J. Define the tensor k in terms of J and the metric h of V as follows:

$$k(v_1, v_2) = h(Jv_1, Jv_2).$$

Since the metric of V is positive definite, so is the quadratic form k defined by h, and therefore, the metric defined by

$$g = \tfrac{1}{2}(h + k)$$

is also positive definite. A computation yields

$$g(Jv_1, Jv_2) = g(v_1, v_2).$$

The 2-form Ω defined by J and g has rank $2n$. Indeed, the coefficients of Ω are given by $F_{AB} = g_{BC}\, F_A{}^C$.

Relative to a J-basis the metric tensor g has $g_{ij^*} = g_{j^*i}$ as its only non-vanishing components as one may easily see from (5.2.8) and (5.2.4). Moreover, since g is a real tensor

$$g_{ij^*} = \overline{g_{i^*j}}.$$

The tensor g on V^c is then said to be *self adjoint*.

More generally, let t be a tensor and denote by J^* the operation o starring the indices of its components (with respect to a J-basis). Then, if $\overline{J^*t} = t$ the tensor t is said to be *self adjoint*. Evidently, this is equivalent to saying that t is a real tensor.

From (5.2.4) one deduces that the only non-vanishing components of the covariant form of the tensor J with respect to a J-basis are

$$F_{ij^*} = \sqrt{-1}\, g_{ij^*},\ F_{j^*i} = -\sqrt{-1}\, g_{j^*i}. \tag{5.2.11}$$

The form Ω then has the following representation

$$\Omega = \sqrt{-1}\, g_{ij^*}\, \omega^i \wedge \omega^{j^*}. \tag{5.2.12}$$

We also consider the tensor $F^A{}_B = g^{AC}\, F_{CB}$. From (5.2.4) and (5.2.11) its only non-vanishing components (with respect to a J-basis) are

$$F^i{}_j = -\sqrt{-1}\, \delta^i_j,\ \ F^{i^*}{}_{j^*} = \sqrt{-1}\, \delta^i_j.$$

Evidently,

$$F^A{}_B = -F_B{}^A$$

and

$$F^A{}_C\, F^C{}_B = -\,\delta^A_B.$$

Thus, the tensor $F^A{}_B$ defines a complex structure $\bar{J}$ on V called the *conjugate* of J.

Let v_1 and v_2 be any orthogonal vectors on the hermitian vector space V. If we insist that v_2 be orthogonal to Jv_1 as well, then, from (5.2.7), v_1, v_2, Jv_1 and Jv_2 are mutually orthogonal.

Let $\{f_i, f_{i^*}\}$, $i = 1, \cdots, n$ where $f_{i^*} = Jf_i$ be a real orthonormal basis of V. Such a basis is assured by the hermitian structure defined by J and g. Then, in terms of the J-basis $\{e_i, e_{i^*}\}$ defined by (5.2.6)

$$g(e_i, e_{j^*}) = \delta_{ij}, \tag{5.2.13}$$

that is $g_{ij^*} = g(e_i, e_{j^*}) = \delta_{ij}$. The form Ω may then be written as

$$\Omega = \sqrt{-1}\sum_{i=1}^{n} \omega^i \wedge \omega^{i^*}. \tag{5.2.14}$$

A differentiable manifold M is said to possess an *almost complex structure* if it carries a real differentiable tensor field J of type (1, 1) (and class k) satisfying

$$J^2 = -I.$$

(By § 1.2, the tensor field J may be considered as a linear endomorphism of the space of tangent vector fields on M). It follows that an almost complex structure is equivalently defined by a *field of subspaces* W^c (of class k and dimension n) of T^c (the complexification of the space of tangent vector fields) such that

$$T^c = W^c \oplus \bar{W}^c. \tag{5.2.15}$$

A manifold with an almost complex structure is said to be an *almost complex manifold.*

Evidently, *an almost complex manifold is even dimensional.*

We now show that *a complex manifold M is almost complex*. Indeed, let U be a coordinate neighborhood of M with the local complex coordinates $z^1, \cdots, z^n$. We have seen that M possesses an underlying real analytic structure and that relative to it $z^1, \cdots, z^n, \bar{z}^1, \cdots, \bar{z}^n$ may be used as local coordinates. Following the notation of § 5.1, we define

$$\frac{\partial}{\partial z^i} = \frac{1}{2}\left(\frac{\partial}{\partial x^i} - \sqrt{-1}\,\frac{\partial}{\partial y^i}\right), \quad \frac{\partial}{\partial \bar{z}^i} = \frac{1}{2}\left(\frac{\partial}{\partial x^i} + \sqrt{-1}\,\frac{\partial}{\partial y^i}\right).$$

Let P be a point of U. Then, the differentials $dz^1, \cdots, dz^n, d\bar{z}^1, \cdots, d\bar{z}^n$ at P define a frame in the complexification $(T_P^c)^*$ of the dual space T_P^* of the tangent space T_P at P and by duality a frame $\{\partial/\partial z^i, \partial/\partial \bar{z}^i\}$ in T_P^c.

If P belongs to the intersection $U \cap U'$ of the coordinate neighborhoods U and U' the differentials (dz^i) and (dz'^i) are related by

$$dz^i = a^i_j\, dz'^j \tag{5.2.16}$$

and their duals $(\partial/\partial z^i)$, $(\partial/\partial z'^i)$ by

$$\frac{\partial}{\partial z'^j} = a^i_j \frac{\partial}{\partial z^i} \tag{5.2.17}$$

where $(a^i_j) \in GL(n, C)$ is the matrix of coefficients

$$a^i_j = \frac{\partial z^i}{\partial z'^j}.$$

It follows that the n vectors $(\partial/\partial z^i)_P$ define a subspace W_P^c of T_P^c and that

$$T_P^c = W_P^c \oplus \bar{W}_P^c,$$

that is, these vectors determine a complex structure on T_P. Hence, at each point $P \in M$ a complex structure is defined in the tangent space at that point. Moreover, at a given point any two frames are related by equations of the form (5.2.3), that is, only those frames $\{X_1, \cdots, X_n, X_{1*}, \cdots, X_{n*}\}$ are allowed which are obtained from the frame

$$\left\{\frac{\partial}{\partial z^1}, \cdots, \frac{\partial}{\partial z^n}, \frac{\partial}{\partial \bar{z}^1}, \cdots, \frac{\partial}{\partial \bar{z}^n}\right\}$$

by

$$X_j = b_j^i \frac{\partial}{\partial z^i}, \quad X_{j*} = b_{j*}^{i*} \frac{\partial}{\partial \bar{z}^i}, \quad b_{j*}^{i*} = \bar{b}_j^i .$$

Hence, the complex structure on M defines a real analytic tensor field J of type (1, 1) on M.

One may easily check that if a differentiable manifold possesses two complex structures, giving rise to the same almost complex structure, they must coincide.

We have seen that a complex manifold is orientable. An almost complex manifold also enjoys this property, this being a consequence of the fact that for every neighborhood U of a point P of the manifold and at every point Q of U there exists a set of real vectors $X_1, \cdots, X_n$ such that $X_1, \cdots, X_n, JX_1, \cdots, JX_n$ are independent vectors; moreover, from (5.2.3) and (5.2.5) any two real bases of this type are related by a matrix of positive determinant. In other words, the existence of a J-basis (cf. 5.2.6) at each point ensures that the almost complex manifold is orientable (cf. § 5.1 for the dual argument).

Let M be an almost complex manifold with the almost complex structure J. The almost complex structure is said to be *integrable* if M can be made into a complex manifold so that in a coordinate neighborhood with the complex coordinates (z^i) operating with J is equivalent to transforming $\partial/\partial z^i$ and $\partial/\partial \bar{z}^i$ into $\sqrt{-1}\ \partial/\partial z^i$ and $-\sqrt{-1}\ \partial/\partial \bar{z}^i$, respectively. It is not difficult to see that if the almost complex structure which is equivalently defined by the tensor field $F^A{}_B$ of type (1, 1) in the (real) local coordinates $(u^A) = (z^i, \bar{z}^i)$ is integrable, then

$$\left(\frac{\partial F^A{}_B}{\partial u^C} - \frac{\partial F^A{}_C}{\partial u^B}\right) F^B{}_D = \left(\frac{\partial F^A{}_B}{\partial u^D} - \frac{\partial F^A{}_D}{\partial u^B}\right) F^B{}_C . \qquad (5.2.18)$$

One merely considers a J-basis with respect to which the functions $F^A{}_B$ are given by (5.2.4).

Conversely, if the almost complex structure given by J is of class $1 + \alpha$ $(0 < \alpha < 1)$, that is, the derivatives are Hölder continuous with exponent α, and if the structure tensor satisfies the (integrability) conditions (5.2.18), it is integrable [*85*]. The proof of this important fact is patterned after that of Newlander and Nirenberg [*84*] who assumed that the structure is of class $2n + \alpha$. Hence, in order that an almost complex structure define a complex structure it is not necessary that it be analytic or even of class ∞. For real analytic manifolds with real analytic $F^A{}_B$ the above result follows from a theorem of Frobenius (cf. I.D.4). For $n = 1$ the problem is equivalent to that of introducing isothermal parameters with respect to the metric

$$ds^2 = |\, dz + \rho d\bar{z} \,|^2,$$

and Chern showed that this is possible even if the structure is of class α.

5.3. Local hermitian geometry

If at each point P of the complex manifold M of complex dimension n the tangent space T_P is endowed with an hermitian metric so that (as functions of local coordinates) the metric tensor g is of class ∞, M is said to be an *hermitian manifold*. Evidently, such a manifold is also Riemannian. On the other hand, since the complex structure is defined by a tensor field J of type (1, 1), if the complex manifold M is given an 'arbitrary' Riemannian metric, a new metric g can be found which commutes with J. The metric g together with the tensor field J is said to define an *hermitian structure* on M (cf. 5.2.8). In this way, it is seen that every complex manifold possesses an hermitian metric. The (local) geometry of an hermitian manifold is called *hermitian geometry*.

In the same way as the bundle of frames with the orthogonal group as structural group is natural for the study of Riemannian geometry, the bundle of unitary frames, that is, the bundle of frames with the unitary group $U(n)$ as structural group, is natural for hermitian geometry. Indeed, by a *unitary frame* at the point $P \in M$ we shall mean a J-basis $\{X_1, \cdots, X_n, X_{1^*}, \cdots, X_{n^*}\}$ at P of the type satisfying (5.2.13), that is

$$X_i \cdot X_{j^*} = \delta_{ij}, \quad i,j = 1, \cdots, n,$$

where $X_i \cdot X_{j^*} = g(X_i, X_{j^*})$.

The collection of all such frames at all points $P \in M$ forms a fibre bundle B over M with $U(n)$ as structural group. A frame at P, that is an element of the fibre over P may be determined by means of a system of local complex coordinates (z^i) at P by the natural basis $\{\partial/\partial z^i\}$, $i = 1, \cdots, n$ of $T_P^{1,0}$ and the group $U(n)$. In the notation of § 1.8, we put

$$X_i = \xi^k_{(i)} \frac{\partial}{\partial z^k}, \quad X_{i^*} = \overline{\xi^k_{(i)}} \frac{\partial}{\partial \bar{z}^k}, \quad i = 1, \cdots, n.$$

Since the vector $X_{i^*} \in T_P^{0,1}$ is the conjugate of $X_i \in T_P^{1,0}$, $\bar{\xi}^k_{(i)} = \xi^{k^*}_{(i^*)}$ where we have written $\bar{\xi}^k_{(i)}$ for $\overline{\xi^k_{(i)}}$. By putting $\xi^{k^*}_{(i)} = \xi^{\ k}_{(i^*)} = 0$ these equations may be written in the abbreviated form

$$X_A = \xi^B_{(A)} \frac{\partial}{\partial z^B}, \quad A, B = 1, \cdots, n, 1^*, \cdots, n^*.$$

The coefficients $\xi^B_{(A)}$ are the elements of a matrix in $GL(n, C)$. However, they are not independent. For, they must satisfy the relation

$$\xi^k_{(i)} \bar{\xi}^l_{(j)} g_{kl^*} = \delta_{ij}, \tag{5.3.1}$$

where $g_{kl^*} = g(\partial/\partial z^k, \partial/\partial \bar{z}^l)$.

Let $(\xi^{(A)}_B)$ denote the inverse matrix of $(\xi^B_{(A)})$. As in § 1.8 it defines $2n$ linearly independent differential forms θ^A in B: In the overlap of the coordinate neighborhoods with the local coordinates $(z^A, \xi^B_{(A)})$ and $(z'^A, \xi'^B_{(A)})$, we have by (1.8.3)

$$\xi'^{(A)}_C = \frac{\partial z^B}{\partial z'^C} \xi^{(A)}_B.$$

Hence, by (5.2.17)

$$\xi'^{(i)}_k = \frac{\partial z^j}{\partial z'^k} \xi^{(i)}_j, \quad \bar{\xi}'^{(i)}_k = \frac{\partial \bar{z}^j}{\partial \bar{z}'^k} \bar{\xi}^{(i)}_j.$$

The $2n$ covariant vector fields $\xi^{(A)}_B$ therefore define $2n$ independent 1-forms $\theta^A = (\theta^i, \theta^{i^*})$ in B with $\theta^{i^*} = \bar{\theta}^i$ $(i = 1, \cdots, n)$. In terms of the local coordinates (z^i), they may be expressed by

$$\alpha^i = \xi^{(i)}_j dz^j, \quad \bar{\alpha}^i = \bar{\xi}^{(i)}_j d\bar{z}^j, \quad \pi^* \alpha^A = \theta^A, \tag{5.3.2}$$

where $\pi : B \to M$ is the projection map.
By analogy they form a 'frame' in T_P^* and for this reason this frame is called a *coframe*.

There are several ways of defining a metrical connection in M. We propose to do this in a manner compatible with the complex and hermitian structures since this approach seems to be natural for hermitian manifolds. Indeed, as in § 1.7 we prescribe $(2n)^2$ linear differential forms $\omega^A_B = \Gamma^A_{BC} dz^C$ in each coordinate neighborhood of a covering in such a way that in the overlap of two coordinate neighborhoods related by holomorphic functions the equations (1.7.5) are satisfied by the n^2 forms ω^i_j given below. We then insist that the $2n^2$ forms ω^i_j, ω^{i*}_{j*} be given by

$$\omega^i_j = \Gamma^i_{jk}\, dz^k, \quad \omega^{i*}_{j*} = \bar{\omega}^i_j \; (\equiv \overline{\omega^i_j})$$

from which it follows that $\bar{\Gamma}^i_{jk} = \Gamma^{i*}_{j*k*}$; the remaining $2n^2$ forms are set equal to zero.

In terms of this connection we take the covariant differential of each of the vectors $\xi^B_{(A)}$ thereby obtaining as in § 1.8 the forms α^B_A. Their images in B will be denoted by $\theta^B{}_A$. By (1.8.6) and (1.8.5)

$$\alpha^j_i = (d\xi^k_{(i)} + \omega^k_l\, \xi^l_{(i)})\, \xi^{(j)}_k, \tag{5.3.3}$$

from which, by (5.3.2)

$$d\theta^i = \theta^k \wedge \theta^i{}_k + \Theta^i \tag{5.3.4}$$

—the torsion forms being given by

$$\Theta^i = -\tfrac{1}{2}\, \xi^{(i)}_j\, \xi^p_{(l)}\, \xi^q_{(m)}\, T_{pq}{}^j\, \theta^l \wedge \theta^m, \quad T_{pq}{}^j = \Gamma^j_{pq} - \Gamma^j_{qp}. \tag{5.3.5}$$

This is the first of the equations of structure. The forms θ^A_B are not independent, but rather, are related by

$$\theta^j_i + \bar{\theta}^i_j = 0, \quad \bar{\theta}^i_j \equiv \overline{\theta^i_j}\,.$$

For, from (5.3.3)

$$\theta^j{}_i + \bar{\theta}^i{}_j = (d\xi^k_{(i)} + \omega^k_l\, \xi^l_{(i)})\, \xi^{(j)}_k + (d\bar{\xi}^k_{(j)} + \bar{\omega}^k_l\, \bar{\xi}^l_{(j)})\, \bar{\xi}^{(i)}_k \tag{5.3.6}$$

and, from (5.3.1)

$$d\xi^k_{(i)}\, \bar{\xi}^l_{(j)}\, g_{kl^*} + \xi^k_{(i)}\, d\bar{\xi}^l_{(j)}\, g_{kl^*} + \xi^k_{(i)}\, \bar{\xi}^l_{(j)}\, dg_{kl^*} = 0. \tag{5.3.7}$$

Since,

$$g_{kl^*} = \sum_{r=1}^{n} \xi^{(r)}_k\, \bar{\xi}^{(r)}_l \tag{5.3.8}$$

(5.3.7) becomes

$$d\xi^k_{(i)}\, \xi^{(j)}_k + d\bar{\xi}^k_{(j)}\, \bar{\xi}^{(i)}_k + \xi^k_{(i)}\, \bar{\xi}^l_{(j)}\, dg_{kl^*} = 0. \tag{5.3.9}$$

Evaluating the differential of the metric tensor g as in § 1.9 we obtain

$$dg_{ij*} = \omega^k{}_i \, g_{kj*} + \bar{\omega}^k{}_j \, g_{ik*}. \tag{5.3.10}$$

This is precisely the condition that the $\omega^i{}_j$ must satisfy in order to define a metrical connection. Hence, for a metrical connection

$$\frac{\partial g_{ij*}}{\partial z^m} dz^m + \frac{\partial g_{ij*}}{\partial \bar{z}^m} d\bar{z}^m = g_{kj*} \, \Gamma^k_{im} \, dz^m + g_{ik*} \, \Gamma^{k*}_{j*m*} \, d\bar{z}^m$$

from which

$$\Gamma^l_{im} = g^{j*l} \frac{\partial g_{ij*}}{\partial z^m}, \quad \Gamma^{l*}_{i*m*} = g^{jl*} \frac{\partial g_{ji*}}{\partial \bar{z}^m}. \tag{5.3.11}$$

Substituting from (5.3.9) into (5.3.6), applying (5.3.11) and observing that

$$g^{jk*} = \sum_{r=1}^{n} \xi^j_{(r)} \, \bar{\xi}^k_{(r)} \tag{5.3.12}$$

we obtain the desired relation.

The second of the equations of structure (1.8.8)

$$d\theta^B_A - \theta^C_A \wedge \theta^B_C = \Theta^B_A$$

splits into

$$d\theta^j{}_i - \theta^k{}_i \wedge \theta^j{}_k = \Theta^j{}_i,$$
$$d\bar{\theta}^j{}_i - \bar{\theta}^k{}_i \wedge \bar{\theta}^j{}_k = \bar{\Theta}^j{}_i, \quad \bar{\Theta}^j{}_i = \Theta^{j*}_{i*} \tag{5.3.13}$$

by virtue of the decomposition $T^c = T^{1,0} \oplus T^{0,1}$.

Denote the curvature forms in the local coordinates $(z^i, \bar{z}^i)$ by $\Omega^j{}_i$, that is, the $\Omega^j{}_i$ are the forms $\Theta^j{}_i$ pulled down to M by means of the cross-section $M \to \{(\partial/\partial z^i)_P, (\partial/\partial \bar{z}^i)_P\}$. Consequently, in much the same way as above, it may be shown that if they are locally given by

$$\Omega^j{}_i = d\omega^j{}_i - \omega^r{}_i \wedge \omega^j{}_r \tag{5.3.14}$$

then, in the bundle of unitary frames, the curvature forms are the $\Theta^j{}_i$.

Since $\omega^j{}_i = \Gamma^j_{ik} \, dz^k$, the equations (5.3.14) become

$$\Omega^j{}_i = \left(\frac{\partial \Gamma^j_{im}}{\partial z^l} - \Gamma^r_{il} \, \Gamma^j_{rm}\right) dz^l \wedge dz^m - \frac{\partial \Gamma^j_{il}}{\partial \bar{z}^m} dz^l \wedge d\bar{z}^m. \tag{5.3.15}$$

Thus, if we put

$$-\Omega^j{}_i = R^j{}_{ilm}\, dz^l \wedge dz^m + R^j{}_{ilm^*}\, dz^l \wedge d\bar{z}^m$$

where

$$R^j{}_{ilm} + R^j{}_{iml} = 0$$

we have

$$2R^j{}_{ilm} = \frac{\partial \Gamma^j_{il}}{\partial z^m} - \frac{\partial \Gamma^j_{im}}{\partial z^l} + \Gamma^r_{il}\Gamma^j_{rm} - \Gamma^r_{im}\Gamma^j_{rl} \tag{5.3.16}$$

and

$$R^j{}_{ilm^*} = \frac{\partial \Gamma^j_{il}}{\partial \bar{z}^m}. \tag{5.3.17}$$

Its only non-vanishing components are

$$R^j{}_{ilm^*},\ R^j{}_{im^*l},\ R^{j^*}{}_{i^*l^*m},\ R^{j^*}{}_{i^*ml^*}.$$

For, substituting (5.3.11) into (5.3.16) and (5.3.17) and applying the relation $d(g^{ij^*}g_{j^*k}) = 0$, we derive

$$R^j{}_{ilm} = 0.$$

Moreover,

$$R^j{}_{ilm^*} = \frac{\partial g^{k^*j}}{\partial \bar{z}^m}\frac{\partial g_{ik^*}}{\partial z^l} + g^{jk^*}\frac{\partial^2 g_{ik^*}}{\partial z^l\, \partial \bar{z}^m}. \tag{5.3.18}$$

Since $\bar{\Gamma}^i_{jk} = \Gamma^{i^*}_{j^*k^*}$ the curvature tensor is self adjoint.

Transvecting (5.3.18) with g_{jr^*} we obtain

$$R_{r^*ilm^*} = g_{jr^*}\frac{\partial g^{k^*j}}{\partial \bar{z}^m}\frac{\partial g_{ik^*}}{\partial z^l} + \frac{\partial^2 g_{ir^*}}{\partial z^l\, \partial \bar{z}^m}. \tag{5.3.19}$$

Hence, the only non-vanishing 'covariant' components of the curvature tensor are

$$R_{ij^*kl^*},\ R_{ij^*k^*l},\ R_{i^*jkl^*},\ R_{i^*jk^*l}.$$

Again, by virtue of the given splitting the Bianchi identities have the form

$$\begin{aligned} d\Theta^j &= \theta^k \wedge \Theta^j{}_k - \Theta^k \wedge \theta^j{}_k, \\ d\Theta^j{}_i &= \theta^k{}_i \wedge \Theta^j{}_k - \Theta^k{}_i \wedge \theta^j{}_k \end{aligned} \tag{5.3.20}$$

together with the conjugate relations.

In a complex coordinate system the first of (5.3.20) are given by

$$R^i{}_{jkl^*} - R^i{}_{kjl^*} = D_{l^*} T_{jk}{}^i \tag{5.3.21}$$

and their conjugates together with the *Jacobi identities*

$$D_l T_{jk}{}^i + D_j T_{kl}{}^i + D_k T_{lj}{}^i - T_{rk}{}^i T_{lj}{}^r - T_{rl}{}^i T_{jk}{}^r - T_{rj}{}^i T_{kl}{}^r = 0 \tag{5.3.22}$$

and their conjugates where as usual D_i denotes covariant differentiation with respect to the connection (5.3.11). From the second Bianchi identity we derive the relations

$$D_m R^i{}_{jkl^*} - D_k R^i{}_{jml^*} = R^i{}_{jrl^*} T_{mk}{}^r \tag{5.3.23}$$

together with their conjugates.

Since the connection is a metrical connection

$$D_k g_{ij^*} = D_{k^*} g_{ij^*} = 0. \tag{5.3.24}$$

Hence, from (5.3.23)

$$D_m R_{i^*jkl^*} - D_k R_{i^*jml^*} = R_{i^*jrl^*} T_{mk}{}^r \tag{5.3.25}$$

together with the conjugate relations.

In terms of the hermitian metric, the torsion tensor has the components

$$\begin{aligned} T_{jk}{}^i &= g^{il^*}\left(\frac{\partial g_{l^*j}}{\partial z^k} - \frac{\partial g_{l^*k}}{\partial z^j}\right), \\ T_{j^*k^*}{}^{i^*} &= g^{i^*l}\left(\frac{\partial g_{lj^*}}{\partial \bar{z}^k} - \frac{\partial g_{lk^*}}{\partial \bar{z}^j}\right). \end{aligned} \tag{5.3.26}$$

Thus, a necessary and sufficient condition that the torsion forms vanish may be given in terms of the hermitian metric tensor g by the system of differential equations

$$\frac{\partial g_{i^*j}}{\partial z^k} = \frac{\partial g_{i^*k}}{\partial z^j}, \quad \frac{\partial g_{ij^*}}{\partial \bar{z}^k} = \frac{\partial g_{ik^*}}{\partial \bar{z}^j}. \tag{5.3.27}$$

In this case, g is said to define a *Kaehler metric*. A complex manifold endowed with this particular metric is called a *Kaehler manifold*.

If the metric of an hermitian manifold is given by

$$g_{ij^*} = \frac{\partial^2 f}{\partial z^i \, \partial \bar{z}^j}$$

(locally) for some real-valued function f, then, clearly, from (5.3.27) it is a Kaehler metric. Conversely, the metric of a Kaehler manifold is locally expressible in this form. For, since the equations (5.3.27) must be satisfied, the equations

$$\frac{\partial \varphi_i}{\partial \bar{z}^j} = g_{ij^*}$$

are completely integrable. If $\tilde{\varphi}_i$ is a solution, the general solution is given by

$$\varphi_i = \tilde{\varphi}_i(z, \bar{z}) + \psi_i(z)$$

where the ψ_i are arbitrary functions of the variables (z). Consider the system of first order equations

$$\frac{\partial f}{\partial z^i} = \tilde{\varphi}_i(z, \bar{z}) + \psi_i(z).$$

The integrability conditions of this system are given by

$$\left(\frac{\partial \tilde{\varphi}_k}{\partial z^i} - \frac{\partial \tilde{\varphi}_i}{\partial z^k}\right) + \left(\frac{\partial \psi_k}{\partial z^i} - \frac{\partial \psi_i}{\partial z^k}\right) = 0.$$

Differentiating these equations with respect to $\bar{z}^j$ we find, after applying the conditions (5.3.27), that functions ψ_i can be chosen satisfying the integrability conditions. That f may be taken to be real is a consequence of the fact that $\bar{f}$ is also a solution of the system.

We remark that an even-dimensional analytic Riemannian manifold M with a locally Kaehlerian metric, that is, whose metric in local complex coordinates satisfies the equations (5.3.27) is not necessarily a Kaehler manifold. For, consider the cartesian product of a circle with a compact 3-dimensional Euclidean space form whose first betti number is zero [*24*]. It can be shown that such a space form exists; in fact, there is only one. This manifold is compact, orientable, and has a locally flat metric. The last property implies that its metric is locally Kaehlerian. (We have invoked the theorem that an even-dimensional locally flat analytic Riemannian manifold is locally Kaehlerian). Since its first betti number is one it cannot be a Kaehler manifold (cf. theorem 5.6.2).

An hermitian metric g is expressible in the local coordinates $(z^i, \bar{z}^i)$ by means of the positive definite quadratic form

$$\begin{aligned} ds^2 &= g_{AB}\, dz^A\, dz^B \\ &= 2g_{ij^*}\, dz^i\, d\bar{z}^j. \end{aligned} \tag{5.3.28}$$

If g is a Kaehler metric, the real 2-form

$$\Omega = \sqrt{-1}\, g_{ij^*}\, dz^i \wedge d\bar{z}^j, \tag{5.3.29}$$

canonically defined by this metric, is closed. Conversely, if Ω is closed, g is a Kaehler metric.

In an hermitian manifold, the 2-form Ω is called the *fundamental form.* We remark that the tensor field g as well as the fundamental form can be given a particularly simple representation in terms of the $2n$ forms $(\alpha^i, \bar{\alpha}^i)$ on M. For, from (5.3.2) and (5.3.8)

$$g = 2\sum_{r=1}^{n} \alpha^r \otimes \bar{\alpha}^r \tag{5.3.30}$$

and

$$\Omega = \sqrt{-1}\sum_{r=1}^{n} \alpha^r \wedge \bar{\alpha}^r. \tag{5.3.31}$$

From the equations (5.3.4) and (5.3.13) we deduce the equations of structure of a Kaehler manifold M:

$$d\theta^i = \theta^k \wedge \theta^i{}_k \tag{5.3.32}$$

and

$$\Theta^i{}_j = d\theta^i{}_j - \theta^k{}_j \wedge \theta^i{}_k \tag{5.3.33}$$

where the 2-forms $\Theta^i{}_j$ define the curvature of the manifold. They are locally expressible in terms of local complex coordinates by

$$\Omega^i{}_j = -R^i{}_{jkl^*}\, dz^k \wedge d\bar{z}^l. \tag{5.3.34}$$

The Ricci tensor of M is given locally by

$$R_{kl^*} = -R^i{}_{ikl^*}, \tag{5.3.35}$$

and so from (5.3.17) it may be expressed explicitly in terms of the metric g by

$$R_{kl^*} = -\frac{\partial^2 \log \det G}{\partial z^k\, \partial \bar{z}^l}, \quad G = (g_{ij^*}). \tag{5.3.36}$$

Now, from (5.3.34)

$$\Omega^i{}_i = R_{kl^*}\, dz^k \wedge d\bar{z}^l$$

and from (5.3.33)

$$\Theta^i_{\ i} = d\theta^i_{\ i}.$$

It follows that $\sqrt{-1}\, d\theta^i_{\ i}$ is a (real) closed 2-form in the bundle of frames over M. Moreover,

$$\frac{\partial R_{ij^*}}{\partial z^k} = \frac{\partial R_{kj^*}}{\partial z^i}, \qquad \frac{\partial R_{i^*j}}{\partial \bar{z}^k} = \frac{\partial R_{k^*j}}{\partial \bar{z}^i}. \tag{5.3.37}$$

Since the operator d is real (that is, it sends real forms into real forms), $\sqrt{-1}\,\theta^i_{\ i}$ defines a real 1-form (which we denote by $2\pi\chi$) on the bundle B of unitary frames. Let $\pi : B \to M$ denote the projection map and put

$$\psi = \frac{1}{2\pi\sqrt{-1}} R_{kl^*}\, dz^k \wedge d\bar{z}^l. \tag{5.3.38}$$

Then, $\pi^*\psi = -\,d\chi$. The 2-form ψ defines the 1st *Chern class* of M (cf. § 6.12).

In contrast with Kaehler geometry there are three distinct contractions of the curvature tensor in an hermitian manifold with non-vanishing torsion. They are called the *Ricci tensors* and are defined as follows:

$$R_{ij^*} = g^{kl^*} R_{il^*kj^*}, \qquad S_{ij^*} = g^{kl^*} R_{ij^*kl^*}, \qquad T_{ij^*} = g^{kl^*} R_{kl^*ij^*}.$$

If the contracted torsion tensor vanishes, that is if $T_{ji}^{\ \ i} = 0$, $T_{ij*} = R_{ij*}$. This is one of two rather natural conditions that can be imposed on the torsion, the other being that the torsion forms be holomorphic. From (5.3.21) we see that the latter condition implies the symmetry relation

$$R^i_{\ jkl^*} = R^i_{\ kjl^*}. \tag{5.3.39}$$

Since the curvature tensor is skew-symmetric in its last two indices the symmetry relation (5.3.39) shows that $S_{ij^*} = R_{ij^*}$.

Now, from (5.3.21) we obtain

$$R_{i^*jkl^*} - R_{kl^*i^*j} = D_{l^*}T_{jki^*} - D_jT_{l^*i^*k} \tag{5.3.40}$$

where $T_{jki^*} = g_{i^*l}\, T_{jk}^{\ \ l}$. Hence, the conditions $\partial T_{jk}^{\ \ i}/\partial \bar{z}^l = 0$ imply the symmetry relations

$$R_{ij^*kl^*} = R_{kl^*ij^*} \tag{5.3.41}$$

as in a Riemannian manifold. We conclude that $S_{ij*} = T_{ij*}$, that is the Ricci curvature tensors coincide as in a Kaehler manifold. That they need not be the same may be seen by the following example [*15*].

Consider the cartesian product of a 1-sphere and a 3-sphere: $M = S^1 \times S^3$. In example 6 of § 5.1 it was shown that M is a complex manifold. A natural metric is given by

$$ds^2 = \frac{1}{\lambda}(dz^1\,d\bar{z}^1 + dz^2\,d\bar{z}^2), \quad \lambda = z^1\bar{z}^1 + z^2\bar{z}^2,$$

so that

$$g_{ij^*} = \frac{1}{\lambda}\,\delta_{ij}, \quad g^{ij^*} = \lambda\delta^{ij}.$$

A computation yields

$$R_{ij^*kl^*} = \frac{1}{\lambda^3}\,(\lambda\delta_{ij}\,\delta_{kl} - \delta_{ij}\,\bar{z}^k\,z^l)$$

from which we obtain

$$R_{ij^*} = \frac{1}{\lambda^2}\,(\lambda\delta_{ij} - \bar{z}^i z^j), \quad S_{ij^*} = \frac{1}{\lambda}\,\delta_{ij}, \quad T_{ij^*} = \frac{2}{\lambda^2}\,(\lambda\delta_{ij} - z^i\bar{z}^j).$$

Summarizing, we see that the curvature tensor defined by a connection with holomorphic torsion has the same symmetry properties as the curvature tensor defined by a Kaehler metric.

The condition that the torsion be holomorphic is a rigidity restriction on the manifold. Indeed, if the manifold is compact, it is actually Kaehlerian [*32*].

One may also consider a connection which carries holomorphic tensor fields into holomorphic tensor fields (cf. § 6.5). Such a connection must satisfy

$$\frac{\partial \Gamma^i_{jk}}{\partial \bar{z}^l} = 0, \quad \frac{\partial \Gamma^{i^*}_{j^*k^*}}{\partial z^l} = 0$$

and, for this reason, the connection is said to be holomorphic. From (5.3.17) it follows that *the curvature tensor of a holomorphic connection must vanish.*

In an hermitian manifold M with non-vanishing torsion, if the Ricci tensor R_{ij^*} defines a positive definite quadratic form, then it defines an hermetian metric g on M. From the second of equations (5.3.20) it follows that the form $\Theta^i{}_i$ is closed, and hence g is a Kaehler metric.

A complex manifold M of complex dimension n is said to be *complex parallelisable* if there are n linearly independent holomorphic vector fields defined everywhere over M (cf. p. 247). In an hermitian manifold, it is not difficult to prove that the vanishing of the curvature tensor is a necessary condition for the manifold to be complex parallelisable. (Hence, the connection of a complex parallelisable manifold is holomorphic.) In Chapter VI it is shown, if the manifold is simply connected, that this condition is also sufficient. Hence, for a complex manifold the existence of a metric with zero curvature is a somewhat weaker property than parallelisability.

5.4. The operators L and Λ

Let M be a complex manifold of complex dimension n and denote by $\wedge^{*c}(M)$ the bundle of exterior differential polynomials with complex values. From § 5.1, a p-form $\alpha \in \wedge^{*c}(M)$ may be represented as a sum

$$\alpha = \alpha_{p,0} + \alpha_{p-1,1} + \dots + \alpha_{0,p}$$

where $\alpha_{q,r}$ is of degree q in the dz^i and of degree r in the conjugate variables. The coefficients of α when expressed in terms of real coordinates are complex-valued functions of class ∞. Thus, there is a canonically defined map

$$\tilde{d}: \wedge^{*c}(M) \to \wedge^{*c}(M)$$

obtained from d by extending the latter to $\wedge^{*c}(M)$ by linearity, that is, if $\alpha = \lambda + \sqrt{-1}\mu$ where λ and μ are real forms, then

$$\tilde{d}\alpha = d\lambda + \sqrt{-1}\, d\mu.$$

Clearly, $\tilde{d}\bar{\alpha} = \overline{\tilde{d}\alpha}$, that is $\tilde{d}$ is a real operator. In the sequel, we shall write d in place of $\tilde{d}$ with no resulting confusion.

The exterior differential operator d maps a form α of bidegree (q, r) into the sum of a form of bidegree $(q + 1, r)$ and one of bidegree $(q, r + 1)$. For, if

$$\alpha = a_{j_1 \dots j_q k_1 \dots k_r}\, dz^{j_1} \wedge \dots \wedge dz^{j_q} \wedge d\bar{z}^{k_1} \wedge \dots \wedge d\bar{z}^{k_r},$$

$$d\alpha = \frac{\partial a_{j_1 \dots j_q k_1 \dots k_r}}{\partial z^i}\, dz^i \wedge dz^{j_1} \wedge \dots \wedge dz^{j_q} \wedge d\bar{z}^{k_1} \wedge \dots \wedge d\bar{z}^{k_r}$$

$$+ \frac{\partial a_{j_1 \dots j_q k_1 \dots k_r}}{\partial \bar{z}^i}\, d\bar{z}^i \wedge dz^{j_1} \wedge \dots \wedge dz^{j_q} \wedge d\bar{z}^{k_1} \wedge \dots \wedge d\bar{z}^{k_r}.$$

The term of bidegree $(q + 1, r)$ will be denoted by $d'\alpha$ and that of bidegree $(q, r + 1)$ by $d''\alpha$. Symbolically we write

$$d = d' + d''$$

and say that d' is of type (1, 0) and d'' of type (0,1). By linearity, we extend d' and d'' to all forms. (An operator on $\wedge^{*c}(M)$ is said to be of *type* (a, b) if it maps a form of bidegree (q, r) into a form of bidegree $(q + a, r + b)$). Both d' and d'' are *complex operators*, that is if α is real, $d'\alpha$ and $d''\alpha$ are complex.

Since

$$0 = dd = d'd' + (d'd'' + d''d') + d''d''$$

it follows, by comparing types, that

$$d'd' = 0, \quad d''d'' = 0$$

and

$$d'd'' + d''d' = 0.$$

We remark that the operators d' and d'' define cohomology theories in the same manner as d gives rise to the de Rham cohomology (cf. § 6.10).

If f is a holomorphic function on M, $d''f$ vanishes. A *holomorphic form* α of degree p is a form of bidegree $(p, 0)$ whose coefficients relative to local complex coordinates are holomorphic functions. This may be expressed simply, by the condition, $d''\alpha = 0$. It follows that *a closed form of bidegree* $(p, 0)$ *is a holomorphic form.*

At this point it is convenient to make a slight change in notation writing θ_i in place of θ^i.

Let $\{\theta_1, \cdots, \theta_n\}$ be a base for the forms of bidegree (1, 0) on M. Then, the conjugate forms $\bar{\theta}_1, \cdots, \bar{\theta}_n$ comprise a base for the forms of bidegree (0, 1). Suppose M has a metric g (locally) expressible in the form

$$g = 2\sum_{i=1}^{n} \theta_i \otimes \bar{\theta}_i.$$

The operator $*$ may then be defined in terms of the given metric. Our procedure is actually the following: As originally defined $*$ was applied to real forms and played an essential role in the definition of the global scalar product on a compact manifold or, on an arbitrary Riemannian manifold when one of the forms has a compact carrier. In order that the properties of the global scalar product be maintained we extend $*$ to complex differential forms by linearity, that is

$$*(\lambda + \sqrt{-1}\,\mu) = *\lambda + \sqrt{-1}\,*\mu.$$

Hence, if M is compact (or, one of α, β has a compact carrier) we define the global scalar product

$$(\alpha, \beta) = \int_M \alpha \wedge *\bar{\beta},$$

so that, in general, (α, β) is complex-valued. However, $(\alpha, \alpha) \geqq 0$, equality holding, if and only if, $\alpha = 0$. Two p-forms α and β are said to be *orthogonal* if $(\alpha, \beta) = 0$. Evidently, if α and β are pure forms of different bidegrees they must be orthogonal.

The dual of a linear operator is defined as in § 2.9.

The operator $*$ maps a form of bidegree (q, r) into a form of bidegree $(n - r, n - q)$. The dual of the exterior differential operator d is the operator δ which maps p-forms into $(p - 1)$-forms. We define operators δ' and δ'' as follows:

$$\delta' = -*d''* \quad \text{and} \quad \delta'' = -*d'*$$

(cf. formula 2.8.7).

Clearly, then, δ' is of type $(-1, 0)$ and δ'' of type $(0, -1)$. Moreover,

$$\delta = \delta' + \delta''$$

For, $\delta = -*d* = -*d'* -*d''*$.

If M is compact or, one of α, β has a compact carrier,

$$(d'\alpha, \beta) = (\alpha, \delta'\beta) \quad \text{and} \quad (d''\alpha, \beta) = (\alpha, \delta''\beta)$$

where α is a p-form and β a $(p + 1)$-form. For,

$$(d'\alpha, \beta) + (d''\alpha, \beta) = (d\alpha, \beta) = (\alpha, \delta\beta) = (\alpha, \delta'\beta) + (\alpha, \delta''\beta).$$

If α is of bidegree (q, r), β is of bidegree $(q + 1, r)$; for, otherwise $d'\alpha$ and β are orthogonal. In this way, it is evident that the desired relations hold. Hence, δ' and δ'' are the duals of d' and d'', respectively.

Evidently,

$$\delta' \delta' = 0, \quad \delta'' \delta'' = 0, \quad \delta' \delta'' + \delta'' \delta' = 0.$$

In terms of the basis forms $\{\theta_i\}$ and $\{\bar{\theta}_i\}$ $(i = 1, \cdots, n)$, the fundamental form Ω is given by

$$\Omega = \sqrt{-1} \sum_{i=1}^{n} \theta_i \wedge \bar{\theta}_i. \tag{5.4.1}$$

We define the operator L on p-forms α of bidegree (q, r) as follows:

$$L\alpha = \alpha \wedge \Omega, \quad p \leqq 2n - 2.$$

Hence, $L\alpha$ is of bidegree $(q + 1, r + 1)$, that is, L is of type $(1, 1)$. For a p'-form β

$$L\alpha \wedge *\beta = \alpha \wedge L*\beta = \alpha \wedge **^{-1} L*\beta = (-1)^{p'} \alpha \wedge ** L*\beta.$$

We define an operator Λ of type $(-1, -1)$ in terms of L as follows:

$$\Lambda = (-1)^p *L*$$

on p-forms. The operator Λ is therefore dual to L and lowers the degree of a form by 2 whereas the operator L raises the degree by 2.

Moreover, if α is of bidegree (q, r), $\Lambda\alpha$ is of bidegree $(q-1, r-1)$. Evidently, $\Lambda\alpha = 0$ for p-forms α of degree less than 2. From (5.4.1)

$$\Lambda = \sqrt{-1}\sum_{k=1}^{n} i(\theta_k)\, i(\bar{\theta}_k) \tag{5.4.2}$$

where $i(\xi)$ is the *interior product operator*, that is, the dual of the operator $\epsilon(\xi)$. Following (3.3.4), we define

$$\epsilon(\xi)\alpha = \xi \wedge \alpha, \quad p < 2n$$

where α is a p-form, and, by (3.3.5)

$$i(\xi) \equiv i(X) = *\epsilon(\xi)*$$

where X is the tangent vector dual to the 1-form ξ.

Since $i(\theta_k)$ is an anti-derivation, $\Lambda\Omega = n$. The operator Λ is not a derivation. For, since a form α of bidegree (q, r) may be expressed as a linear combination of the forms $\theta_{j_1} \wedge \cdots \wedge \theta_{j_q} \wedge \bar{\theta}_{k_1} \wedge \cdots \wedge \bar{\theta}_{k_r}$, and Λ is linear, one need only examine the effect of Λ on such forms. Indeed, since $i(\theta_l)$ is an anti-derivation

$$\begin{aligned} &i(\theta_l)\,(\theta_{j_1} \wedge \ldots \wedge \theta_{j_q} \wedge \bar{\theta}_{k_1} \wedge \ldots \wedge \bar{\theta}_{k_r}) \\ &\quad = \begin{cases} 0, & l \neq j_1, \ldots, j_q; \\ \theta_{j_2} \wedge \ldots \wedge \theta_{j_q} \wedge \bar{\theta}_{k_1} \wedge \ldots \wedge \bar{\theta}_{k_r}, & l = j_1; \end{cases} \end{aligned} \tag{5.4.3}$$

a similar statement holds for $i(\bar{\theta}_l)$. Hence,

$$\begin{aligned} &i(\bar{\theta}_l)i(\theta_l)(\theta_{j_1} \wedge \ldots \wedge \theta_{j_q} \wedge \bar{\theta}_{k_1} \wedge \ldots \wedge \bar{\theta}_{k_r}) \\ &\quad = (-1)^{q-1}\,\theta_{j_2} \wedge \ldots \wedge \theta_{j_q} \wedge \bar{\theta}_{k_2} \wedge \ldots \wedge \bar{\theta}_{k_r} \end{aligned}$$

for $l = j_1 = k_1$ and is zero for $l \neq j_1, \cdots j_q, k_1, \cdots, k_r$. Thus,

$$\begin{aligned} &\Lambda(\theta_{j_1} \wedge \cdots \wedge \theta_{j_q} \wedge \bar{\theta}_{k_1} \wedge \cdots \wedge \bar{\theta}_{k_r}) \\ &= \sqrt{-1}\sum_{l=1}^{n} i(\theta_l)i(\bar{\theta}_l)\,(\theta_{j_1} \wedge \ldots \wedge \theta_{j_q} \wedge \bar{\theta}_{k_1} \wedge \ldots \wedge \bar{\theta}_{k_r}) \\ &= (-1)^q\ \sqrt{-1}\sum_{l=1}^{n}(-1)^{l+1}\,\theta_{j_1} \wedge \ldots \wedge \theta_{j_{l-1}} \wedge \theta_{j_{l+1}} \wedge \ldots \wedge \theta_{j_q} \wedge \bar{\theta}_{k_1} \wedge \ldots \\ &\qquad \wedge \bar{\theta}_{k_{l-1}} \wedge \bar{\theta}_{k_{l+1}} \wedge \ldots \wedge \bar{\theta}_{k_r}. \end{aligned}$$

In particular,

$$\Lambda(\theta_{j_1} \wedge \dots \wedge \theta_{j_q} \wedge \bar{\theta}_{k_1} \wedge \dots \wedge \bar{\theta}_{k_r} \wedge \Omega)$$

$$= (-1)^r \sqrt{-1} \sum_{i=1}^{n} \Lambda(\theta_{j_1} \wedge \dots \wedge \theta_{j_q} \wedge \theta_i \wedge \bar{\theta}_{k_1} \wedge \dots \wedge \bar{\theta}_{k_r} \wedge \bar{\theta}_i)$$

$$= (-1)^r \sqrt{-1} \Big\{ \sum_{i=1}^{n} [\sqrt{-1} \sum_{l=1}^{n} (-1)^{l+q+1}$$

$$\theta_{j_1} \wedge \dots \wedge \theta_{j_{l-1}} \wedge \theta_{j_{l+1}} \wedge \dots \wedge \theta_{j_q} \wedge \theta_i \wedge \bar{\theta}_{k_1} \wedge \dots \wedge \bar{\theta}_{k_{l-1}} \wedge \bar{\theta}_{k_{l+1}} \wedge \dots \wedge \bar{\theta}_{k_r} \wedge \bar{\theta}_i]$$

$$+ \sqrt{-1} (-1)^{2q+r-3} (n-p) (\theta_{j_1} \wedge \dots \wedge \theta_{j_q} \wedge \bar{\theta}_{k_1} \wedge \dots \wedge \bar{\theta}_{k_r}) \Big\}$$

$$= \Lambda(\theta_{j_1} \wedge \dots \wedge \theta_{j_q} \wedge \bar{\theta}_{k_1} \wedge \dots \wedge \bar{\theta}_{k_r}) \wedge \Omega$$

$$+ (n-p)\, \theta_{j_1} \wedge \dots \wedge \theta_{j_q} \wedge \bar{\theta}_{k_1} \wedge \dots \wedge \bar{\theta}_{k_r}.$$

Thus, for any p-form α, $\Lambda(\alpha \wedge \Omega) = \Lambda\alpha \wedge \Omega + (n-p)\alpha$. This result will prove useful in the sequel.

Consider the space C_n of n complex variables with complex coordinates $z^1, \cdots, z^n$ and metric

$$ds^2 = 2\sum_{i=1}^{n} dz^i\, d\bar{z}^i. \tag{5.4.4}$$

Let $\alpha = a_{j_1 \dots j_q k_1 \dots k_r}\, dz^{j_1} \wedge \cdots \wedge dz^{j_q} \wedge d\bar{z}^{k_1} \wedge \cdots \wedge d\bar{z}^{k_r}$ and denote by ∂_i the operator which replaces each coefficient $a_{j_1 \dots j_q k_1 \dots k_r}$ by the coefficient of dz^i in $da_{j_1 \dots j_q k_1 \dots k_r}$. In a similar way we define the operator $\bar{\partial}_i$. The forms $\partial_i\alpha$ and $\bar{\partial}_i\alpha$ are each of bidegree (q, r). Moreover, the operators ∂_i and $-\bar{\partial}_i$ are duals, that is, $(\partial_i\alpha, \beta) = -(\alpha, \bar{\partial}_i\beta)$. If we put $\theta_i = dz^i$, then

$$d' = \sum_i \epsilon(\theta_i)\partial_i, \quad d'' = \sum_i \epsilon(\bar{\theta}_i)\bar{\partial}_i \tag{5.4.5}$$

and, since δ' and δ'' are dual to d' and d'', respectively,

$$\delta' = -\sum_j \bar{\partial}_j\, i(\theta_j), \quad \delta'' = -\sum_j \partial_j\, i(\bar{\theta}_j). \tag{5.4.6}$$

Consider, for example, the linear differential form $\alpha = a_i dz^i + b_i d\bar{z}^i$. Then, since $\{dz^i\}$ and $\{\partial/\partial z^i\}$ are dual bases

$$\delta\alpha = \delta'\alpha + \delta''\alpha = -\sum_i \left(\frac{\partial a_i}{\partial \bar{z}^i} + \frac{\partial b_i}{\partial z^i}\right).$$

Lemma 5.4.1. *In* C_n

$$\Lambda d' - d'\Lambda = -\sqrt{-1}\,\delta''$$

and (5.4.7)

$$\Lambda d'' - d''\Lambda = \sqrt{-1}\,\delta'.$$

In the first place, it is easily checked that

$$i(\theta_k)d + di(\theta_k) = \partial_k$$

and

$$i(\bar{\theta}_k)d + di(\bar{\theta}_k) = \bar{\partial}_k.$$

Pre-multiplying the first of these equations by $i(\bar{\theta}_k)$ and post-multiplying the second by $i(\theta_k)$ one obtains after subtracting and summing with respect to k

$$\Lambda d - d\Lambda = \sqrt{-1}(\delta' - \delta'')$$

since $i(\theta_k)$ commutes with ∂_k. The desired formulae are obtained by separating the components of different types.

5.5. Kaehler manifolds

Let M be a complex manifold with an hermitian metric g. Then, in general, there does not exist at each point P of M a local complex coordinate system which is *geodesic*, that is a local coordinate system (z^i) with the property that g is equal to $2\Sigma_i\, dz^i \otimes d\bar{z}^i$ modulo terms of higher order. (Two tensors coincide up to the order k at $P \in M$ if their coefficients, as well as their partial derivatives up to the order k, coincide at P. A complex geodesic coordinate system at P should have the property that g coincide with $2\Sigma_i\, dz^i \otimes d\bar{z}^i$ up to the order 1 at P.)

We seek a condition to ensure that such local coordinates exist.

Let $\{\theta_1, \cdots, \theta_n\}$ be a base for the forms of bidegree (1, 0) on M with the property that g may be expressed in the form

$$g = 2\sum_i \theta_i \otimes \bar{\theta}_i \tag{5.5.1}$$

(cf. 5.3.30).

Our problem is to find n 1-forms ω_i of bidegree (1, 0) such that

(i) $\omega_i(P) = \theta_i(P)$, $i = 1, \cdots, n$;

(ii) $g = 2\,\Sigma_i\, \omega_i \otimes \bar{\omega}_i$ modulo terms of higher order; and

(iii) $d\omega_i(P) = 0$, $i = 1, \cdots, n$.

This latter condition is the requirement that in the sought after coordinates, the coefficients of connection vanish at P, that is, in terms of the metric tensor g, $dg_{ij*}(P) = 0$ (cf. 5.3.3, 5.3.32 and 5.3.10).

Let (z^i) be a system of local complex coordinates at P such that $z^i(P) = 0$, $i = 1, \cdots, n$ and $\theta_i(P) = dz^i(P)$. We put

$$\omega_i = \theta_i - \sum_{j,k} a_{ijk}\, z^j\, \theta_k - \sum_{j,k} b_{ijk}\, \bar{z}^j\, \theta_k \tag{5.5.2}$$

and look for the relations satisfied by the coefficients a_{ijk} and b_{ijk} in order that (i), (ii), and (iii) hold. For condition (ii) to hold it is necessary and sufficient that

$$a_{ijk} + \bar{b}_{kji} = 0. \tag{5.5.3}$$

Now, put

$$d\theta_i = \tfrac{1}{2}\sum_{j,k} c_{ijk}\, \theta_j \wedge \theta_k + \sum_{j,k} c'_{ijk}\, \bar{\theta}_j \wedge \theta_k,$$

$$c_{ijk} + c_{ikj} = 0.$$

Then, (iii) is satisfied, if and only if

$$\tfrac{1}{2}(a_{ijk} - a_{ikj}) = c_{ijk} \quad \text{and} \quad b_{ijk} = c'_{ijk}. \tag{5.5.4}$$

Substituting in (5.5.3), we derive

$$c_{ijk} = \bar{c}'_{jki} - \bar{c}'_{kji}.$$

These are the necessary conditions that a complex geodesic local coordinate system exists at P.

Conversely, assume that there exist c_{ijk}, c'_{ijk} satisfying $c_{ijk} = \bar{c}'_{jki} - \bar{c}'_{kji}$. If we put $a_{ijk} = -\bar{c}'_{kji}$ and $b_{ijk} = c'_{ijk}$ the relations (5.5.3) and (5.5.4) are satisfied. If we define the forms θ_i by (5.5.2), the conditions (i), (ii), and (iii) for a complex geodesic local coordinate system are satisfied.

We recall that an hermitian metric is a Kaehler metric if the associated 2-form $\Omega = \sqrt{-1}\,\Sigma_i \theta_i \wedge \bar{\theta}_i$ is closed and, in this case, M is a Kaehler manifold. Hence, *at each point of a Kaehler manifold there exists a system of local complex coordinates which is geodesic.* This property of the Kaehler metric leads to many significant topological properties of compact Kaehler manifolds which we now pursue.

5.6. Topology of a Kaehler manifold

The formulae (5.4.7) hold in a Kaehler manifold as one easily sees by choosing a complex geodesic coordinate system (z^i) at a point P. Indeed, for C_n we may take $g = 2 \sum_i dz^i \otimes d\bar{z}^i$. Since the metric of a Kaehler manifold has this form modulo terms of higher order, and since only first order terms enter into the derivation of the formulae (5.4.7) they must also hold in a Kaehler manifold.

Lemma 5.6.1. *In a Kaehler manifold*

$$\Lambda d' - d'\Lambda = -\sqrt{-1}\,\delta''$$

and (5.6.1)

$$\Lambda d'' - d''\Lambda = \sqrt{-1}\,\delta'.$$

These formulae are of fundamental importance in determining the basic topological properties of compact Kaehler manifolds.

Lemma 5.6.2. *In a Kaehler manifold the operators Λ and δ commute. Hence, by comparing types Λ commutes with δ' and δ''.*

Clearly, the operators L and d commute. Hence,

$$*d**^{-1}L* = *L**^{-1}d*,$$

that is

$$\delta\Lambda = \Lambda\delta.$$

Several interesting consequences may be derived from lemmas 5.6.1 and 5.6.2 for a complex manifold with a Kaehler metric. To begin with we have

Lemma 5.6.3. *In a Kaehler manifold*

$$d'\,\delta'' + \delta''\,d' = 0 \quad \text{and} \quad d''\,\delta' + \delta'\,d'' = 0. \tag{5.6.2}$$

The proof is immediate from lemma 5.6.1.

Lemma 5.6.4. *In a Kaehler manifold*

$$d'\,\delta' + \delta'\,d' = d''\,\delta'' + \delta''\,d''.$$

For, from lemma 5.6.1 the expression $-\sqrt{-1}(d'\Lambda d''-d''\Lambda d'+d''d'\Lambda - \Lambda d'd'')$ is equal to $d''\,\delta'' + \delta''\,d''$ from the first relation and to $d'\,\delta' + \delta'\,d'$ from the second.

Lemma 5.6.5. *In a Kaehler manifold the Laplace-Beltrami operator* $\Delta = d\,\delta + \delta\,d$ *has the expressions*

$$\Delta = 2(d'\,\delta' + \delta'\,d') = 2(d''\,\delta'' + \delta''\,d''). \tag{5.6.3}$$

For,

$$\begin{aligned}\Delta &= d\delta + \delta d \\ &= (d' + d'')(\delta' + \delta'') + (\delta' + \delta'')(d' + d'') \\ &= (d'\,\delta' + \delta'\,d') + (d''\,\delta'' + \delta''\,d'')\end{aligned}$$

by lemma 5.6.3. Applying lemma 5.6.4, the result follows.

A complex p-form α is called *harmonic* if $\Delta\alpha$ vanishes.

Since a p-form may be written as a sum of forms of bidegree (q, r) with $q + r = p$ we have:

Lemma 5.6.6. *A p-form is harmonic, if and only if its various terms of bidegree (q, r) with $q + r = p$ are harmonic.*

This follows from the fact that Δ is an operator of type $(0, 0)$. Indeed, d' is of type $(1, 0)$ and δ' of type $(-1, 0)$. Moreover, a p-form is zero, if and only if its various terms of bidegree (q, r) vanish.

Lemma 5.6.7. *In a Kaehler manifold Δ commutes with L and Λ. Hence, if α is a harmonic form so are $L\alpha$ and $\Lambda\alpha$.*

This follows easily from lemmas 5.6.1 and 5.6.2 since $\delta'\,\delta'' + \delta''\,\delta' = 0$ and $*\Delta = \Delta *$.

Lemma 5.6.8. *In a Kaehler manifold the forms $\Omega^p = \Omega \wedge \cdots \wedge \Omega$ (p times) for every integer $p \leq n$ are harmonic of degree $2p$.*

The proof is by induction. In the first place, $\Delta\Omega = 0$ since the manifold is Kaehlerian. For, by lemma 5.6.1, $\delta'\Omega = \delta''\Omega = 0$ since $d'\Omega = d''\Omega = 0$ and $\Lambda\Omega = n$. Now,

$$\Delta(\Omega^p) = \Delta(L\Omega^{p-1}) = L\Delta(\Omega^{p-1}) = 0.$$

Lemma 5.6.9. *The cohomology groups $H^{2p}(M, C)$ of a compact Kaehler manifold M with complex coefficients C are different from zero for $p = 0, 1, \cdots, n$.*

Indeed, by the results of Chapter II, $H^q(M, C)$ is isomorphic with the space of the (complex) harmonic forms of degree q on M. Since the constant functions are harmonic of degree 0, the lemma is proved for $p = 0$. The proof is completed by applying the previous lemma and showing that $\Omega^p \neq 0$ for $p < n$. In fact, we need only show that $\Omega^n \neq 0$, and this is so, since Ω^n defines an orientation of M (cf. § 5.1).

Theorem 5.6.1. *A holomorphic form on a Kaehler manifold is harmonic.*

For, if α is a holomorphic form, it is of bidegree $(p, 0)$; moreover $d''\alpha$ vanishes. Now, since δ'' is an operator of type $(0, -1)$, $\delta''\alpha$ is a form of bidegree $(p, -1)$, that is $\delta''\alpha = 0$. It follows that

$$\Delta\alpha = 2(d''\,\delta'' + \delta''\,d'')\alpha = 0.$$

Corollary. *A holomorphic form on a compact Kaehler manifold is closed.*

Conversely, *a harmonic form of bidegree $(p, 0)$ on a compact hermitian manifold is holomorphic.* For, a harmonic form is closed and a closed form of bidegree $(p, 0)$ is holomorphic.

The term of bidegree $(p, 0)$ of a harmonic p-form α is holomorphic. Similarly, the conjugate of the term of bidegree $(0, p)$ is holomorphic. For, let

$$\alpha = \sum_{k=0}^{p} \alpha_{p-k,k}\,,$$

the subscripts indicating the bidegree. Then, since α is harmonic and the manifold is compact

$$\sum_{k=0}^{p} d''\alpha_{p-k,k} = 0.$$

Since the terms on the left side of this equation are of different bidegrees they must vanish individually. In particular,

$$d''\alpha_{p,0} = 0.$$

Similarly, $d'\alpha_{0,p} = 0$ implies $\overline{d'\alpha_{0,p}} = d''\overline{\alpha_{0,p}} = 0$.

Let $\wedge_H^p$ be the linear space of complex harmonic forms of degree p. Then, by lemma 5.6.6, $\wedge_H^p$ is the direct sum of the subspaces $\wedge_H^{q,r}$ of the harmonic forms of bidegree (q, r) with $q + r = p$. The pth betti number $b_p(M)$ of the Kaehler manifold M is equal to the sum

$$\sum_{q+r=p} b_{q,r} \tag{5.6.4}$$

where $b_{q,r}$ is the complex dimension of $\wedge_H^{q,r}$. Now, if $\alpha \in \wedge_H^{q,r}$, its conjugate $\bar{\alpha} \in \wedge_H^{r,q}$, and conversely. For,

$$\tfrac{1}{2}\,\overline{\Delta\alpha} = \overline{d'\,\delta'\,\alpha + \delta'\,d'\,\alpha} = \overline{d'\,\delta'\,\alpha} + \overline{\delta'\,d'\,\alpha} = d''\,\delta''\,\bar{\alpha} + \delta''\,d''\,\bar{\alpha} = \tfrac{1}{2}\,\Delta\bar{\alpha}.$$

Hence,

$$b_{q,r} = b_{r,q}, \tag{5.6.5}$$

and, since $\alpha + \bar{\alpha}$ is real, (5.6.4) is also the (real) dimension of the space of real harmonic forms of degree p.

Since

$$b_p = b_{p,0} + \cdots + b_{0,p},$$

we have shown that

$$2b_{p,0} \leqq b_p \text{ for } p \neq 0.$$

Hence, *the number of holomorphic p-forms is majorized by half the* p^{th} *betti number*.

Moreover, from (5.6.5) we may also conclude that $b_p(M)$ is even if p is odd. Summarizing, we have:

Theorem 5.6.2. *The* p^{th} *betti number of a compact Kaehler manifold is even if p is odd. The first betti number is twice the dimension of the space of holomorphic 1-forms sometimes called* abelian differentials of the first kind. *The even-dimensional betti numbers* b_p $(p \leqq 2n)$ *are different from zero.*

The last part follows from lemma 5.6.9.

The number $\sum_{q=0}^{n} (-1)^q b_{0,q}$ is an important invariant of the complex structure called the *arithmetic genus*.

In the next section it is shown that for $p \leqq n - 1$, $b_p \leqq b_{p+2}$.

Since the first betti number of the Riemann sphere S^2 is zero there are no holomorphic 1-forms on S^2.

Consider the torus (cf. § 5.8) with the complex structure induced by C. Since $b_1 = 2$, the differential dz is (apart from a constant factor) the only holomorphic differential on the torus.

Let M be a compact (connected) Riemann surface. Put $b_1 = 2g$; the integer g is called the *genus* of M. It is equal to the number of independent abelian differentials of the first kind on M. Since there are $2g$ independent 1-cycles and g independent abelian differentials the periods of an abelian differential may not be arbitrarily prescribed on a basis of 1-cycles. However, it may be shown that a unique abelian differential exists with prescribed real parts of the periods.

Let α be a p-form. Then, by (II.B.4) and lemma 5.6.5

$$\begin{aligned}\alpha &= d\delta G\alpha + \delta dG\alpha + H[\alpha] \\ &= 2(d'\,\delta'\,G\alpha + \delta'\,d'\,G\alpha) + H[\alpha] \\ &= 2(d''\,\delta''\,G\alpha + \delta''\,d''\,G\alpha) + H[\alpha]\end{aligned}$$

where the operators H and G are the complex extensions of the corresponding real operators. Moreover, since the Green's operator G commutes with d and δ it commutes with d', d'', δ', δ'' as one sees by comparing types.

Since Δ commutes with d, it also commutes with d' and d'' as one sees by comparing types. This result is very important since it relates harmonic forms with the cohomology theories arising from d' and d''.

5.7. Effective forms on an hermitian manifold

There is a special class of forms defined as the zeros of the operator Λ on the (linear)space of harmonic forms. They are called effective harmonic forms and the dimension of the space determined by them is a topological invariant. More precisely, the number e_p of linearly independent effective harmonic forms of degree p on a compact Kaehler manifold M is equal to the difference $b_p - b_{p-2}$ for $p \leqq n + 1$ where dim $M = 2n$. This important result hinges on a relation measuring the defect of the operator $L^k\Lambda$ from ΛL^k where $L^k\alpha = \alpha \wedge \Omega^k$. The fact that these operators do not commute is crucial for the determination of the invariants e_p.

Lemma 5.7.1. *For any p-form* α *on an hermitian manifold M*

$$(\Lambda L^k - L^k\Lambda)\alpha = k(n - p - k + 1)L^{k-1}\,\alpha.$$

It was shown in § 5.4 that

$$\Lambda\, L\alpha = L\Lambda\alpha + (n - p)\,\alpha.$$

Hence, proceeding by induction on the integer k

$$\begin{aligned}\Lambda L^{k+1}\alpha = \Lambda L^k(L\alpha) &= L^k\Lambda(L\alpha) + k(n - p - 2 - k + 1)L^k\alpha \\ &= L^k[L\Lambda\alpha + (n - p)\alpha] + k(n - p - k - 1)L^k\alpha \\ &= L^{k+1}\Lambda\alpha + (k + 1)\,(n - p - k)L^k\,\alpha.\end{aligned}$$

This completes the proof.

In the remainder of this section a subscript on a given form will indicate its degree; thus deg $\alpha_p = p$.

A form α is said to be *effective* if it is a zero of the operator Λ, that is, if $\Lambda\alpha = 0$. Since Λ annihilates $\wedge^p(T^{c*})$ for $p = 0,1$ the elements of these spaces are effective.

Lemma 5.7.2. *If α_p is an effective form, then, for any $s \geqq 0$*

$$(-1)^{k+1}\Lambda^k L^{k+s}\alpha_p = (s+1)\cdots(s+k)(s-n+p)\cdots(s-n+p+k-1)L^s\alpha_p.$$

This follows inductively from the preceding lemma.

Corollary. *There are no effective p-forms for $p \geqq n+1$.*

This is an immediate consequence, if we take $k = n+1$ and $s \geqq n-p+1$.

Theorem 5.7.1. *Every p-form $\alpha_p (p \leqq n+1)$ on an hermitian manifold of complex dimension n has a unique representation as a sum*

$$\alpha_p = \sum_{k=0}^{r} L^k \varphi_{p-2k} \tag{5.7.1}$$

where the φ_{p-2k}, $0 \leqq k \leqq r$ are effective forms and $r = [\frac{p}{2}]$.

The theorem is trivial for $p = 0,1$. Proceeding inductively, assume its validity for $p \leqq n-1$. Then, to any p-form β_p is associated a unique p-form α_p such that

$$\Lambda L \alpha_p = \beta_p, \quad p \leqq n-1. \tag{5.7.2}$$

For,

$$\beta_p = \sum_{k=0}^{r} L^k \psi_{p-2k}$$

where the forms ψ_{p-2k} are effective. Now, by (5.7.1) and lemma 5.7.1

$$\Lambda L \alpha_p = \sum_{k=0}^{r} \Lambda L^{k+1} \varphi_{p-2k}$$

$$= \sum_{k=0}^{r} (k+1)(n-p-k) L^k \varphi_{p-2k}.$$

Since $p \leqq n-1$, $n-p+k \neq 0$. Consequently, in order that (5.7.2) hold, it is sufficient to take

$$\varphi_{p-2k} = \frac{\psi_{p-2k}}{(k+1)(n-p+k)}, \quad k = 0, 1, \cdots, r,$$

and by uniqueness, this is also necessary. Now, let β_{p+2} be an arbitrary $(p+2)$-form and put $\Lambda\, \beta_{p+2} = \beta_p$ in (5.7.2).

Then, the form $\chi_{p+2} = \beta_{p+2} - L\alpha_p$ is effective, and

$$\beta_{p+2} = \chi_{p+2} + L\alpha_p$$
$$= \chi_{p+2} + \sum_{k=0}^{r} L^{k+1}\,\varphi_{p-2k}$$

is the representation sought for β_{p+2} thereby completing the induction. The uniqueness is evident from that of α_p. For, let

$$\beta_{p+2} = \chi'_{p+2} + L\alpha'_p$$

be another decomposition for β_{p+2}. Then, $(\chi'_{p+2} - \chi_{p+2}) + L(\alpha'_p - \alpha_p) = 0$. Applying the operator Λ to this relation we obtain $\Lambda\, L\alpha'_p = \Lambda\, L\alpha_p$ since $\chi'_{p+2} - \chi_{p+2}$ is effective. Applying (5.7.2), we conclude that $\alpha'_p = \alpha_p$ from which $\chi'_{p+2} = \chi_{p+2}$.

Corollary 5.7.1. *ΛL is an automorphism of $\wedge^p(T^{c*})$ for $p \leqq n - 1$.*

For, if $\alpha_p \in \wedge^p(T^{c*})$, $\Lambda L\alpha_p \in \wedge^p(T^{c*})$. Conversely, by (5.7.2) for any β_p there is an α_p such that $\Lambda L\alpha_p = \beta_p$. Moreover, $\Lambda L\alpha_p = 0$ implies $\alpha_p = 0$.

Corollary 5.7.2. *L is an isomorphism of $\wedge^p(T^{c*})$ into $\wedge^{p+2}(T^{c*})$ for $p \leqq n - 1$.*

Indeed, $L\alpha_p = 0$ implies $\Lambda\, L\alpha_p = 0$ from which by the preceding corollary α_p must vanish.

Assume now that M is a Kaehler manifold. Then, since Δ commutes with the operator L (cf. lemma 5.6.7) we may conclude

Corollary 5.7.3. *Every harmonic p-form $\alpha_p (p \leqq n + 1)$ on a Kaehler manifold may be uniquely represented as a sum*

$$\alpha_p = \sum_{k=0}^{r} L^k\,\varphi_{p-2k}$$

where the $\varphi_{p-2k} (0 \leqq k \leqq r)$ are effective harmonic forms and $r = [p/2]$.

Let M be a compact Kaehler manifold. Then, from lemma 5.6.7 and corollary 5.7.2, it follows that

$$b_p(M) \leqq b_{p+2}(M), \quad p \leqq n - 1. \tag{5.7.3}$$

Corollary 5.7.4. *The betti numbers b_p for $p \leqq n - 1$ of a compact Kaehler manifold satisfy the monotonicity condition* (5.7.3). *Moreover, $b_{2s} \neq 0$ for $s \leqq n$.*

The difference $b_p - b_{p-2}$ may be measured in terms of the number e_p of effective harmonic forms of degree p, $p \leqq n+1$ and is given by the following

Theorem 5.7.2. *On a compact Kaehler manifold*

$$e_p = b_p - b_{p-2}$$

for $p \leqq n+1$.

To see this, denote by $\tilde{\wedge}^p_H$ the linear subspace of $\wedge^p_H$ of effective harmonic p-forms. Then, by corollary 5.7.3

$$\wedge^p_H = \tilde{\wedge}^p_H \oplus L\tilde{\wedge}^{p-2}_H \oplus \ldots \oplus L^r \tilde{\wedge}^{p-2r}_H \tag{5.7.4}$$

where $r = [p/2]$, and

$$\wedge^{p+2}_H = \tilde{\wedge}^{p+2}_H \oplus L\tilde{\wedge}^p_H \oplus \ldots \oplus L^r \tilde{\wedge}^{p+2-2r}_H \tag{5.7.5}$$

where $r = [p/2] + 1$.

Applying the operator L to the relation (5.7.4) we obtain

$$L\wedge^p_H = L\tilde{\wedge}^p_H \oplus \ldots \oplus L^{r+1}\tilde{\wedge}^{p-2r}_H, \qquad r = \left[\frac{p}{2}\right]. \tag{5.7.6}$$

Combining (5.7.5) and (5.7.6)

$$\wedge^{p+2}_H = \tilde{\wedge}^{p+2}_H \oplus L\wedge^p_H.$$

Since L is an isomorphism from $\wedge^p(T^{c*})$ into $\wedge^{p+2}(T^{c*})$ ($p \leqq n-1$) and since Δ commutes with L, $\dim L\wedge^p_H = \dim \wedge^p_H$. Hence,

$$\dim \wedge^{p+2}_H = \dim \tilde{\wedge}^{p+2}_H + \dim \wedge^p_H$$

that is $b_{p+2} = e_{p+2} + b_p$, $p \leqq n-1$ or $b_p - b_{p-2} = e_p$ for $p \leqq n+1$.

5.8. Holomorphic maps. Induced structures

Let M and M' be complex manifolds. A differentiable map $f: M \to M'$ is said to be a *holomorphic map* if the induced dual map $f^*: \wedge^{*c}(M') \to \wedge^{*c}(M)$ sends forms of bidegree $(1, 0)$ into forms of bidegree $(1, 0)$. Under the circumstances, f^* preserves types, that is, it maps forms of bidegree (q, r) on M' into forms of bidegree (q, r) on M. For, since f^* is a ring homomorphism we need only examine its effect on the decomposable forms (cf. § 1.5).

If $f : M \to M'$ and $g: M' \to M''$ are holomorphic maps, so is the composed map $g \cdot f : M \to M''$. By a *holomorphic isomorphism* $f : M \to M'$ is meant a 1-1 holomorphic map f together with a differentiable map $g : M' \to M$ such that both $f \cdot g$ and $g \cdot f$ are the identity maps on M' and M, respectively. If f is a holomorphic isomorphism, it follows that the inverse map g is also a holomorphic isomorphism.

Lemma 5.8.1. *Let M be a complex manifold and f a complex-valued differentiable function on M. In order that f be a holomorphic map of M into C (considered as a complex manifold), it is necessary and sufficient that f be a holomorphic function.*

Since dz is a base for the forms of bidegree (1, 0) on C, in order that f be a holomorphic map, it is necessary and sufficient that $f^*(dz) = df$ be of bidegree (1, 0). Hence, since $df = d'f + d''f$, it is necessary and sufficient that $d''f$ vanish.

Lemma 5.8.2. *The induced dual map of a holomorphic map sends holomorphic forms into holomorphic forms.*

Let $f: M \to M'$ be a holomorphic map and α a form of bidegree $(p, 0)$ on M'. Then, since f^* preserves bidegrees, $f^*(\alpha)$ is a form of bidegree $(p, 0)$ on M. Hence, since f^* and d commute, so do f^* and d''. Thus, if α is holomorphic, so is $f^*(\alpha)$.

Proposition 5.8.3. *Let $\tilde{M}$ be a covering space of the complex manifold M and π the canonical projection of $\tilde{M}$ onto M. (We denote this covering space by $(\tilde{M}, \pi)$.) Then, there exists a unique complex structure on $\tilde{M}$ with respect to which π is a holomorphic map.*

For, let $\{V_\alpha\}$ be an open covering of $\tilde{M}$ such that for every α the restriction π_α of π to V_α is a homeomorphism of V_α onto $\pi(V_\alpha)$. Such a covering of $\tilde{M}$ always exists. To each α is associated a complex structure operator J_α on V_α in terms of which $\pi_\alpha : V_\alpha \to M$ is holomorphic. To see this, we need only define $\pi_{\alpha*} \cdot J_\alpha = J \cdot \pi_{\alpha*}$. On the intersection $V_\alpha \cap V_\beta$, the complex structure operators J_α and J_β coincide since $\pi_\beta^{-1} \cdot \pi_\alpha$ is the identity map on $V_\alpha \cap V_\beta$, and as such is holomorphic. Thus, the operator on $\tilde{M}$ having the J_α as its restrictions defines a complex structure on $\tilde{M}$. With respect to this complex structure on $\tilde{M}$ the projection π is evidently a holomorphic map. The uniqueness is clear.

Corollary. *Let $(\tilde{M}, \pi)$ be a covering space of the Kaehler manifold M. Then, $(\tilde{M}, \pi)$ has a canonically defined Kaehler structure.*

For, let Ω be the Kaehler 2-form of M canonically defined by the Kaehler metric ds^2 of M. Let π^* denote the induced dual map of π.

Then, since $\pi^*(ds^2)$ is positive and hermitian and

$$d(\pi^*\Omega) = \pi^*(d\Omega) = 0,$$

the result follows. $\pi^*(ds^2)$ is positive since the Jacobian of the map is different from zero.

Conversely, suppose that the covering space $(\tilde{M}, \pi)$ of the manifold M has a complex structure. Moreover, assume that every point of M has an open connected neighborhood U such that each component of $\pi^{-1}(U)$ is open in $\tilde{M}$, that is, the union of disjoint open sets V_α on each of which π induces a homeomorphism π_α of V_α onto U in such a way that for any α and β, $\pi_\beta^{-1} \cdot \pi_\alpha$ is a holomorphic isomorphism of V_α onto V_β with respect to the complex structures induced on V_α and V_β by that of $\tilde{M}$. Then, U has a complex structure induced by the maps π_α—the complex structure being independent of the choice of α. We conclude that M has a complex structure called the *quotient complex structure* of that on $\tilde{M}$ by the relation of equivalence $\pi(\tilde{P}) = \pi(\tilde{P}')$, $\tilde{P}$ and $\tilde{P}'$ being points of $\tilde{M}$.

If $(\tilde{M}, \pi)$ has a Kaehlerian structure, then by exactly the same argument as given above M has a canonically defined Kaehlerian structure.

Consider the important case where the manifold M is the quotient space $\tilde{M}/G$ of the complex manifold $\tilde{M}$ by the relation of equivalence determined by a *properly discontinuous group* G of homeomorphisms of $\tilde{M}$ onto $\tilde{M}$ without fixed points. In other words, by the relation for which the equivalence class of the point $\tilde{P} \in \tilde{M}$ is the set of transforms $g(\tilde{P})$ of $\tilde{P}$ by the elements g of G such that every point of $\tilde{M}$ has a neighborhood V with the property (A): $V \cap gV$ is empty for all $g \in G$ other than the identity. Then, $\tilde{M}$ is a covering space of $M = \tilde{M}/G$. Indeed, any point $P \in M$ has a neighborhood U such that $\pi^{-1}(U)$ is the union of disjoint open sets V_α on each of which π induces a homeomorphism π_α of V_α onto U. To see this, take a point $\tilde{Q} \in \tilde{M}$ such that $P = \pi(\tilde{Q})$; then, take $U = \pi(V)$ where V is a neighborhood of $\tilde{Q}$ with the property (A). Moreover, for the neighborhoods V_α take the transforms gV of V for all $g \in G$.

In order that $M = \tilde{M}/G$ have a complex structure it is necessary and sufficient that G be a group of holomorphic isomorphisms.

If $\tilde{M}$ has a Kaehlerian structure the above condition is also necessary and sufficient for M to have a Kaehlerian structure.

5.9. Examples of Kaehler manifolds

1. A complex manifold of complex dimension 1 is usually called a *Riemann surface*. Let S be a Riemann surface with an hermitian metric

$ds^2 = \rho^2 dz\, d\bar{z}$ where ρ is a real, positive function (of class ∞) of the local coordinates $x, y (z = x + iy)$, $i = \sqrt{-1}$. The fundamental 2-form $\Omega = (i/2)\, \rho^2 dz \wedge d\bar{z}$ is the element of area of S. Clearly, $d\Omega = 0$ since dim $S = 2$. The real unit tangent vectors which are given by

$$e(\varphi) = \frac{1}{\rho}\left(e^{i\varphi}\frac{\partial}{\partial z} + e^{-i\varphi}\frac{\partial}{\partial \bar{z}}\right)$$

determine a sub-bundle $\tilde{B}$ of the tangent bundle called the *circle bundle.* We define a differential form ω of bidegree (1,0) by the formula

$$\omega = e^{-i\varphi}\rho\, dz.$$

Evidently, $\langle e(\varphi), \omega\rangle = 1$. Conversely, ω is uniquely determined by the conditions: (i) it is of bidegree (1,0) and (ii) its inner product with the vectors of $\tilde{B}$ is 1. Consider the 1-form θ on $\tilde{B}$ defined as follows:

$$\theta = -\, d\varphi + i(d' - d'')\log \rho.$$

One may easily check that θ is real and satisfies the differential equation

$$d\omega = i\theta \wedge \omega.$$

In fact, θ is the only real-valued linear differential form satisfying this differential equation with the property that $\theta \equiv -\, d\varphi$ (mod $(dz, d\bar{z})$). Hence, θ is globally defined in $\tilde{B}$, independent of the choice of local coordinates. Moreover,

$$d\theta = -\, 2i\frac{\partial^2 \log \rho}{\partial z\, \partial \bar{z}}\, dz \wedge d\bar{z}.$$

Now, the Gaussian curvature K of S is given by

$$K = -\frac{4}{\rho^2}\frac{\partial^2 \log \rho}{\partial z\, \partial \bar{z}},$$

from which

$$d\theta = K\Omega.$$

It is known that a compact Riemann surface can be given an hermitian metric of constant curvature and that such surfaces may be classified according to whether K is positive, negative or zero.

Incidentally, besides the Riemann sphere ($K > 0$) and the torus ($K = 0$) any other compact Riemann surface can be considered as the quotient space of the unit disc by some Fuchsian group.

2. Consider C_n with the metric

$$ds^2 = 2\sum_{i=1}^{n} dz^i \, d\bar{z}^i.$$

The fundamental 2-form in this case is given by

$$\Omega = \sqrt{-1}\sum_{i=1}^{n} dz^i \wedge d\bar{z}^i.$$

Clearly, this form is closed, and so the metric defines a Kaehler structure on C_n.

3. Let Γ be a discrete subgroup of maximal rank of the additive group of C_n and denote by T_n the quotient space C_n/Γ; Γ is actually the discrete additive group (over R) generated by $2n$ independent vectors. It is clear that Γ is a properly discontinuous group without fixed points. As a topological space, C_n/Γ is homeomorphic with the product of a torus of dimension $2n$ and a vector space over R. However, C_n/Γ is compact since Γ has rank $2n$, and so it is isomorphic as a topological group with the torus. Since the complex structure on C_n is invariant under Γ (cf. § 5.8) one is able to define a complex structure (and one only) on the quotient space T_n. With this complex structure the manifold $T_n = C_n/\Gamma$ is called a *complex multi-torus*.

Let π denote the natural projection of C_n onto T_n. Then, π is a holomorphic map. The metric of C_n defined in example 2 is invariant by the translations of Γ. We are therefore able to define a metric on T_n in such a way that π is locally an isometry. Since the property of a complex manifold which ensures that it be Kaehlerian is a local property, T_n is a Kaehler manifold.

We describe the homology properties of the multi-torus T_n: The projection π induces a canonical isomorphism π^* of the space of differential forms on T_n onto the space of differential forms on C_n invariant by the translations of Γ. Since the isomorphism π^* commutes with the operators d and δ, π^* defines an isomorphism of the space $\wedge^{*c}_H(T_n)$ of the harmonic forms on T_n onto $\wedge^{*c}_1(C_n)$—the vector subspace of $\wedge^{*c}(C_n)$ generated by $\{dz^A\}$ and their exterior products. For, the elements of $\wedge^{*c}_1(C_n)$ are harmonic and invariant by the translations of Γ. Conversely, every form α on C_n may be expressed as

$$\alpha = a_{i_1\ldots i_q\, j_1\ldots j_r}\, dz^{i_1} \wedge \cdots \wedge dz^{i_q} \wedge d\bar{z}^{j_1} \wedge \cdots \wedge d\bar{z}^{j_r}$$

where the coefficients are complex-valued functions. If α is the image by π^* of a harmonic form on T_n it is harmonic and invariant by Γ, that is, its coefficients $a_{i_1\cdots i_q j_1\cdots j_r}$ are harmonic functions which are invariant by Γ. Consequently, these functions are the images by π^* of harmonic functions on T_n. But a harmonic function on a compact manifold is a constant function, and so $\alpha \in \wedge_1^{*c}(C_n)$.

4. On a bounded open set M contained in C_n there exists a well-defined 2-form invariant by the group of complex automorphisms of M. This is a consequence of the theory of Bergman. One can construct canonically from this form a 2-form Ω having the Kaehler property, namely, $d\Omega = 0$ [72].

5. Complex projective n-space P_n: By identifying pairs of antipodal points of the sphere

$$\sum_{i=0}^{n} z^i \bar{z}^i = 1$$

contained in C_{n+1} we obtain P_n. For every index j, let U_j be the open subspace of P_n defined by $t^j \neq 0$ where $t^0, t^1, \cdots, t^n$ denote the homogeneous coordinates of the points of P_n. The map

$$(t^0, t^1, \cdots, t^n) \rightarrow (z_j^0, z_j^1, \cdots, \hat{z}_j^j, \cdots, z_j^n), \quad z_j^i = \frac{t^i}{t^j}$$

is a holomorphic isomorphism of U_j onto C_n. It is easily checked that these maps for $j = 0, 1, \cdots, n$ define a complex structure on P_n.

Consider the functions $\varphi_j = \sum_{i=0}^{n} z_j^i \bar{z}_j^i$ defined in each open set U_j of the covering. On $U_j \cap U_k$ we have

$$z_j^i = z_k^i / z_k^j \quad (k \text{ not summed})$$

and

$$\varphi_k = \sum_{i=0}^{n} z_k^i \bar{z}_k^i = \sum_{i=0}^{n} (z_j^i \bar{z}_j^i) z_k^j \bar{z}_k^j = \varphi_j z_k^j \bar{z}_k^j \quad (j, k \text{ not summed})$$

where z_k^j is a holomorphic function in U_k, and hence in $U_j \cap U_k$. The φ_j define a real closed form Ω of bidegree (1,1) on P_n. Indeed, in $U_j \cap U_k$

$$d'd''(\log \varphi_j - \log \varphi_k) = 0.$$

Hence, Ω is given by

$$\Omega = \sqrt{-1}\, d'd'' \log \varphi_i$$

in each open set U_i. In particular in U_0

$$\Omega = \sqrt{-1}\, d'd'' \log \varphi_0. \tag{5.9.1}$$

Clearly, Ω is a closed 2-form, and since

$$\varphi_0 = \sum_{i=0}^{n} z_0^i \bar{z}_0^i = 1 + \sum_{i=1}^{n} z_0^i \bar{z}_0^i,$$

$$\Omega = \sqrt{-1}\, \frac{\Sigma\, dz_0^i \wedge d\bar{z}_0^i + \Sigma\, |z_0^i|^2\, \Sigma\, dz_0^i \wedge d\bar{z}_0^i - \Sigma\, \bar{z}_0^i dz_0^i \wedge \Sigma\, z_0^i d\bar{z}_0^i}{(1 + \Sigma\, |z_0^i|^2)^2}.$$

The associated metric tensor g (sometimes called the *Fubini metric*) is given by

$$ds^2 = 2\, \frac{\Sigma\, |dz_0^i|^2 + \Sigma\, |z_0^i|^2\, \Sigma\, |dz_0^i|^2 - |\Sigma\, \bar{z}_0^i\, dz_0^i|^2}{(1 + \Sigma\, |z_0^i|^2)^2}$$

or, more explicitly by

$$g_{ij^*} = \frac{\delta_{ij}}{\varphi_0} - \frac{z_0^i \bar{z}_0^j}{\varphi_0^2}.$$

We remark that the fundamental form Ω of any Kaehler manifold may be written in the form (5.9.1). For, by § 5.3, since the metric tensor g is (locally) expressible as

$$g_{ij^*} = \frac{\partial^2 f}{\partial z^i \partial \bar{z}^j}$$

for some real-valued function f,

$$\Omega = \sqrt{-1}\, \frac{\partial^2 f}{\partial z^i\, \partial \bar{z}^j}\, dz^i \wedge d\bar{z}^j = \sqrt{-1}\, d'd''f.$$

6. Let M be a Kaehler manifold and M' a complex manifold holomorphically imbedded (that is, without singularities) in M. The metric g on M induces an hermitian metric on M'. The associated 2-form Ω' on M' coincides with the form induced by Ω and is therefore closed. In this way, the induced complex structure on M' is Kaehlerian (cf. § 5.8).

7. Let $G(n, k)$ denote the *Grassman manifold* of k-dimensional projective subspaces of P_n [26]. It can be shown that it is a non-singular irreducible rational variety in a P_N for sufficiently large N. Moreover, its odd-dimensional betti numbers vanish whereas b_{2p} is the number of partitions of $p = a_0 + a_1 + \cdots + a_k$ (a_i: integers) such that $0 \leqq a_0 \leqq a_1 \leqq \cdots \leqq a_k \leqq n - k$.

Example 6 in § 5.1 cannot be given a Kaehler structure except for $S^1 \times S^1$ since in all other cases b_2 is zero. It may be shown by employing the algebra of Cayley numbers (cf. V.B.7) that the 6-sphere S^6 possesses an almost complex structure. However, since $b_2(S^6) = 0$, S^6 does not have a Kaehlerian structure.

Besides S^2, the only sphere which may carry a complex structure is S^6. However, it can be shown that the almost complex structure defined by the Cayley numbers is not integrable.

EXERCISES

A. Holomorphic functions [*50*]

1. Let S be an open subset of C_n. In order that $f \in F$ (the algebra of differentiable functions on S) be a holomorphic function it is necessary and sufficient that

$$\left(\frac{\partial}{\partial x^i} + \sqrt{-1}\,\frac{\partial}{\partial y^i}\right) f = 0, \quad i = 1, \cdots, n$$

where $z^i = x^i + \sqrt{-1}y^i$. Put $f = u + \sqrt{-1}v$. Then,

$$\frac{\partial u}{\partial x^i} = \frac{\partial v}{\partial y^i} \quad \text{and} \quad \frac{\partial v}{\partial x^i} = -\frac{\partial u}{\partial y^i}, \quad i = 1, \cdots, n.$$

These are the *Cauchy-Riemann* equations. Prove that the holomorphic functions on S are those functions which may be expanded in a convergent power series in the neighborhood of every point of S.

If f is a holomorphic function and $a = (a^1, \cdots, a^n) \in S$, then, for every $b = (b^1, \cdots, b^n) \in C_n$, the function

$$g(z) = f(a + bz)$$

is a holomorphic function in a neighborhood of $z = 0 \in C$.

2. (a) Let f be a holomorphic function on the complex manifold M. If, for every point P with local coordinates $(z^1, \cdots, z^n)$ in a neighborhood of P_0 with the local coordinates $(a^1, \cdots, a^n)$, $|f(z^1, \cdots, z^n)| \leqq |f(a^1, \cdots, a^n)|$, then $f(z^1, \cdots, z^n) = f(a^1, \cdots, a^n)$ for all P in a neighborhood of P_0. Hence, if M is compact (and connected), a holomorphic function is necessarily a constant.

(b) A compact connected submanifold of C_n is a point.

3. Show that a holomorphic function on a (connected) complex manifold M which vanishes on some non-empty open subset must vanish everywhere on M.

4. Let α be a holomorphic 1-form on the Riemann sphere S^2. Then, in C_1—the complex plane, $\alpha = f(z)dz$ where $f(z)$ is an entire function. By employing the map given by $1/z$ at ∞ show that $f(1/z)1/z^2$ has a pole at the origin unless $f(z) = 0$. In this way, we obtain a direct proof of the fact that S^2 is of genus 0.

B. Almost complex manifolds [50]

1. Let X and Y be any two vector fields of type (0,1) on the almost complex manifold M. Then, in order that M be complex it is necessary that $[X,Y]$ be of type (0,1). Denote by $T^{1,0}$ and $T^{0,1}$ the spaces of tangent vector fields of types (1,0) and (0,1), respectively, on M.

2. On an almost complex manifold M the following conditions are equivalent:

(a) $[T^{0,1}, T^{0,1}] \subset T^{0,1}$;

(b) $d \wedge^{q,r} \subset \wedge^{q+1,r} \oplus \wedge^{q,r+1}$ for every q and r;

(c) $h(X,Y) \equiv [X,Y] + J[JX,Y] + J[X,JY] - [JX,JY] = 0$ for any vector fields X and Y where J is the almost complex structure operator of M.

Hence, in order that M be complex it is necessary that $h(X,Y) = 0$, for any X and Y. Show that the condition (c) is equivalent to (5.2.18).

3. $h(X,Y)$ is a tensor of type (1,2) with the properties:

(i) $h(X + Y,Z) = h(X,Z) + h(Y,Z)$,

(ii) $h(X,Y) = - h(Y,X)$,

(iii) $h(X,fY) = f\,h(X,Y)$

for any $X,Y,Z \in T$ and $f \in F$.

4. If $\dim M = 2$, M is complex.

Hint: $h(X,JX) = 0$ for all X.

5. Let G be a $2n$-dimensional Lie group, L the Lie algebra of left invariant vector fields on G and J an almost complex structure on G. If the tensor field J of type (1,1) on G is left invariant, that is, if J is a *left invariant almost complex structure*, then $JL = L$. The integrability condition may consequently be expressed as $h(X,Y) = 0$ for any $X,Y \in L$. Since every bi-invariant (that is, both left and right invariant) tensor field on a Lie group is analytic it follows that every left invariant almost complex structure on an abelian Lie group defines a complex structure on the underlying manifold. (It is known that a bi-invariant almost complex structure on any Lie group is integrable.)

6. Show that any two complex structures on a differentiable manifold which define the same almost complex structure coincide.

7. Let **C** denote the algebra of *Cayley numbers*: It has a basis $\{I, e_0, e_1, \cdots, e_6\}$ where I is the unit element and the multiplication table is

$$e_i^2 = -I, \quad e_j \cdot e_i = -e_i \cdot e_j \ (i \neq j), \quad i,j = 0,1, \cdots, 6,$$
$$e_0 \cdot e_1 = e_2, \quad e_0 \cdot e_3 = e_4, \quad e_0 \cdot e_5 = -e_6,$$
$$e_1 \cdot e_4 = e_5, \quad e_1 \cdot e_3 = e_6, \quad e_2 \cdot e_3 = e_5, \quad e_2 \cdot e_4 = -e_6,$$

the other $e_i \cdot e_j$ being given by permuting the indices cyclically. The algebra **C** is non-associative.

Any element of **C** may be written as

$$xI + X, \quad x \in R$$

where

$$X = \sum_{i=0}^{6} x^i e_i, \quad x^i \in R, \quad i = 0,1, \cdots, 6.$$

If $x = 0$, the element is called a *purely imaginary Cayley number*. These numbers form a 7-dimensional subspace $E^7 \subset$ **C**. The product $X \cdot Y$ of $X = \sum_{i=0}^{6} x^i e_i \in E^7$ and $Y = \sum_{i=0}^{6} y^i e_i \in E^7$ may be expressed in the form

$$X \cdot Y = -\langle X, Y \rangle I + X \times Y$$

where

$$\langle X, Y \rangle = \sum_{i=0}^{6} x^i y^i$$

is the scalar product in E^7, and

$$X \times Y = \sum_{i \neq j} x^i y^j e_i \cdot e_j$$

is the *vector product* of X and Y. The vector product has the properties:

(i) $(aX_1 + bX_2) \times Y = a(X_1 \times Y) + b(X_2 \times Y)$,

(i)′ $X \times (cY_1 + dY_2) = c(X \times Y_1) + d(X \times Y_2)$

for any $a, b, c, d \in R$;

(ii) $\langle X, X \times Y \rangle = \langle Y, X \times Y \rangle = 0$ and

(iii) $X \times Y = -Y \times X$.

Consider the unit 6-sphere S^6 in E^7:

$$S^6 = \{X \in E^7 \mid \langle X, X \rangle = 1\}.$$

Let g denote the (canonically) induced metric on S^6. The tangent space T_X at $X \in S^6$ may be identified with a subspace of E^7.

Define the endomorphism

$$J_X : T_X \to T_X$$

by

$$J_X Y = X \times Y, \quad Y \in T_X.$$

It has the properties:

(i) $J_X^2 = -$ identity;

(ii) $g(J_X Y, J_X Z) = g(Y,Z)$

for $Y,Z \in T_X$.

Property (i) implies that S^6 has an almost complex structure whereas (ii) says that the metric on S^6 is hermitian. Under the circumstances, S^6 is said to possess an *almost hermitian structure.*

8. Consider the 3-dimensional subspace $E^3 \subset E^7$ spanned by the vectors $e_0,e_1,e_2 \in E^7$. $S^6 \cap E^3$ is a 2-sphere S^2. Show that S^2 is an *invariant submanifold* of S^6, that is, for any $X \in S^2$ the tangent space T_X to S^2 at X is invariant under J_X.

C. Hermitian manifolds [50]

1. Let M be a Riemannian manifold with metric tensor g. Show that there exists a mapping

$$X \to D_X$$

of T into the space of endomorphisms of T with the properties:

(a) $Zg(X,Y) - g(D_Z X,Y) - g(X,D_Z Y) = 0$

(parallel translation is an isometry);

(b) $D_X Y - D_Y X = [X,Y]$

(torsion is zero)

for any $X,Y,Z \in T$.

Hint: Assume the existence of this map and show that

$$\begin{aligned} 2g(X,D_Z Y) = Zg(X,Y) - Xg(Y,Z) + Yg(Z,X) \\ + g(Y,[X,Z]) - g(X,[Y,Z]) - g(Z,[Y,X]) \end{aligned}$$

for any X,Y and $Z \in T$. Conversely, this relation defines for every $Y,Z \in T$ an element $D_Z Y \in T$. The map $Z \to D_Z$ is thus unique. For every $Z \in T$, D_Z is called the operation of *covariant differentiation with respect to Z.*

2. Establish the identities:

(i) $D_{X+Y} = D_X + D_Y$,

(ii) $D_{fX}Y = f(D_X Y)$,

(iii) $D_X(Y + Z) = D_X Y + D_X Z$,

(iv) $D_X(fY) = (Xf)Y + f(D_X Y)$,

(v) $\overline{D_X Y} = D_{\bar{X}} \bar{Y}$ (if M is almost complex)

for all $X, Y, Z \in T$ and $f \in F$—the algebra of differentiable functions on M;

(vi) $$g\left(D_{\frac{\partial}{\partial u^i}} \frac{\partial}{\partial u^j}, \frac{\partial}{\partial u^k}\right) = g_{mk}\Gamma^m_{ij}$$

where the Γ^m_{kj} are the coefficients of the Levi Civita connection.

From (ii) it follows that for any point P, $D_X Y(P)$ depends only on $X(P)$ and Y, that is, if $X_1(P) = X_2(P)$, then $D_{X_1} Y(P) = D_{X_2} Y(P)$.

3. A p-form α on M may be considered as an alternating multilinear form on the F-module T with values in F, that is $\alpha(X_1, \cdots, X_p) \in F$ for any $X_1, \cdots, X_p \in T$. To a p-form α on M we may associate a p-form $D_X\alpha$ on M called the *covariant derivative of* α *with respect to* X by putting

$$(D_X\alpha)(Y_1, \cdots, Y_p) = X\alpha(Y_1, \cdots, Y_p) - \sum_{i=1}^{p} \alpha(Y_1, \cdots, D_X Y_i, \cdots, Y_p).$$

Show that the map

$$D_X : \wedge^p(M) \to \wedge^p(M)$$

so defined is a derivation.

The map D_X may be extended in the obvious way to tensors on M of type $(0,p)$ which are not necessarily skew-symmetric. Hence, the covariant derivative of the metric tensor g with respect to the vector field X vanishes, that is

$$D_X g = 0$$

for all $X \in T$.

4. Establish the equivalence of the following statements for an hermitian manifold with metric g whose complex structure is defined by J:

(a) $D_X(JY) = J(D_X Y)$,

(b) $D_X\Omega = 0$ where $\Omega(X,Y) = g(JX,Y)$,

(c) $d\Omega = 0$

for any $X,Y \in T$.

Hint: In a Riemannian parallelisable manifold, the map

$$\alpha \to \hat{d}\alpha = \sum_{i=1}^{n} \epsilon(\theta^i) D_{X_i}\alpha$$

where $\{X_i\}$ and $\{\theta^i\}$ are dual bases is an anti-derivation. Show that d and $\hat{d}$ agree on $\wedge^0(M)$ and $\wedge^1(M)$, and hence on $\wedge(M)$.

If any of these conditions is satisfied, the manifold is Kaehlerian and Ω is the fundamental form defining the Kaehlerian structure. Note that

$$g(X,JY) + g(JX,Y) = 0.$$

Incidentally, from the formula

$$d\alpha = \sum_{i=1}^{n} \epsilon(\theta^i) D_{X_i}\alpha$$

we may derive the formula

$$(d\alpha)(Y_1, \cdots, Y_{p+1}) = \sum_{i=1}^{p+1} (-1)^{i+1} Y_i\alpha(Y_1, \cdots, \hat{Y}_i, \cdots, Y_{p+1})$$

$$+ \sum_{i<j} (-1)^{i+j} \alpha([Y_i, Y_j], Y_1, \cdots, \hat{Y}_i, \cdots, \hat{Y}_j \cdots, Y_{p+1})$$

Hence, $(d\alpha)(X,Y) = X\alpha(Y) - Y\alpha(X) - \alpha([X,Y])$ (cf. formula (3.5.2)).

5. If M is Kaehlerian, show that $D_X\wedge^{q,r}(M) \subset \wedge^{q,r}(M)$ for every pair of integers (q,r) and any $X \in T$.

6. Let M be a complex manifold, J the linear endomorphism of T defining the complex structure of M and Ω a real form of bidegree (1,1) on M. Then,

$$\Omega(JX,Y) + \Omega(X,JY) = 0$$

from which

$$\Omega(X,Y) = \Omega(JX,JY)$$

for any $X,Y \in T$. Show that the 'metric' g defined by

$$g(X,Y) = \Omega(X,JY)$$

is symmetric, hermitian and real; hence if Ω is closed and g is positive definite, the metric is Kaehlerian.

D. The 2-form Ω

1. The form

$$\Omega = \sqrt{-1}\, d'd''f$$

where f is a real-valued function of class ∞ on the complex manifold M is real, closed and of bidegree (1,1). Let $\{U_i\}$ be an open covering of M. For each i

let f_i be a real-valued function of class ∞ with no zeros in U_i. If, for each pair of integers (i,j) there exists a holomorphic function h_{ij} on $U_i \cap U_j$ such that

$$f_i = f_j \, h_{ij} \, \overline{h_{ij}},$$

then, there exists a real closed form Ω of bidegree (1,1) on M such that

$$\Omega = \sqrt{-1}\, d'd'' \log f_i$$

on each open set U_i.

2. Let $\{U_i\}$ be an open covering of M by coordinate neighborhoods with complex coordinates (z^i) and Ψ a real $2n$-form of maximal rank $2n$ on M. Then, the restriction Ψ_i of Ψ to each U_i is given by

$$\Psi_i = f_i \, dz^1 \wedge \cdots \wedge dz^n \wedge d\bar{z}^1 \wedge \cdots \wedge d\bar{z}^n,$$

where f_i is either a real or purely imaginary function with no zeros in U_i; moreover, on $U_i \cap U_j$

$$f_i = f_j \, h_{ij} \, \overline{h_{ij}}$$

where h_{ij} is a holomorphic function on $U_i \cap U_j$. Show that Ψ determines a real, closed 2-form of bidegree (1,1) and maximal rank on M.

Bergman has shown that on every bounded open subset S of C_n there exists a well-defined real form of degree $2n$ invariant under the complex automorphisms of S and independent of the imbedding. With respect to this form we may construct a 2-form Ω on S whose associated metric is Kaehlerian.

E. The fundamental commutativity formulae. Topology of Kaehler manifolds [*50, 72*]

1. Let M be an hermitian manifold with metric g. Assume that $T^{1,0}$ is a free F-module; this is certainly the case if M is holomorphically isomorphic with an open subset of C_n. Let $\{X_1, \cdots, X_n\}$ be a basis of $T^{1,0}$; then, $\{\bar{X}_1, \cdots, \bar{X}_n\}$ is a basis of $T^{0,1}$. By employing the Schmidt orthonormalization process the X_i, $i = 1, \cdots, n$ may be chosen so that

$$g(X_i, \bar{X}_j) = \delta_{ij}$$

(cf. equations (5.2.13) and (5.3.1)). Consider this basis of $T^c = T^{1,0} \oplus T^{0,1}$ and denote by $\{\theta^i, \bar{\theta}^i\}$, $i = 1, \cdots, n$ the dual basis. Then,

$$g = \sum_{i=1}^{n} (\theta^i \otimes \bar{\theta}^i + \bar{\theta}^i \otimes \theta^i)$$

and

$$\Omega = \sqrt{-1} \sum_{i=1}^{n} \theta^i \wedge \bar{\theta}^i.$$

Establish the formulae

$$d' = \sum_{i=1}^{n} \epsilon(\theta^i) D_{X_i}, \quad d'' = \sum_{i=1}^{n} \epsilon(\bar{\theta}^i) D_{\bar{X}_i}$$

and

$$\delta' = -\sum_{j=1}^{n} i(\theta^j) D_{\bar{X}_j}, \quad \delta'' = -\sum_{j=1}^{n} i(\bar{\theta}^j) D_{X_j}.$$

Hint: Employ C.4.

2. Using the above formulae for d', d'', δ', and δ'' as well as formula (5.4.2) derive the fundamental lemma 5.6.1.

3. Establish the formulae

$$\delta' L - L\delta' = \sqrt{-1}\, d''.$$

and

$$\delta'' L - L\delta'' = -\sqrt{-1}\, d'.$$

These relations are the duals of those in lemma 5.6.1.

4. For a complex manifold M, $\wedge^{*c}(M)$ is a direct sum of the subspaces $\wedge^{q,r}$, that is any $\alpha \in \wedge^{*c}(M)$ may be uniquely expressed as a sum of pure forms $\alpha_{q,r}$ of bidegrees (q,r), respectively. Consider the map

$$P_{q,r} : \wedge^{*c}(M) \to \wedge^{q,r}$$

sending α into $\alpha_{q,r}$. If M is Kaehlerian denote by A the algebra of operators generated by $*$, d, L, and $P_{q,r}$. Show that Δ belongs to the center of A. If M is compact prove that the operators H and G associated with the underlying Riemannian structure also belong to the center. In particular, Δ, H and G commute with d', d'', δ', δ'', and Λ.

5. Prove that the harmonic part $H[\alpha]$ of a pure form α of bidegree (q,r) on a compact hermitian manifold is itself of bidegree (q,r) (cf. II.B.3).

6. Let $D^{q,r}(M)$ denote the quotient space of the space of d-closed forms of bidegree (q,r) on the compact Kaehler manifold M by the space of exact forms of bidegree (q,r). Prove that $D^p(M)$ is the direct sum of the spaces $D^{q,r}(M)$ with $q + r = p$. (Note that this decomposition is independent of the Kaehler metric.)

The map $\alpha \to \bar{\alpha}$ induces an isomorphism of $D^{q,r}(M)$ onto $D^{r,q}(M)$. Hence, $b_{q,r} = b_{r,q}$ where $b_{q,r} = \dim D^{q,r}(M)$.

In terms of the complex structure on $D^p_R(M)$ (the pth cohomology space constructed from the subspace of real forms) induced by that of M, it may be shown once again that b_p is even for p odd.

Hint: Extend the complex structure J of M to p-forms on M and prove that $\tilde{J}^2 = (-1)^p I$ where $\tilde{J}$ denotes the induced map on $\wedge^p$; then, prove that $\tilde{J}$ and Δ commute.

CHAPTER VI

CURVATURE AND HOMOLOGY OF KAEHLER MANIFOLDS

It is a classical theorem that compact Riemann surfaces belong to one of three classes (cf. example 1, § 5.9). However, for several complex variables the situation is not quite so simple. In any case, there is the following generalization, namely, if M is a compact Kaehler manifold of constant holomorphic curvature k (cf. § 6.1), its universal covering space is either complex projective space $P_n(k > 0)$, the interior of a unit sphere $B_n(k < 0)$, or the space C_n of n complex variables $(k = 0)$. These spaces are of interest in algebraic geometry; indeed, they provide a source of examples of algebraic varieties. In analogy with the real case (cf. § 3.1) a (compact) Kaehler manifold of constant holomorphic curvature is called elliptic, if $k > 0$, hyperbolic, if $k < 0$ and parabolic if $k = 0$. By an application of the results of Chapter V it is shown that an elliptic space is homologically equivalent to complex projective space. It is, in fact known, in this case, that M is actually P_n itself. If the manifold M is parabolic it can be represented as the quotient space C_n/D where D is a discrete group of motions in C_n, namely, the fundamental group. The group Γ in example 5, § 5.1 is a normal subgroup of D of finite index with $2n$ independent generators. The complex torus $T_n = C_n/\Gamma$ is then a covering space of M.

On the 1-dimensional (complex) torus T_1 there is essentially only one holomorphic differential, namely, dz in contrast with the Riemann sphere on which none exist (cf. § 5.6). In higher dimensions there is the analogous situation, that is, on T_n there are n independent holomorphic pfaffian forms whereas in the elliptic case there are no holomorphic 1-forms. More generally, on a compact Kaehler manifold of positive definite Ricci curvature, there do not exist holomorphic p-forms $(0 < p \leqq n)$ [58].

The reader is referred to § 5.9, example 3 for a description of the complex torus. Now, the torus has 'zero curvature' and this fact is decisive from a geometrical standpoint in describing its homology. More generally, a compact hermitian manifold M of zero curvature has as its universal covering space $\tilde{M}$ a complex Lie group. If D (the fundamental group) is a discrete group of covering transformations of M whose elements are isometries acting without fixed points, then M is homeomorphic with $\tilde{M}/D$. If M is simply connected, a necessary condition for zero curvature is complex parallelisability by means of a parallel field of orthonormal frames, that is, the existence of n globally defined linearly independent holomorphic vector fields which are parallel with respect to the connection defined in § 5.3. On the other hand, a complex parallelisable manifold has a natural hermitian metric of zero curvature. The existence of a metric with zero curvature is consequently a weaker property than parallelisability. The problem of determining those manifolds with a locally flat hermitian metric is considered. It is shown that a compact hermitian manifold of zero curvature is homeomorphic with a quotient space of a complex Lie group modulo a discrete subgroup. It is Kaehlerian, if and only if, it is a multi-torus [*69*].

The hyperbolic spaces will be considered from the point of view of the problem of imbedding into a locally flat space. Our interest lies in the local properties of a manifold for which a holomorphic imbedding which induces the metric is possible. If the Ricci curvature is positive, it is not possible to define such an imbedding. On the other hand, negative Ricci curvature is not sufficient to guarantee this. For, one need only consider the classical hyperbolic space defined by the metric $g(z, \bar{z}) = (1 - z\bar{z})^{-2}$ in the unit circle $|z| < 1$. Such imbeddings consequently appear rather remote and can only occur if the Ricci curvature is not positive [*5*].

Whereas positive Ricci curvature yields information on homology, negative curvature is of interest in the study of groups of transformations (cf. Chap. III). Chapter VII is concerned essentially with the study of groups of holomorphic and conformal homeomorphisms of Kaehler manifolds, and so some of the results for negative curvature are postponed until then. In any case, the elliptic and parabolic spaces are particularly interesting from our point of view in that their homology properties may be described by the methods of Chapters III and V.

For negative curvature no holomorphic contravariant tensor fields of bidegree $(p, 0)$ can exist. Hence, in particular (as already observed), the manifold is not complex parallelisable. A generalization may be obtained by assuming that the 1st Chern class is negative definite (cf. VII.A.4).

The Gauss-Bonnet formula is also particularly interesting from our point of view. In fact, if M is a compact Kaehler manifold on which there are 'sufficiently many' holomorphic pfaffian forms, then $(-1)^n \chi(M) \geqq 0$ where $\chi(M)$ is the Euler characteristic. An example is provided by T_n for which it is clear that $\chi(T_n) = 0$ [8].

Denote by the pair (M, g) a Kaehler manifold with metric g and underlying complex manifold M. Consider the Kaehler manifolds (M, g) and (M, g'). If the connections ω and ω' canonically defined by g and g', respectively, are projectively related, a certain tensor w (the complex analogue of the Weyl projective curvature tensor) is an invariant of these connections. Its vanishing is of interest. For, if $w = 0$, the manifold (M, g) (or (M, g')) has constant holomorphic curvature. Conversely, for a manifold of constant holomorphic curvature, $w = 0$. In this way, constant holomorphic curvature is seen to be the complex analogue of constant curvature in a Riemannian manifold [33]. (A Kaehler manifold of constant curvature is of zero curvature). The homological structure of elliptic space is, as previously mentioned, identical with that of P_n. However, the betti numbers of P_n are retained even for deviations from projective flatness [7].

An important application of the results of Chapter III is sketched in § 6.14 where the so-called vanishing theorems of Kodaira are obtained. These theorems are of interest in the applications of sheaf theory to complex manifolds since it is important to know when certain cohomology groups vanish.

6.1. Holomorphic curvature

Let M be a Kaehler manifold of constant curvature K whose complex dimension is n. Then, from (1.10.4) the curvature tensor is given by

$$R_{ABCD} = K(g_{BC}g_{AD} - g_{AC}g_{BD}).$$

(The same systems of indices as in Chapter V are maintained throughout.) In terms of local complex coordinates these equations take the form

$$R_{ij^*kl^*} = Kg_{j^*k}\,g_{il^*}$$

from which

$$-R^j{}_{ikl^*} = K\delta^j_k\,g_{il^*}.$$

Substitution of this last set of equations into (5.3.39) gives

$$K\delta^j_k\,g_{il^*} = K\delta^j_i\,g_{kl^*},$$

that is

$$Kg_{kl^*} = nKg_{kl^*}.$$

Hence,

Theorem 6.1.1. *A Kaehler manifold of constant curvature is locally flat provided $n > 1$.*

If, instead of insisting that all sectional curvatures at a given point are equal, we require that only those determined by any two orthogonal vectors in the tangent space at each point are equal, the same conclusion prevails, since the bundle of orthogonal frames suffices to determine the Riemannian geometry. For complex manifolds, however, it is natural to consider only those 2-dimensional subspaces of the tangent space defined by a vector and its image by the linear endomorphism J giving the complex structure. Indeed, to each tangent vector X_P at a point P of the hermitian manifold M, one may associate the tangent vector $(JX)_P$ at P orthogonal to X_P. The section determined by these vectors will be called a *holomorphic section* since it is defined by the complex structure. We shall denote the sectional curvature defined by the holomorphic section determined by the vector X_P by $R(P, X)$ and call it the *holomorphic sectional curvature* defined by X_P.

We seek a formula in local complex coordinates for $R(P, X)$. To begin with, if

$$X = \xi^A \frac{\partial}{\partial z^A} = \xi^i \frac{\partial}{\partial z^i} + \xi^{i^*} \frac{\partial}{\partial \bar{z}^i},$$

then, from (5.2.4)

$$JX = \eta^A \frac{\partial}{\partial z^A} = \sqrt{-1}\, \xi^i \frac{\partial}{\partial z^i} - \sqrt{-1}\, \xi^{i^*} \frac{\partial}{\partial \bar{z}^i}.$$

Hence, from (1.10.4)

$$R(P,X) = \frac{R_{ABCD}\, \xi^A \eta^B \xi^C \eta^D}{(g_{BC}\, g_{AD} - g_{BD}\, g_{AC})\, \xi^A \eta^B \xi^C \eta^D}$$

where $\eta^i = \sqrt{-1}\, \xi^i$ and $\eta^{i^*} = -\sqrt{-1}\, \xi^{i^*}$. Now, it is easy to see that

$$R_{ABCD}\, \xi^A \eta^B \xi^C \eta^D = -4R_{ij^*kl^*}\, \xi^i \xi^{j^*} \xi^k \xi^{l^*}$$

and

$$(g_{BC}\, g_{AD} - g_{BD}\, g_{AC})\, \xi^A \eta^B \xi^C \eta^D = -4g_{ij^*}\, g_{kl^*}\, \xi^i \xi^{j^*} \xi^k \xi^{l^*}.$$

Consequently,

$$R(P,X) = \frac{R_{ij^*kl^*}\, \xi^i \xi^{j^*} \xi^k \xi^{*l}}{g_{ij^*}\, g_{kl^*}\, \xi^i \xi^{j^*} \xi^k \xi^{l^*}}$$

which, by reasons of symmetry, may be expressed in the form

$$R(P,X) = \frac{2R_{ij^*kl^*}\, \xi^i \xi^{j^*} \xi^k \xi^{l^*}}{(g_{ij^*}\, g_{kl^*} + g_{il^*}\, g_{kj^*})\, \xi^i \xi^{j^*} \xi^k \xi^{l^*}}.$$

Suppose that $R(P, X)$ is independent of the tangent vector X chosen to define it. Then, the curvature tensor at P has the representation

$$R_{ij^*kl^*} = \frac{k}{2}(g_{ij^*}\, g_{kl^*} + g_{il^*}\, g_{kj^*}) \tag{6.1.1}$$

where $k = k(P)$ denotes the common value of $R(P, X)$ for all tangent vectors X at P. For, by assumption, the equation

$$\left[R_{ij^*kl^*} - \frac{k}{2}(g_{ij^*}\, g_{kl^*} + g_{il^*}\, g_{kj^*})\right] \xi^i\, \xi^{j^*}\, \xi^k\, \xi^{l^*} = 0$$

is satisfied by the $2n$ independent variables (ξ^i, ξ^{i^*}). Hence, since both sides of (6.1.1) are symmetric in the pairs (i, k) and (j, l) we have the desired conclusion.

Theorem 6.1.2. *If the holomorphic sectional curvatures at each point of a Kaehler manifold are independent of the holomorphic sections passing through the point, they are constant over the manifold.*

We wish to show that the function k appearing in (6.1.1) is a constant. By assumption, the curvature tensor has this form at each point of M. Transvecting (6.1.1) with g^{kl^*} we derive

$$R_{ij^*} = \frac{n+1}{2}\, k g_{ij^*}\,, \tag{6.1.2}$$

that is M is a '(Kaehler-) Einstein' space (cf. § 6.4). Hence, from (5.3.29) and (5.3.38) the 1st Chern class of M is given by

$$\psi = -\frac{n+1}{4\pi}\, k\Omega.$$

Since ψ is closed,

$$dk \wedge \Omega = 0,$$

from which by corollary 5.7.2, dk must vanish for $n \geqq 2$.

If at each point of a Kaehler manifold the holomorphic sectional curvature is independent of the tangent vector defining it, the manifold is said to have *constant holomorphic curvature*.

Theorem 6.1.3. *P_n may be given a metric g in terms of which it is a manifold of constant holomorphic curvature.*

Indeed, we give to P_n the Fubini metric g of example 5, § 5.9:

$$g_{ij^*} = \frac{\delta_{ij}}{\varphi_0} - \frac{z_0^i\, \bar{z}_0^j}{{\varphi_0}^2}, \quad \varphi_0 = \sum_{i=0}^{n} z_0^i\, \bar{z}_0^i \tag{6.1.3}$$

in the coordinate neighborhood U_0.

At the origin of this system of local complex coordinates $g_{ij^*} = \delta_{ij}$. Hence, from (5.3.19), a straightforward computation yields

$$R_{ij^*kl^*} = \delta_{ij}\ \delta_{kl} + \delta_{il}\ \delta_{kj}\ ,$$

and so from the covering of P_n given in § 5.9, since

$$d'\,d''\log\varphi_0 = d'\,d''\log\varphi_j$$

for every index $j = 1, \cdots, n$, the curvature tensor has this form everywhere. In other words, since there exists a transitive Lie group of holomorphic homeomorphisms preserving the metric, the curvature tensor has the prescribed form everywhere.

Corollary. *The holomorphic sectional curvature with respect to g of complex projective space is positive.*

An application of theorem 3.2.4 in conjunction with theorem 5.7.2 yields the betti numbers of a compact Kaehler manifold with the Fubini metric (6.1.3) and, in particular, those of P_n.

Theorem 6.1.4. *The betti numbers b_p of a compact Kaehler manifold M of positive constant holomorphic curvature vanish if p is odd and are equal to 1 if p is even:*

$$b_{2r} = 1, \quad b_{2r+1} = 0, \quad 0 \leqq r \leqq n.$$

To see this, let β be an effective harmonic p-form on M. Then, $\bar\beta$ is a harmonic p-form, and since

$$\begin{aligned} \Lambda\bar\beta &= (-1)^p * L * \bar\beta \\ &= \overline{\Lambda\beta} = 0 \end{aligned}$$

(cf. § 5.4 for the definition of $*$ for complex differential forms), it is also effective. It follows that

$$\alpha = \beta + \bar\beta$$

is a real effective harmonic p-form. Now, put $\alpha = a_{A_1\dots A_p}\,dz^{A_1}\wedge\cdots\wedge dz^{A_p}$ and compute $F(\alpha)$ (cf. formula (3.2.10)). In the first place, from (6.1.2)

$$\begin{aligned} R_{AB}a^{AA_2\dots A_p}a^B{}_{A_2\dots A_p} &= 2(R_{ij^*}a^{ikA_3\dots A_p}a^{j^*}{}_{kA_3\dots A_p} \\ &\quad + R_{ij^*}a^{ik^*A_3\dots A_p}a^{j^*}{}_{k^*A_3\dots A_p}) \qquad (6.1.4) \\ &= (n+1)\,k(a^{ikA_3\dots A_p}a_{ikA_3\dots A_p} + a^{ik^*A_3\dots A_p}a_{ik^*A_3\dots A_p}) \end{aligned}$$

and from (6.1.1)

$$R_{ABCD}a^{ABA_3\ldots A_p}a^{CD}{}_{A_3\ldots A_p} = 4R_{ij^*kl^*}a^{ij^*A_3\ldots A_p}a^{kl^*}{}_{A_3\ldots A_p}$$

$$= 2k(g_{ij^*}g_{kl^*}a^{ij^*A_3\ldots A_p}a^{kl^*}{}_{A_3\ldots A_p} - a^{ij^*A_3\ldots A_p}a_{ij^*A_3\ldots A_p}). \tag{6.1.5}$$

Next, we derive an explicit formula for $\Lambda\alpha$ in local complex coordinates (z^i). From (5.4.2) and (5.3.12)

$$\Lambda\alpha = \sqrt{-1}\sum_{r=1}^{n}\xi^j_{(r)}\xi^{k*}_{(r)}\, i\Big(\frac{\partial}{\partial\bar{z}^k}\Big)\, i\Big(\frac{\partial}{\partial z^j}\Big)\alpha$$

$$= \sqrt{-1}\, g^{jk^*}\, i\Big(\frac{\partial}{\partial\bar{z}^k}\Big)\, i\Big(\frac{\partial}{\partial z^j}\Big)\, a_{j_1\ldots j_q k_1\ldots k_r}dz^{j_1}\wedge\ldots\wedge dz^{j_q}\wedge d\bar{z}^{k_1}\wedge\ldots\wedge d\bar{z}^{k_r}.$$

Hence, since the interior product operator is an anti-derivation

$$\Lambda\alpha = \sum_{q+r=p}\sqrt{-1}\,(-1)^{q-1}g^{jk^*}a_{jj_1\ldots j_{q-1}k^*k^*_1\ldots k^*_{r-1}} \cdot dz^{j_1}\wedge\ldots\wedge dz^{j_{q-1}}\wedge d\bar{z}^{k_1}\wedge\ldots\wedge d\bar{z}^{k_{r-1}}. \tag{6.1.6}$$

Returning to equation (6.1.5), we conclude that

$$R_{ABCD}a^{ABA_3\ldots A_p}a^{CD}{}_{A_3\ldots A_p} = -2ka^{ij^*A_3\ldots A_p}a_{ij^*A_3\ldots A_p}.$$

Combining (6.1.4) and (6.1.5), the quadratic form

$$F(\alpha) = (n+1)\,ka^{ijA_3\ldots A_p}a_{ijA_3\ldots A_p} + (n-p+2)\,ka^{ij^*A_3\ldots A_p}a_{ij^*A_3\ldots A_p}$$

$$> 0, \quad 0 < p \leqq n,$$

that is, there are no non-trivial real effective harmonic p-forms for $p \leqq n$. Hence, by theorem 5.7.2

$$b_{p-2} = b_p, \quad p \leqq n+1.$$

Now, by theorem 3.2.1, since the Ricci curvature is positive definite (by virtue of the fact that k is positive), b_1 vanishes. Thus

$$b_{2r+1} = 0, \quad 2r \leqq n.$$

On the other hand, since M is connected, $b_0 = 1$, and so

$$b_{2r} = 1, \quad 2r \leqq n+1.$$

The desired conclusion then follows by Poincaré duality.

Corollary 1. *The betti numbers of P_n are*

$$b_{2r} = 1, \quad b_{2r+1} = 0, \quad 0 \leqq r \leqq n.$$

Since P_n is connected, it is only necessary to show that P_n is compact. The following proof is instructive: In C_{n+1} with the canonical metric g define the sphere

$$S^{2n+1} = \{e_0 \in C_{n+1} \,|\, g(e_0, e_{0*}) = 1\}.$$

Consider the equivalence relation

$$e_0' \sim e_0$$

defined by

$$e_0' = e^{i\varphi} e_0$$

where φ is a real-valued function. P_n is thus the quotient space of S^{2n+1} by this equivalence relation. In fact, P_n may be identified with the quotient space $U(n+1)/U(n) \times U(1)$. To see this, consider the unitary frame (e_A, e_{A^*}), $A = 0, 1, \cdots, n$ obtained by adjoining to e_0, n vectors e_i in such a way that the frames obtained from (e_A, e_{A^*}) by a transformation of $U(n+1)$ are unitary. Since the frames obtained from (e_i, e_{i^*}), $i = 1, \cdots, n$ by means of the group $U(n)$ are unitary, P_n has the given representation. That P_n is compact now follows immediately from the fact that the unitary group is compact [*27*].

Incidentally, this gives another proof that P_n is a Kaehler manifold. For, by the compactness of $U(n+1)$ we may construct an invariant hermitian metric by 'averaging' over $U(n+1)$. The fundamental form Ω is thus invariant. Hence, since $U(n+1)/U(n) \times U(1)$ is a symmetric space, that is, the curvature tensor associated with this metric has vanishing covariant derivative, Ω is closed (cf. VI.E for the definition of a symmetric space). We have invoked the theorem that an invariant form in a symmetric space is closed. (That P_n is a symmetric space follows directly from the fact that with the Fubini metric it is a manifold of constant holomorphic curvature). The reader is referred to VI.E for further details.

Corollary 2. *There are no holomorphic p-forms, $0 < p \leqq n$ on P_n. In degree 0 the holomorphic forms are constant functions.*

Indeed, by (5.6.4) the pth betti number

$$b_p = \sum_{q+r=p} b_{q,r}\,.$$

Since the even-dimensional betti numbers are each one and $b_{q,r} = b_{r,q}$ we conclude that

$$b_{0,0} = b_{1,1} = \cdots = b_{n,n} = 1$$

with all remaining $b_{q,r}$ zero. In particular,

$$b_{p,0} = 0 \text{ for } p \neq 0.$$

By employing the methods of theorem 3.2.7, it can be shown that a 4-dimensional δ-pinched compact Kaehler manifold is homologically P_2 provided δ is strictly greater than zero (strictly positive curvature). The reader is referred to VI.D for details. Hence, $S^2 \times S^2$ considered as a Kaehler manifold cannot be provided with a metric of strictly positive curvature. In fact, it is still an open question as to whether $S^2 \times S^2$ can be given a Riemannian structure of strictly positive curvature. For more recent results the reader is referred to [90] and [94].

The n-sphere, complex projective n-space, quaternionic projective n-space and the Cayley plane are the only known examples of compact, simply connected manifolds which may be endowed with a Riemannian structure of strictly positive curvature [*1*].

6.2. The effect of positive Ricci curvature

Since the Ricci curvature associated with the Fubini metric of P_n is positive it is natural to ask if corollary 2 of the previous section can be extended to any compact Kaehler-Einstein manifold with positive Ricci curvature. An examination of the proof of theorem 6.1.4 reveals more, however. For, if β is a holomorphic form of degree p,

$$\alpha = \beta + \bar{\beta}$$

is a real p-form; in fact, α is harmonic since β and $\bar{\beta}$ are harmonic. Hence, since α is the sum of a form of bidegree $(p, 0)$ and one of bidegree $(0, p)$ it follows from the symmetry properties of the curvature tensor that

$$\begin{aligned} F(\alpha) &= R_{AB} a^{AA_2 \ldots A_p} a^{B}{}_{A_2 \ldots A_p} \\ &= 2R_{ij^*} a^{ii_2 \ldots i_p} a^{j^*}{}_{i_2 \ldots i_p} . \end{aligned}$$

Let M be a compact Kaehler manifold of positive definite Ricci curvature. Then, by theorem 3.2.4, since α is harmonic, and $F(\alpha)$ is positive definite, α must vanish.

We have proved

Theorem 6.2.1. *On a compact Kaehler manifold of positive definite Ricci curvature, a holomorphic form of degree p, $0 < p \leqq n$ is necessarily zero* [*4*, *58*].

6.3. Deviation from constant holomorphic curvature

In this section a class of compact spaces having the same homology structure as P_n and of which P_n is itself a member is considered. They have one common local property, namely, their Ricci curvatures are positive. Aside from this their local structures can be quite different—their classification being made complete, however, by means of a condition on the projective curvature tensors associated with these spaces. They need not have constant holomorphic curvature. If instead, a measure W of their deviation from this property is given, and if the function W associated with a space M satisfies a certain inequality depending on the Ricci curvature of the space, M is a member of the class.

Consider the Kaehler manifolds (M, g) and (M, g') of complex dimension n. If the matrices of connection forms ω and ω' canonically defined by g and g', respectively are projectively related their coefficients of connection are related by

$$\Gamma'^{i}_{jk} = \Gamma^{i}_{jk} + p_j\, \delta^i_k + p_k\, \delta^i_j \tag{6.3.1}$$

(cf. § 3.11). Since

$$\begin{aligned} R'^{i}{}_{jkl^*} &= \frac{\partial \Gamma'^{i}_{jk}}{\partial \bar{z}^l} \\ &= \frac{\partial}{\partial \bar{z}^l}(\Gamma^{i}_{jk} + p_j\, \delta^i_k + p_k\, \delta^i_j), \end{aligned}$$

$$R'^{i}{}_{jkl^*} = R^{i}{}_{jkl^*} + \delta^i_k\, D_{l^*} p_j + \delta^i_j\, D_{l^*} p_k. \tag{6.3.2}$$

It follows easily that the tensor w with components

$$W^{i}{}_{jkl^*} = R^{i}{}_{jkl^*} + \frac{1}{n+1}(R_{jl^*}\, \delta^i_k + R_{kl^*}\, \delta^i_j) \tag{6.3.3}$$

is an invariant of the connections ω and ω'. For this reason we shall call it the *projective curvature tensor* of (M, g) (or (M, g')). It is to be noted that w vanishes for $n = 1$. Its vanishing for the dimensions $n > 1$ is of some interest. For, if $w = 0$, the curvature of M (relative to g or g') has the representation

$$R^i_{jkl^*} = -\frac{1}{n+1}(R_{jl^*}\,\delta^i_k + R_{kl^*}\,\delta^i_j)$$

from which

$$R_{ij^*kl^*} = \frac{1}{n+1}(R_{il^*}\,g_{kj^*} + R_{kl^*}\,g_{ij^*}). \tag{6.3.4}$$

Applying the symmetry relation (5.3.41) we obtain

$$R_{il^*}\,g_{kj^*} + R_{kl^*}\,g_{ij^*} = R_{kj^*}\,g_{il^*} + R_{ij^*}\,g_{kl^*}$$

which after transvection with g^{il^*} may be written as

$$R_{kj^*} = \frac{R}{2n}\,g_{kj^*}.$$

Substituting for the Ricci curvature in (6.3.4) results in

$$R_{ij^*kl^*} = \frac{R}{2n(n+1)}(g_{il^*}\,g_{kj^*} + g_{kl^*}\,g_{ij^*}). \tag{6.3.5}$$

Thus (M, g) (or (M, g')) is a manifold of constant holomorphic curvature.

Conversely, assume that (M, g) is a manifold of constant holomorphic curvature. Then, its curvature has the representation (6.3.5). Substituting for the curvature from (6.3.5) into (6.3.3) we conclude that the tensor w vanishes.

Hence, *a necessary and sufficient condition that a Kaehler manifold be of constant holomorphic curvature is given by the vanishing of the projective curvature tensor* w.

It is known (cf. theorem 6.1.4) that a compact Kaehler manifold of positive constant holomorphic curvature ($w = 0$) is homologically equivalent with complex projective space. It is of some interest to inquire into the effect on homology in the case where w does not vanish. Under suitable restrictions we shall see that the betti numbers of P_n are retained. Indeed, the homology structure of a compact manifold of positive constant holomorphic curvature is preserved under a variation of the metric preserving the signature of the Ricci curvature and the inequality (6.3.7) given below. To this end, we introduce a function

$$2\,W = \sup_t \frac{|\,W_{ij^*kl^*}\,t^{ij^*}\,t^{kl^*}\,|}{\langle t,t\rangle} \tag{6.3.6}$$

where $W_{ij^*kl^*} = -\,g_{rj^*}W^r{}_{ikl^*}$, the least upper bound being taken over all skew-symmetric tensors of type $\binom{1\,1}{0\,0}$.

Theorem 6.3.1. *In a compact Kaehler manifold M of complex dimension n with positive definite Ricci curvature, if*

$$\left(1 - \frac{p-1}{n+1}\right) \lambda_0 > (p-1)\, W$$

for all $p = 1, \cdots, n$ where

$$\lambda_0 = \inf_{\xi} \frac{\langle Q\xi, \xi\rangle}{\langle \xi, \xi\rangle}$$

the greatest lower bound being taken over all (non-trivial) forms of degree 1, M is homologically equivalent with P_n [7].

The idea of the proof, as in theorem 6.1.4, is to show that under the circumstances there can be no non-trivial effective harmonic p-forms on M for $p \leqq n$. Once this is accomplished the result follows by Poincaré duality.

Let $\alpha = a_{A_1 \ldots A_p} dz^{A_1} \wedge \cdots \wedge dz^{A_p}$ be a real effective harmonic p-form on M. Then, from (3.2.10), (6.3.3), and (6.3.6),

$$\begin{aligned}
\tfrac{1}{2}F(\alpha) &= R_{ij^*}\, a^{iA_2\ldots A_p} a^{j^*}{}_{A_2\ldots A_p} + (p-1) R_{ij^*kl^*}\, a^{ij^*A_3\ldots A_p} a^{kl^*}{}_{A_3\ldots A_p} \\
&= R_{ij^*}\, a^{ikA_3\ldots A_p} a^{j^*}{}_{kA_3\ldots A_p} + \left(1 - \frac{p-1}{n+1}\right) R_{ij^*}\, a^{ik^*A_3\ldots A_p} a^{j^*}{}_{k^*A_3\ldots A_p} \\
&\quad + (p-1) W_{ij^*kl^*}\, a^{ij^*A_3\ldots A_p} a^{kl^*}{}_{A_3\ldots A_p} \\
&\geqq R_{ij^*}\, a^{ikA_3\ldots A_p} a^{j^*}{}_{kA_3\ldots A_p} + \left[\left(1 - \frac{p-1}{n+1}\right)\lambda_0 - (p-1)W\right] \cdot \\
&\quad \cdot a^{ij^*A_3\ldots A_p} a_{ij^*A_3\ldots A_p} .
\end{aligned}$$

Since $\lambda_0 > 0$ the desired conclusion follows.

Corollary. *Under the conditions of the theorem, if*

$$W < \frac{2\lambda_0}{(n-1)(n+1)}, \tag{6.3.7}$$

M is homologically equivalent with complex projective n-space.

6.4. Kaehler-Einstein spaces

In a manifold of constant holomorphic curvature k, the general sectional curvature K is dependent, in a certain sense, upon the value of the constant k. In fact, if $k > 0$ (< 0), so is K; moreover, the ratio of the smallest (largest) to the largest (smallest) value of K is $\frac{1}{4}$ provided $k > 0$ (< 0). To see this, let $K = K(X, Y)$ denote the sectional curvature determined by the vector fields $X = \xi^A \partial/\partial z^A$ and $Y = \eta^A \partial/\partial z^A$. Then,

$$K = \frac{R_{ABCD}\, \xi^A\, \eta^B\, \xi^C\, \eta^D}{(g_{BC}\, g_{AD} - g_{BD}\, g_{AC})\, \xi^A\, \eta^B\, \xi^C\, \eta^D}$$

$$= \frac{R_{ij^*kl^*}\, (\xi^i\, \eta^{j^*} - \xi^{j^*}\, \eta^i)\, (\xi^k\, \eta^{l^*} - \xi^{l^*}\, \eta^k)}{(g_{ij^*}\, \xi^i\, \eta^{j^*} + g_{ij^*}\, \eta^i\, \xi^{j^*})^2 - 4g_{ij^*}\, \xi^i\, \xi^{j^*}\, g_{kl^*}\, \eta^k\, \eta^{l^*}}$$

$$= \frac{k(g_{ij^*}\, g_{kl^*} + g_{il^*}\, g_{kj^*})\, (\xi^i\, \eta^{j^*} - \xi^{j^*}\, \eta^i)\, (\xi^k\, \eta^{l^*} - \xi^{l^*}\, \eta^k)}{[\langle X,Y\rangle + \langle Y,X\rangle]^2 - 4\langle X,X\rangle\, \langle Y,Y\rangle}$$

$$= k\, \frac{\langle X,Y\rangle^2 + \langle Y,X\rangle^2 - \langle X,Y\rangle\, \langle Y,X\rangle - \langle X,X\rangle\, \langle Y,Y\rangle}{\langle X,Y\rangle^2 + 2\langle X,Y\rangle\, \langle Y,X\rangle + \langle Y,X\rangle^2 - 4\langle X,X\rangle\, \langle YY\rangle}$$

where $\langle X,Y\rangle = g_{ij^*}\xi^i\eta^{j^*}$ denotes the (local) scalar product of the vector fields $\xi^i\partial/\partial z^i$ and $\eta^{i^*}\partial/\partial \bar{z}^i$ in that order.

If we put

$$\frac{\langle X,Y\rangle}{[\langle X,X\rangle\, \langle Y,Y\rangle]^{1/2}} = re^{i\theta}, \quad i = \sqrt{-1},$$

then

$$K = k\, \frac{1 + r^2 - 2r^2 \cos 2\theta}{4 - 2r^2 - 2r^2 \cos 2\theta} = k\left(1 - \frac{3}{4}\, \frac{1 - r^2}{1 - r^2 \cos^2\theta}\right).$$

Hence, since $0 \leqq r \leqq 1$,

$$0 \leqq \frac{1 - r^2}{1 - r^2 \cos^2\theta} \leqq 1,$$

from which we conclude that

$$0 < \tfrac{1}{4}k \leqq K \leqq k \tag{6.4.1}$$

if k is positive, and if k is negative

$$k \leqq K \leqq \tfrac{1}{4}k < 0. \tag{6.4.2}$$

Theorem 6.4.1. *The general sectional curvature K in a manifold of constant holomorphic curvature k satisfies the inequalities* (6.4.1) *for $k > 0$ and* (6.4.2) *for $k < 0$ where the upper limit in* (6.4.1) *and the lower limit in* (6.4.2) *are attained when the section is holomorphic* [5].

Thus, for $k > 0$ the manifold is $\frac{1}{4}$-pinched. This result should be compared with theorem 3.2.7.

From 1.10.4 it is seen that the Ricci curvature κ in the direction of the tangent vector X is given by

$$\kappa = \frac{\langle QX, X \rangle}{\langle X, X \rangle}$$

Therefore, in analogy with § 1.10 a Kaehler manifold for which the Ricci directions are indeterminate is called a *Kaehler-Einstein manifold* and the Ricci curvature is given by

$$R_{ij^*} = \kappa g_{ij^*}$$

or, in terms of the fundamental form Ω and the 2-form ψ determining the 1st Chern class, ψ is proportional to Ω, that is

$$\psi = -\frac{\kappa}{2\pi}\Omega.$$

Since ψ is closed, $d\kappa \wedge \Omega = 0$. Thus, if $n > 1$, κ is a constant.

6.5. Holomorphic tensor fields

We have seen that there exist no (non-trivial) holomorphic p-forms on a compact Kaehler manifold with positive Ricci curvature. In this section this result is generalized to tensor fields of type $\binom{p\ 0}{q\ 0}$ as follows: Denote by $\lambda_{\max}$ and $\lambda_{\min}$ the algebraically largest and smallest eigenvalues of the Ricci operator $\underset{\sim}{Q}$ (cf. § 3.2), respectively. Then, for a holomorphic tensor field t of type $\binom{p\ 0}{q\ 0}$, if

$$q\lambda_{\min} - p\lambda_{\max} \geqq 0$$

everywhere and is strictly positive at least at one point, t must vanish, that is no such tensor fields exist.

The idea of the proof is based on a part of the 'Bochner lemma' (cf. VI.F) which for our purposes is easily established. (The non-orientable case is more difficult to prove. The applications of this lemma made by Bochner and others have led to many important results on the homology properties of Riemannian manifolds). We shall state it as

Proposition 6.5.1. *Let f be a function of class 2 on a compact and orientable Riemannian manifold M. Then, if $\Delta f \geqq 0$ ($\leqq 0$) on M, Δf vanishes identically.*

Corollary. *If $\Delta f \geqq 0$ ($\leqq 0$) on M, then f is a constant.*

The proof is an easy application of lemma 3.2.1. For,

$$\int_M \Delta f * 1 = \int_M \delta df * 1 = 0.$$

In order to establish the above result, we put f equal to the 'square length' of the tensor field t. But first, *a tensor field of type $\binom{p\ 0}{q\ 0}$ is said to be holomorphic if its components (with respect to a given system of local complex coordinates) are holomorphic functions.* This notion is evidently an invariant of the complex structure. Since the Γ^i_{jk} and $\bar{\Gamma}^i_{jk}$ are the only non-vanishing coefficients of connection, the tensor field

$$t = t^{i_1 \cdots i_p}{}_{j_1 \cdots j_q} \frac{\partial}{\partial z^{i_1}} \otimes \cdots \otimes \frac{\partial}{\partial z^{i_p}} \otimes dz^{j_1} \otimes \cdots \otimes dz^{j_q}$$

of type $\binom{p\ 0}{q\ 0}$ is holomorphic, if and only if, the covariant derivatives of t with respect to $\bar{z}^i$ for all $i = 1, \cdots, n$ are zero.

Consider the tensor field $t + \bar{t}$. If t is holomorphic,

$$D_{k^*}\, t^{i_1 \cdots i_p}{}_{j_1 \cdots j_q} = 0.$$

Applying the interchange formula (1.7.21) it follows that

$$\begin{aligned} D_{l^*} D_k t^{i_1 \cdots i_p}{}_{j_1 \cdots j_q} &= \sum_{\mu=1}^{p} t^{i_1 \cdots i_{\mu-1} r i_{\mu+1} \cdots i_p}{}_{j_1 \cdots j_q} R^{i_\mu}{}_{rkl^*} \\ &\quad - \sum_{\nu=1}^{q} t^{i_1 \cdots i_p}{}_{j_1 \cdots j_{\nu-1} s j_{\nu+1} \cdots j_q} R^s{}_{j_\nu k l^*}. \end{aligned} \tag{6.5.1}$$

Transvecting (6.5.1) with g^{kl^*} we obtain

$$\begin{aligned} g^{kl^*} D_{l^*} D_k t^{i_1 \cdots i_p}{}_{j_1 \cdots j_q} &= \sum_{\nu=1}^{q} t^{i_1 \cdots i_p}{}_{j_1 \cdots j_{\nu-1} s j_{\nu+1} \cdots j_q} R^s{}_{j_\nu} \\ &\quad - \sum_{\mu=1}^{p} t^{i_1 \cdots i_{\mu-1} r i_{\mu+1} \cdots i_p}{}_{j_1 \cdots j_q} R^{i_\mu}{}_r. \end{aligned} \tag{6.5.2}$$

Now, put

$$f = g_{i_1 r_1^*} \cdots g_{i_p r_p^*} g^{j_1 s_1^*} \cdots g^{j_q s_q^*} t^{i_1 \cdots i_p}{}_{j_1 \cdots j_q} \overline{t^{r_1 \cdots r_p}{}_{s_1 \cdots s_q}}\ ;$$

then, since $t + \bar{t}$ is self adjoint, f is a real-valued function, and since the operator Δ is real, Δf is real-valued. Moreover,

$$-\tfrac{1}{2}\Delta f = \tfrac{1}{2}g^{AB}D_B D_A f = g^{ij^*}D_{j^*}D_i f$$

$$= g_{i_1 r_1^*}\cdots g_{i_p r_p^*}g^{j_1 s_1^*}\cdots g^{j_q s_q^*}g^{kl^*}D_k t^{i_1\cdots i_p}{}_{j_1\cdots j_q}\overline{D_l t^{r_1\cdots r_p}{}_{s_1\cdots s_q}} + G(t) \qquad (6.5.3)$$

where

$$G(t) = g_{i_1 r_1^*}\cdots g_{i_p r_p^*}g^{j_1 s_1^*}\cdots g^{j_q s_q^*}g^{kl^*}D_{l^*}D_k t^{i_1\cdots i_p}{}_{j_1\cdots j_q}\overline{t^{r_1\cdots r_p}{}_{s_1\cdots s_q}}.$$

Expanding $G(t)$ by (6.5.2) gives

$$G(t) = \Big[\sum_{\nu=1}^{q} t^{i_1\cdots i_p}{}_{j_1\cdots j_{\nu-1}sj_{\nu+1}\cdots j_q}R^s{}_{j_\nu} - \sum_{\mu=1}^{p} t^{i_1\cdots i_{\mu-1}ri_{\mu+1}\cdots i_p}{}_{j_1\cdots j_q}R^{i_\mu}{}_r\Big]\cdot \qquad (6.5.4)$$

$$\cdot\, t_{i_1\cdots i_p}{}^{j_1\cdots j_q}$$

Since the first term on the right in 6.5.3 is non-negative we may conclude that $\Delta f \leqq 0$, provided we assume that the function G is non-negative. Hence, as a consequence of proposition 6.5.1

Theorem 6.5.1. *Let t be a holomorphic tensor field of type $\binom{p\ 0}{q\ 0}$. Then, a necessary and sufficient condition that the (self adjoint) tensorfield $t + \bar{t}$ on a compact Kaehler manifold be parallel is given by the inequality*

$$G(t) \geqq 0.$$

On the other hand, if $G(t)$ is positive somewhere, t must vanish, that is there exists no holomorphic tensor field of the prescribed type [11].

An analysis of the expression (6.5.4) for the function G yields without difficulty

Corollary 1. *Let t be a holomorphic tensor field of type $\binom{p\ 0}{q\ 0}$ on the compact Kaehler manifold M. Then, if*

$$q\lambda_{\min} \geqq p\lambda_{\max},$$

t is a parallel tensor field. If strict inequality holds at least at one point of M, t must vanish.

If M is an Einstein space,

$$\lambda_{\min} = \lambda_{\max}.$$

Denoting the common value by λ we obtain

Corollary 2. *There exist no holomorphic tensor fields t of type* $\binom{p\ 0}{q\ 0}$ *on a compact Kaehler-Einstein manifold in each of the cases:*

(i) $q > p$ *and* $\lambda > 0$,

(ii) $q < p$ *and* $\lambda < 0$.

In either case, for $\lambda = 0$, *t is a parallel tensor field.*

6.6. Complex parallelisable manifolds

Let S be a compact Riemann surface (cf. example 1, § 5.8). The genus g of S is defined as half the first betti number of S, that is $b_1(S) = 2g$. By theorem 5.6.2, g is the number of independent abelian differentials of the first kind on S.

We have seen (cf. § 5.6) that there are no holomorphic differentials on the Riemann sphere. On the other hand, there is essentially only one holomorphic differential on the (complex) torus. On the multi-torus T_n there exist n abelian differentials of the first kind, there being, of course, no analogue for the n-sphere, $n > 2$. The Riemann sphere has positive curvature and this accounts (from a local point of view) for the distinction made in terms of holomorphic differentials between it and the torus whose curvature is zero.

Since the torus is locally flat (its metric being induced by the flat metric of C_n) the above facts make it clear that it is complex parallelisable. Indeed, there is no distinction between vectors and covectors in a manifold whose metric is locally flat. On the other hand, a complex parallelisable manifold can be given an hermitian metric in terms of which it may be locally isometrically imbedded in a flat space provided the holomorphic vector fields generate an abelian Lie algebra and, in this case, the manifold is Kaehlerian.

Theorem 6.6.1. *Let M be a complex parallelisable manifold of complex dimension n. Then, by definition, there exists n (globally defined) linearly independent holomorphic vector fields* $X_1, \cdots, X_n$ *on M. If the Lie algebra they generate is abelian, M is Kaehlerian and the metric canonically defined by the* X_i, $i = 1, \cdots, n$, *is locally flat* [10].

Let θ^r, $r = 1, \cdots, n$ denote the 1-forms dual to $X_1, \cdots, X_n$. Thus, they form a basis of the space of covectors of bidegree $(1, 0)$. In terms of these pfaffian forms, the fundamental form Ω has the expression

$$\Omega = \sqrt{-1} \sum_{r=1}^{n} \theta^r \wedge \bar{\theta}^r.$$

If we put (cf. V. B.1)

$$[X_j, X_k] = c^i_{jk} X_i$$

then, by (3.5.3) and (3.5.4)

$$d'\theta^i = -\frac{1}{2} c^i_{jk} \theta^j \wedge \theta^k.$$

Hence, since the Lie algebra of holomorphic vector fields is abelian, the θ^i are d'-closed for all $i = 1, \cdots, n$. (Referring to the proof of theorem 6.7.3, we see that they are also d''-closed.) This being the case for the conjugate forms as well

$$d\Omega = \sqrt{-1} \sum_{r=1}^{n} (d\theta^r \wedge \bar{\theta}^r - \theta^r \wedge d\bar{\theta}^r) = 0,$$

that is, M with the metric $2 \Sigma \theta^r \otimes \bar{\theta}^r$ is a Kaehler manifold. Moreover, the fact that the θ^i are closed allows us to conclude that M is locally flat. To see this, consider the second of the equations of structure (5.3.33):

$$\Theta^i_{\ j} = d\theta^i_{\ j} - \theta^k_{\ j} \wedge \theta^i_{\ k}.$$

Taking the exterior product of these equations by θ^j (actually $\pi^*\theta^j$, cf. § 5.3) and summing with respect to the index j, we obtain

$$\theta^j \wedge \Theta^i_{\ j} = 0, \quad i = 1, \cdots, n.$$

Indeed, from the first of the equations of structure (5.3.32), $\theta^j \wedge \theta^k_{\ j}$ vanishes since the θ^i are closed. Moreover,

$$\theta^j \wedge d\theta^k_{\ j} = -d(\theta^j \wedge \theta^k_{\ j}) = 0.$$

If we pull the forms $\Theta^i_{\ j}$ down to M and apply the equations (5.3.34), we obtain

$$R^i_{\ jkl^*} dz^k \wedge d\bar{z}^l \wedge dz^j = 0,$$

and so, since the curvature tensor is symmetric in j and k, it must vanish.

In § 6.9, it is shown that if M is compact, it is a complex torus.

6.7. Zero curvature

In this section we examine the effect of zero curvature on the properties of hermitian manifolds—the curvature being defined as in § 5.3.

Theorem 6.7.1. *The curvature of an hermitian manifold vanishes, if and only if, it is possible to choose a parallel field of orthonormal holomorphic frames in a neighborhood of each point of the manifold.*

By a *field of frames* on the manifold M or an open subset S of M is meant a cross section in the (principal) bundle of frames over M or S, respectively. The field is said to be *parallel* if each of the vector fields is parallel.

We first prove the sufficiency. If the curvature is zero, the system of equations $\omega^j{}_i = 0$ is completely integrable. Therefore, in a suitably chosen coordinate neighborhood U of each point P it is possible to introduce a field of orthonormal frames $P, \{e_1, \cdots, e_n, \bar{e}_1, \cdots, \bar{e}_n\}$ which are parallel and are uniquely determined by the initially chosen frame at P. For, by § 1.9 the vector fields e_i satisfy the differential system

$$de_i = \omega^j{}_i \, e_j.$$

(The metric being locally flat, the e_i may be thought of as covectors.) Of course, we also have the conjugate relations. Since the e_i are of bidegree (1,0), the condition $de_i = 0$ implies $d''e_i = 0$, that is the e_i are holomorphic vector fields. Hence, the condition that the curvature is zero implies the existence of a field of parallel orthonormal holomorphic frames in U.

Conversely, with respect to any parallel field of orthonormal frames the equations

$$\omega^k{}_i \wedge \omega^j{}_k - d\omega^j{}_i = R^j{}_{ikl^*} \, dz^k \wedge d\bar{z}^l$$

imply $R^j{}_{ikl^*} = 0$. The curvature tensor must therefore vanish for all frames.

Let us call the neighborhoods U of the theorem admissible neighborhoods. Parallel displacement of a frame at P along any path in such a neighborhood U of P is independent of the path since the system $\omega^i{}_j = 0$ has a unique solution through U coinciding with the given frame at $P \in U$. (In the remainder of this section we shall write $U(P)$ in place of U).

Now, given any two points P_0 and P_1 of the manifold and a path C joining them, there is a neighborhood $U(C) = \bigcup_{Q \in C} U(Q)$ of C such that the displacement of a frame from P_0 to P_1 is the same along any path from P_0 to P_1 in $U(C)$. We call $U(C)$ an admissible neighborhood. Let C_0 and C_1 be any two homotopic paths joining P_0 to P_1 and denote by $\{C_t\}$ $(0 \leqq t \leqq 1)$ the class of curves defining the homotopy. Let S be the subset of the unit interval I corresponding to those paths C_s for which parallel displacement of a frame from P_0 to P_1 is identical with that along C_0. Hence, $0 \in S$. That S is an open subset of I is clear. We show that S is closed. If $S \neq I$, it has a least upper bound s'. Consequently, since $U(C_{s'})$ is of finite width we have a contradiction. For, S is both open and closed, and so $S = I$. We have proved

Theorem 6.7.2. *In an hermitian manifold of zero curvature, parallel displacement along a path depends only on the homotopy class of the path [16].*

Corollary. *A simply connected hermitian manifold of zero curvature is (complex) parallelisable by means of parallel orthonormal frames.*

It is shown next that a complex parallelisable manifold has a canonically defined hermitian metric g with respect to which the curvature vanishes. Indeed, in the notation of theorem 6.6.1 let

$$X_r = \xi^{\,i}_{(r)} \frac{\partial}{\partial z^i}$$

with respect to the system (z^i) of local complex coordinates. In terms of the inverse matrix $(\xi^{(r)}_i)$ of $(\xi^{\,i}_{(r)})$ the n 1-forms

$$\theta^r = \xi^{(r)}_i \, dz^i$$

define a basis of the space of covectors of bidegree (1,0). We define the metric g by means of the matrix of coefficients

$$g_{jl^*} = \sum_{r=1}^{n} \xi^{(r)}_j \overline{\xi^{(r)}_l}.$$

Since $\xi^{\,i}_{(r)} \xi^{(r)}_j = \delta^i_j$ and

$$\frac{\partial \xi^{\,i}_{(r)}}{\partial \bar{z}^j} = 0, \quad r = 1, \cdots, n,$$

$$\xi^{\,i}_{(r)} \frac{\partial \xi^{(r)}_j}{\partial \bar{z}^k} = 0.$$

Hence, since $\langle X_r, \theta^s \rangle = \delta^s_r$,

$$\frac{\partial \xi^{(r)}_j}{\partial \bar{z}^k} = 0.$$

In terms of the metric g, the connection defined in § 5.3 is given by the coefficients

$$\begin{aligned} \Gamma^i_{jk} &= g^{il^*} \frac{\partial g_{jl^*}}{\partial z^k} \\ &= \sum_{r=1}^{n} \xi^{\,i}_{(r)} \frac{\partial \xi^{(r)}_{j}}{\partial z^k}. \end{aligned} \tag{6.7.1}$$

Differentiating with respect to $\bar{z}^l$ we conclude that $R^i{}_{jkl^*} = 0$.

Theorem 6.7.3. *A complex parallelisable manifold has a natural hermitian metric of zero curvature.*

Since

$$D_j\, \xi^{\,i}_{(r)} = \frac{\partial \xi^{\,i}_{(r)}}{\partial z^j} + \xi^{\,k}_{(r)}\, \Gamma^i_{kj}$$

(D_j denoting covariant differentiation with respect to the given connection), it follows that

$$\xi^{(r)}_{l}\, D_j\, \xi^{i}_{(r)} = 0.$$

Multiplying these equations by $\xi^{l}_{(s)}$ and taking account of the relations

$$\xi^{\,l}_{(s)}\, \xi^{(r)}_{l} = \delta^r_s$$

we conclude that

$$D_j\, \xi^{\,i}_{(r)} = 0, \quad r = 1, \cdots, n.$$

Thus, we have

Corollary. *A complex parallelisable manifold has a natural hermitian metric with respect to which the given field of frames is parallel.*

The results of this section are interpreted in VI.G.

6.8. Compact complex parallelisable manifolds

Let M be a compact complex parallelisable manifold. Since the curvature of M (defined by the connection (6.7.1)) vanishes, the connection is holomorphic; hence, so is the torsion, that is, in the notation of § 5.3

$$d''\,\Omega^i = 0, \quad i = 1, \cdots, n$$

where the Ω^i are the forms Θ^i pulled down to M by means of the cross section $M \to \{(\partial/\partial z^j)_P, (\partial/\partial \bar{z}^j)_P\}$. Denoting the components of the torsion tensor by $T_{jk}{}^i$ as in § 5.3, put

$$f = T_{jk}{}^i \, \bar{T}^{jk}{}_i, \quad T^{jk}{}_i = g^{jr^*} g^{ks^*} g_{it^*} T_{r^*s^*}{}^{t^*};$$

then, f is a real-valued function.

Lemma 6.8.1. *The $T_{jk}{}^i$ are the constants of structure of a local Lie group.*

For,

$$-\tfrac{1}{2}\delta df = g^{rs^*} D_{s^*} D_r f$$

$$= g^{rs^*}(\bar{T}^{jk}{}_i \, D_{s^*} D_r \, T_{jk}{}^i + D_r \, T_{jk}{}^i D_{s^*} \bar{T}^{jk}{}_i).$$

Hence, since the curvature is zero, an application of the interchange formula (1.7.21) gives

$$-\tfrac{1}{2}\delta df = g^{rs^*} \, D_r T_{jk}{}^i \, \overline{D_s T^{jk}{}_i}.$$

Therefore, by proposition 6.5.1,

$$\Delta f = \delta df \equiv 0,$$

from which we conclude that the $D_r T_{jk}{}^i$ vanish. Consequently, from (5.3.22) they satisfy the Jacobi identities

$$T_{rk}{}^i T_{lj}{}^r + T_{rl}{}^i T_{jk}{}^r + T_{rj}{}^i T_{kl}{}^r = 0. \tag{6.8.1}$$

Since M is complex parallelisable, it follows from the proof of theorem 6.7.3 that there exists n linearly independent holomorphic pfaffian forms $\theta^1, \cdots, \theta^n$ defined everywhere on M. Therefore, their exterior products $\theta^i \wedge \theta^j$ $(i < j)$ are also holomorphic and linearly independent everywhere (cf. lemma (6.10.1)). Moreover, since there are $n(n-1)/2$ such products they form a basis of the space of pure forms of bidegree (2,0).

It is now shown that $d\theta^i$ is a holomorphic 2-form, $i = 1, \cdots, n$. Indeed, θ^i is of bidegree (1,0), and so since $d\theta^i = d'\theta^i$ (by virtue of the fact that the θ^i are holomorphic), $d\theta^i$ is a pure form of bidegree (2,0). On the other hand, $d''d\theta^i = d''d'\theta^i = 0$ since $d'd'' + d''d' = 0$.

We conclude that the $d\theta^i$ may be expressed linearly (with complex coefficients) in terms of the products $\theta^j \wedge \theta^k$, and since M is compact these coefficients (as holomorphic functions) are necessarily constants. That the coefficients are proportional to the $T_{jk}{}^i$ is easily seen from equations (5.3.3) - (5.3.5) by restricting to parallel orthonormal frames (cf. proof of theorem 6.7.1). Consequently,

$$d\theta^i = -\tfrac{1}{2}T_{jk}{}^i\theta^j \wedge \theta^k, \quad T_{jk}{}^i + T_{kj}{}^i = 0. \tag{6.8.2}$$

Equations (6.8.1) and (6.8.2) imply that the θ^i $(i = 1, \cdots, n)$ define a local Lie group. This group cannot, in general, be extended to the whole of M. For this reason we consider the universal covering space $\tilde{M}$ of M. For, $\tilde{M}$ is simply connected and has a naturally induced hermitian metric of zero curvature (cf. theorem 6.7.2, cor., and prop. 5.8.3). We prove

Theorem 6.8.1. *The universal covering space $\tilde{M}$ of a compact complex parallelisable manifold M is a complex Lie group* [69].

In the first place, since the projection $\pi: \tilde{M} \to M$ is a holomorphic map, $\tilde{M}$ has a naturally induced complex structure (cf. prop. 5.8.3). On the other hand, π is a local homeomorphism; hence, it is (1-1). Consequently, the n forms

$$\tilde{\theta}^i = \pi^*(\theta^i)$$

are linearly independent and holomorphic, the latter property being due to the fact that π is holomorphic (cf. lemma 5.8.2). Moreover,

$$\begin{aligned} d\tilde{\theta}^i &= d(\pi^*\theta^i) = \pi^*(d\theta^i) \\ &= \pi^*(-\tfrac{1}{2}T_{jk}{}^i\,\theta^j \wedge \theta^k) \\ &= -\tfrac{1}{2}T_{jk}{}^i\,\pi^*\theta^j \wedge \pi^*\theta^k \\ &= -\tfrac{1}{2}T_{jk}{}^i\tilde{\theta}^j \wedge \tilde{\theta}^k. \end{aligned}$$

Hence, the $\tilde{\theta}^i$ define a local Lie group. The $\tilde{\theta}^i$ being independent we define the (hermitian) metric

$$\tilde{g} = 2\sum_{i=1}^{n} \tilde{\theta}^i \otimes \overline{\tilde{\theta}^i}$$

on $\tilde{M}$. That this metric is not, in general Kaehlerian follows from the fact that the $\tilde{\theta}^i$ are not necessarily d'-closed.

With respect to this metric, $\tilde{M}$ may be shown to be complete (cf. § 7.7). To see this, since M is compact it is complete with respect to the metric

$$g = 2\sum_{i=1}^{n} \theta^i \otimes \overline{\theta^i}.$$

The completeness of $\tilde{M}$ now follows from that of M.

For,

$$\tilde{g} = \pi^* g.$$

Hence, $(\tilde{M}, \tilde{g})$ is a 'hermitian covering space' of (M, g), that is, the holomorphic projection map π induces the metric of $\tilde{M}$.

The universal covering space $\tilde{M}$ of a compact complex parallelisable manifold therefore has the properties:

(i) there are n independent abelian differentials of the first kind on $\tilde{M}$;

(ii) they satisfy the equations of Maurer-Cartan;

(iii) $\tilde{M}$ is simply connected, and

(iv) $\tilde{M}$ is complete (with respect to $\tilde{g}$).

Under the circumstances, $\tilde{M}$ can be given a group structure in such a way that multiplication in the group is holomorphic. Moreover, the abelian differentials are left invariant pfaffian forms. We conclude therefore that $\tilde{M}$ is a complex Lie group.

That compactness is essential to the argument may be seen from the following example:

Let $M = C_2 - 0$. Define the holomorphic pfaffian forms θ^1 and θ^2 on M as follows:

$$\theta^1 = z^1 z^2 \, dz^1, \quad \theta^2 = z^1 z^2 \, dz^2.$$

Denote by X_1 and X_2 their duals in T^c. The components of the torsion tensor with respect to this basis are given by (6.8.2), namely,

$$T_{12}{}^1 = \frac{1}{z^1 (z^2)^2}, \quad T_{12}{}^2 = -\frac{1}{(z^1)^2 z^2}.$$

Although X_1 and X_2 form parallel frames, these components are not constant.

6.9. A topological characterization of compact complex parallelisable manifolds

In this section, a compact complex parallelisable manifold M is characterized as the quotient space of a complex Lie group. In fact, it is shown that M is holomorphically isomorphic with $\tilde{M}/D$ where D is the fundamental group of $\tilde{M}$. As a consequence of this, it follows that M is Kaehlerian, if and only if, it is a multi-torus.

Let D be the fundamental group of the universal covering space $(\tilde{M}, \pi)$ of the compact complex parallelisable manifold M, that is, the group of those homeomorphisms σ of $\tilde{M}$ with itself such that $\pi \cdot \sigma = \pi$ for every element $\sigma \in D$. Then,

$$\sigma^* \cdot \pi^* = \pi^*$$

where σ^* is the induced dual map on $\wedge^{*c}(\tilde{M})$. Hence,

$$\tilde{\theta}^i = \pi^*(\theta^i) = \sigma^*\pi^*(\theta^i) = \sigma^*(\tilde{\theta}^i),$$

that is the $\tilde{\theta}^i$ are invariant under D. It follows that σ is a left translation of $\tilde{M}$, and so D may be considered as a discrete subgroup of the complex Lie group $\tilde{M}$. With this identification of D, M is holomorphically isomorphic with $\tilde{M}/D$. Thus,

Theorem 6.9.1. *A compact complex parallelisable manifold is holomorphically isomorphic with a complex quotient space of a complex Lie group modulo a discrete subgroup* [*69*].

Corollary. *A compact complex parallelisable manifold is Kaehlerian, if and only if, it is a complex multi-torus.*

A complex torus is compact, Kaehlerian, and complex parallelisable (cf. example 3, § 5.9). Conversely, if $M = G/D$ is Kaehlerian, the left invariant pfaffian forms on the complex Lie group G must be closed. It follows that G is abelian. Therefore, M is a complex torus.

Theorem 6.9.1 may be strengthened by virtue of theorem 6.7.1. For, zero curvature alone implies that the $T_{jk}{}^i$ satisfy the equations of Maurer-Cartan. It follows that the θ^i are the left invariant pfaffian forms of a local Lie group.

Theorem 6.9.2. *A compact hermitian manifold of zero curvature is holomorphically isomorphic with a complex quotient space of a complex Lie group modulo a discrete subgroup.*

Corollary. *A compact hermitian manifold M of zero curvature cannot be simply connected.*

For, otherwise the left invariant pfaffian forms on M are closed. Thus, M is an abelian Lie group, and hence is a complex torus. This, of course is impossible.

6.10. d''-cohomology

We have seen that d'' is a differential operator on the graded module $\wedge^{*c}(M)$ (cf. § 5.4) where M is a complex manifold. In this way, since $d''^2 = 0$, it is possible to define a cohomology theory analogous to the de Rham cohomology (d-cohomology) of a differentiable manifold. The reason for considering cohomology with the differential operator d'' is clear. Indeed, it yields information regarding holomorphic forms.

We remark that in this section to every statement regarding the operator d'' there is a corresponding statement for the operator d'. Thus, there is a corresponding cohomology theory defined by d'.

Lemma 6.10.1. *For every form α of bidegree (q, r) and any β*

$$d''(\alpha \wedge \beta) = d''\alpha \wedge \beta + (-1)^{q+r} \alpha \wedge d''\beta.$$

To see this, it is only necessary to apply the operator d to $\alpha \wedge \beta$ and compare the bidegrees in the resulting expansion.

Let $\wedge^{q,r}$ denote the linear space of forms of bidegree (q, r) on M. Consider the sequence of maps

$$\cdots \to \wedge^{q,r-1} \xrightarrow{d''_{q,r-1}} \wedge^{q,r} \xrightarrow{d''_{q,r}} \wedge^{q,r+1} \to \cdots$$

where for the moment we write $d''_{q,r} = d'' \mid \wedge^{q,r}$. Now, put

$$H_{\bar{z}}^{q,r}(M) = \frac{\text{kernel } d''_{q,r}}{\text{image } d''_{q,r-1}};$$

then,

Proposition 6.10.1.

$$H_{\bar{z}}^{p,0}(M) = \text{kernel } d''_{p,0}.$$

For, if $\alpha \in \text{image } d''_{q,r-1}$ it must come from a form of bidegree $(q, r-1)$. Let α be a form of bidegree $(p, 0)$. Then, its image by $d''_{q,r-1}$ must be 0.

Corollary. *$H_{\bar{z}}^{p,0}(M)$ is the linear space of holomorphic p-forms.*

Now, by lemma 6.10.1 if α and β are holomorphic forms, so is $\alpha \wedge \beta$. Define

$$H_{\bar{z}}(M) = \sum_{p=0}^{n} H_{\bar{z}}^{p,0}(M);$$

then, by the remark just made, $H_{\bar{z}}(M)$ has a ring structure.

It is now shown that the d''-cohomology ring of a compact complex parallelisable manifold M depends only upon the local structure of its universal covering space.

Indeed, every holomorphic p-form α on M has a unique representation

$$\alpha = a_{i_1 \ldots i_p} \theta^{i_1} \wedge \cdots \wedge \theta^{i_p}$$

where the coefficients are holomorphic functions. Since M is compact, the coefficients must be constants. Hence, $\pi^*(\alpha)$ is a left invariant holomorphic p-form on $(\tilde{M}, \pi)$—the universal covering space of M. On the other hand, a left invariant p-form on $\tilde{M}$ has constant coefficients. Thus, π^* defines a ring isomorphism from the exterior algebra of holomorphic forms on M onto the exterior algebra of left invariant differential forms on $\tilde{M}$. Moreover, since

$$\pi^* d'' = d'' \pi^*,$$

π^* induces an isomorphism between their cohomology rings. Now, since the cohomology ring of a compact (connected) Lie group is isomorphic with the cohomology ring of its Lie algebra L [*48*], we conclude that the d''-cohomology ring of M is isomorphic with the cohomology ring of L. We have proved

Theorem 6.10.1. *The d''-cohomology ring of a compact complex parallelisable manifold is isomorphic with the cohomology ring of the (complex) Lie algebra of its universal covering space* [*69*].

6.11. Complex imbedding

In this section, the problem of imbedding a Kaehler manifold M holomorphically into a locally flat space is considered. More precisely, we are interested in establishing necessary conditions for such imbeddings to be possible. Moreover, only locally isometric imbeddings are considered. If the Ricci curvature of M is positive, M cannot be so imbedded. On the other hand, negative Ricci curvature is not sufficient as we shall see by considering the classical hyperbolic space defined by means of the metric

$$ds^2 = \frac{dz d\bar{z}}{(1 - z\bar{z})^2} \tag{6.11.1}$$

in the unit circle $|z| < 1$. The possiblity of such imbeddings thus appears to be rather remote.

Let U be a coordinate neighborhood on the complex manifold M

with local complex coordinates $(z^1, \cdots, z^n)$ and assume the existence on U of $N \geqq n$ holomorphic 1-forms

$$\alpha^r = a_i^{(r)}\, dz^i, \quad r = 1, \cdots, N \tag{6.11.2}$$

independent at each point, satisfying the further conditions

$$d'\alpha^r = 0, \quad r = 1, \cdots, N. \tag{6.11.3}$$

Since $d = d' + d''$ we have assumed the existence on U of N closed forms α^r. Thus, the real 2-form

$$\Omega = \sqrt{-1} \sum_{r=1}^{N} \alpha^r \wedge \overline{\alpha^r} \tag{6.11.4}$$

on U is closed and, since it is of maximal rank, the differential forms α^r define a (locally) Kaehlerian metric on U. This metric is not globally defined, that is, we do not assume the existence of N (globally defined) holomorphic 1-forms on M but rather on the coordinate neighborhood U.

The conditions (6.11.3) are the integrability conditions of the system of differential equations

$$a_i^{(r)} = \frac{\partial f^r}{\partial z^i}, \quad i = 1, \cdots, n; \quad r = 1, \cdots, N \tag{6.11.5}$$

where the f^r are holomorphic functions.

Consider the map $f: U \rightarrow C_N$ defined in terms of local coordinates by

$$w^r = f^r(z^1, \cdots, z^n), \quad r = 1, \cdots, N. \tag{6.11.6}$$

Since Ω is of maximal rank, this map is (1-1). Hence, the metric g of U:

$$g_{ij^*} = \sum_{r=1}^{N} a_i^{(r)} \overline{a_j^{(r)}}, \quad i, j = 1, \cdots, n \tag{6.11.7}$$

is induced by the flat Kaehler metric

$$d\sigma^2 = \sum_{r=1}^{N} dw^r\, \overline{dw^r} \tag{6.11.8}$$

of C_N. For,

$$\begin{aligned}\sum_{r=1}^{N} dw^r\, \overline{dw^r} &= \sum_{r=1}^{N} \frac{\partial f^r}{\partial z^i} \overline{\frac{\partial f^r}{\partial z^j}}\, dz^i\, d\bar{z}^j \\ &= \sum_{r=1}^{N} a_i^{(r)} \overline{a_j^{(r)}}\, dz^i\, d\bar{z}^j \\ &= g_{ij^*}\, dz^i\, d\bar{z}^j.\end{aligned}$$

Computing the Ricci curvature with respect to the metric (6.11.7), we obtain

$$R_{ij^*} = -g^{kl^*}\sum_{r=1}^{N} D_i a_k^{(r)} \overline{D_j a_l^{(r)}}. \tag{6.11.9}$$

For, from (5.3.19) the Ricci curvature is given by

$$-R_{ij^*} = g^{kl^*}\left(\frac{\partial^2 g_{ij^*}}{\partial z^k \partial \bar{z}^l} - g^{r^*s}\frac{\partial g_{ir^*}}{\partial z^k}\frac{\partial g_{j^*s}}{\partial \bar{z}^l}\right).$$

Substituting for g_{ij^*} from (6.11.7) and applying (5.3.11), the desired formula for R_{ij^*} follows.

Clearly, then, the Ricci curvature defines a negative semi-definite quadratic form since

$$R_{ij^*}\xi^i\xi^{j^*} = -g^{kl^*}\sum_{r=1}^{N}(\xi^i D_i a_k^{(r)})(\xi^{j^*} D_{j^*} a_{l^*}^{(r)}) \leqq 0.$$

We have proved

Theorem 6.11.1. *Let M be a Kaehler manifold locally holomorphically isometrically imbedded in C_N with the flat metric* (6.11.8). *Then, its Ricci curvature is non-positive* [5].

If M is compact we may draw the following conclusion from cor. 1, theorem 6.5.1.

Theorem 6.11.2. *If the Ricci curvature is strictly negative there are no holomorphic contravariant tensor fields of bidegree $(p, 0)$; otherwise, a tensor field of this type must be a parallel tensor field. In particular, for negative Ricci curvature there are no holomorphic vector fields on M.*

Since a complex torus T_n is locally flat (with respect to the metric of (6.7.1)), we may draw the obvious conclusion:

Corollary. *If a compact Kaehler manifold can be locally holomorphically (1-1) imbedded in some T_n, and if its metric g can be obtained directly from the imbedding, a holomorphic contravariant tensor field of bidegree $(p, 0)$ (if it exists) must be parallel with respect to the connection* (6.7.1) *of the metric g.*

If the Ricci curvature is negative, local imbeddings of the type considered in theorem 6.11.1 are not always possible. The hyperbolic space defined by the metric (6.11.1) in the unit circle shows that this is the case. This is a consequence of the following

Proposition 6.11.2. *Let U be a coordinate neighborhood of complex dimension 1 endowed with the metric*

$$ds^2 = g(z,\bar{z})\, dz\, d\bar{z}$$

where the function g has the special form

$$g(z,\bar{z}) = \sum_{\rho=1}^{\infty} a_\rho\, z^\rho\, \bar{z}^\rho. \tag{6.11.10}$$

If U can be holomorphically, isometrically mapped into $C_N(N \geqq 1)$ with the flat metric (6.11.8), then, the power series (6.11.10) is a polynomial.

For, since U is holomorphically, isometrically imbedded in C_N, the imbedding is given by the functions

$$w^r = f^r(z) = \sum_{\rho=1}^{\infty} b^r{}_\rho\, z^\rho, \quad r = 1, \cdots, N$$

with the property

$$\sum_{r=1}^{N} \left| \frac{df^r(z)}{dz} \right|^2 = \sum_{\rho=1}^{\infty} a_\rho\, z^\rho\, \bar{z}^\rho. \tag{6.11.11}$$

Hence,

$$\rho\sigma \sum_{r=1}^{N} b^r{}_\rho\, \overline{b^r}_\sigma = \delta_{\rho\sigma}\, a_{\rho-1}. \tag{6.11.12}$$

Now, for each ρ the sequence of numbers

$$B_\rho = \{b^r{}_\rho\}, \quad r = 1, \cdots, N$$

is a vector in C_N. But, by (6.11.12) any two are orthogonal; hence, at most N of them can be different from zero. We conclude that at most N^2 of the $b^r{}_\rho$ are different from zero, that is the mapping functions $f^r(z)$ are polynomials. Comparing (6.11.10) with (6.11.11), $g(z, \bar{z})$ must be a (finite) polynomial.

Consider the metric

$$g(z,\bar{z}) = \frac{1}{(1 - z\bar{z})^2} \tag{6.11.13}$$

in the unit circle $| z | < 1$. Hence, from the proposition just proved, the interior of the disc $| z | < 1$ cannot be isometrically imbedded in some C_N with the flat metric. It is not difficult to see that the Ricci curvature of g is given by

$$R(z,\bar{z}) = -\frac{2}{(1 - z\bar{z})^2} < 0. \tag{6.11.14}$$

From (6.11.13) and (6.11.14) we obtain immediately that the scalar curvature is -2. Thus, g has constant negative curvature, that is g is a hyperbolic metric.

Another example is afforded by the higher dimensional analogue, namely, the interior of the unit ball $\Sigma_{i=1}^n \mid z^i \mid^2 < 1$ with the hyperbolic metric

$$ds^2 = \frac{\Sigma \mid dz^i \mid^2 - \Sigma \mid z^i \mid^2 \Sigma \mid dz^j \mid^2 + \mid \Sigma \bar{z}^i \, dz^i \mid^2}{(1 - \Sigma \mid z^i \mid^2)^2}.$$

6.12. Euler characteristic

In the previous section we considered manifolds M on which $N \geqq n$ holomorphic functions $f^r(r = 1, \cdots, N)$ are 'locally' defined. More precisely, in a coordinate neighborhood U of M we assumed the existence of N independent holomorphic 1-forms α^r satisfying $d'\alpha^r = 0$. Now, in this section, we assume that on the complex manifold M there exists $N \geqq n$ 'globally' defined holomorphic differentials

$$\alpha^r = a_i^{(r)} \, dz^i, \quad r = 1, \cdots, N, \quad \text{rank}\,(a_i^{(r)}) = n \text{ everywhere},$$

which are simultaneously d'-closed. The fundamental form

$$\Omega = \sqrt{-1} \sum_{r=1}^{N} \alpha^r \wedge \overline{\alpha^r}$$

of M is then closed and of maximal rank. The distinction made here is that we now have a globally defined Kaehler metric

$$ds^2 = 2 \sum_{r=1}^{N} \alpha^r \otimes \overline{\alpha^r}.$$

In terms of the curvature of this metric, and by means of the generalized Gauss-Bonnet theorem, if M is compact

$$(-1)^n \chi(M) \geqq 0$$

where $\chi(M)$ denotes the Euler characteristic of M. Moreover, $\chi(M)$ vanishes, if and only if the nth Chern class vanishes. Incidentally, the vanishing of $\chi(M)$ is a necessary and sufficient condition for the existence of a continuous vector field with no zeros (on M).

A representative c_r of the $(n-r+1)^{\text{st}}$ *Chern class* of an hermitian manifold is given in terms of the curvature forms $\Theta^i_{\ j}$ by means of the formula [21]

$$c_r = \frac{1}{(2\pi\sqrt{-1})^{n-r+1}\,(n-r+1)!}\,\delta^{j_1\ldots j_{n-r+1}}_{i_1\ldots i_{n-r+1}}\,\Theta^{i_1}_{j_1}\wedge\ldots\wedge\Theta^{i_{n-r+1}}_{j_{n-r+1}}. \tag{6.12.1}$$

The theorem invoked above may be stated as follows:

The Euler characteristic of a compact hermitian manifold M is given by the Gauss-Bonnet formula

$$\chi(M) = \int_M c_1. \tag{6.12.2}$$

As in § 6.11, in each coordinate neighborhood U there exists N holomorphic functions f^r such that

$$a^{(r)}_i = \frac{\partial f^r}{\partial z^i}, \quad i = 1, \cdots, n; \quad r = 1, \cdots, N \tag{6.12.3}$$

by means of which M is mapped locally, (1-1) into C_N. Moreover, the metric g of M defined by the matrix of coefficients

$$g_{ij^*} = \sum_{r=1}^{N} a^{(r)}_i\,\overline{a^{(r)}_j} \tag{6.12.4}$$

is induced by the flat Kaehler metric

$$d\sigma^2 = \sum_{r=1}^{N} dw^r\,\overline{dw^r}$$

of C_N where

$$w^r(z) = \int^z a^{(r)}_i\,dz^i$$

is the rth abelian integral of the first kind on M.

To compute the curvature tensor of the metric g we proceed as follows: In the first place, from (5.3.19) the only non-vanishing components are given by

$$R_{ij^*kl^*} = -\frac{\partial^2 g_{ij^*}}{\partial z^k\,\partial \bar z^l} + g^{r^*s}\,\frac{\partial g_{ir^*}}{\partial z^k}\,\frac{\partial g_{j^*s}}{\partial \bar z^l}. \tag{6.12.5}$$

From (6.12.4), since the functions $a^{(r)}_i$, $r = 1, \cdots, N$; $i = 1, \cdots, n$ are holomorphic,

$$\frac{\partial g_{ij^*}}{\partial z^k} = \sum_{r=1}^{N} \frac{\partial a^{(r)}_i}{\partial z^k}\,\overline{a^{(r)}_j}$$

and

$$\frac{\partial^2 g_{ij^*}}{\partial z^k \partial \bar{z}^l} = \sum_{r=1}^{N} \frac{\partial a_i^{(r)}}{\partial z^k} \overline{\frac{\partial a_j^{(r)}}{\partial z^l}}.$$

Substituting in (6.12.5) and making use of the fact that

$$g^{r^*s} \frac{\partial g_{j^*s}}{\partial \bar{z}^l} = \overline{\Gamma_{jl}^r}$$

we obtain

$$R_{ij^*kl^*} = -\sum_{r=1}^{N} \frac{\partial a_i^{(r)}}{\partial z^k} \overline{\frac{\partial a_j^{(r)}}{\partial z^l}} + \sum_{r=1}^{N} \frac{\partial a_i^{(r)}}{\partial z^k} a_{t^*}^{(r)} \overline{\Gamma_{jl}^t}. \tag{6.12.6}$$

Now, since

$$D_k a_i^{(r)} = \frac{\partial a_i^{(r)}}{\partial z^k} - a_m^{(r)} \Gamma_{ik}^m$$

and

$$\overline{D_l a_j^{(r)}} = \overline{\frac{\partial a_j^{(r)}}{\partial z^l}} - \overline{a_p^{(r)}} \overline{\Gamma_{jl}^p},$$

$$D_k a_i^{(r)} \overline{D_l a_j^{(r)}} = \frac{\partial a_i^{(r)}}{\partial z^k} \overline{\frac{\partial a_j^{(r)}}{\partial z^l}} - \left(\frac{\partial a_i^{(r)}}{\partial z^k} \overline{a_p^{(r)}} \overline{\Gamma_{jl}^p} + \overline{\frac{\partial a_j^{(r)}}{\partial z^l}} a_m^{(r)} \Gamma_{ik}^m \right) + a_m^{(r)} \overline{a_p^{(r)}} \Gamma_{ik}^m \overline{\Gamma_{jl}^p}$$

$$= \frac{\partial a_i^{(r)}}{\partial z^k} \overline{\frac{\partial a_j^{(r)}}{\partial z^l}} - \frac{\partial a_i^{(r)}}{\partial z^k} \overline{a_p^{(r)}} \overline{\Gamma_{jl}^p} - a_m^{(r)} \Gamma_{ik}^m \overline{D_l a_j^{(r)}}. \tag{6.12.7}$$

But

$$D_{l^*} g_{mj^*} = \sum_{r=1}^{N} D_{l^*}(a_m^{(r)} \overline{a_j^{(r)}}) = \sum_{r=1}^{N} \overline{a_j^{(r)}} D_{l^*} a_m^{(r)} + a_m^{(r)} \overline{D_l a_j^{(r)}}),$$

from which we conclude that

$$\sum_{r=1}^{N} a_m^{(r)} \overline{D_l a_j^{(r)}} = -\sum_{r=1}^{N} \overline{a_j^{(r)}} D_{l^*} a_m^{(r)} = 0.$$

Summing (6.12.7) with respect to r and comparing the result with (6.12.6) we obtain

$$R_{ij^*kl^*} = -\sum_{r=1}^{N} D_k a_i^{(r)} \overline{D_l a_j^{(r)}}.$$

Thus,

$$\Omega_{ij*} \equiv R_{ij*kl*} dz^k \wedge d\bar{z}^l = - \sum_{r=1}^{N} D_k a_i^{(r)} \overline{D_l a_j^{(r)}} \, dz^k \wedge d\bar{z}^l.$$

where the $\Omega^k{}_i$ are the forms $\Theta^k{}_i$ pulled down to M. (The Ω_{ij*} are defined by the above relations.)

From 6.12.1 we deduce that

$$\begin{aligned} c_1 &= \frac{1}{(2\pi\sqrt{-1})^n \, n!} \delta^{j_1 \ldots j_n}_{i_1 \ldots i_n} \Theta^{i_1}_{j_1} \wedge \ldots \wedge \Theta^{i_n}_{j_n} \\ &= \frac{(-1)^n}{(2\pi\sqrt{-1})^n} \det(\Omega_{ij*}) \\ &= \frac{1}{(2\pi\sqrt{-1})^n} \det\left(\sum_{r=1}^{N} D_k a_i^{(r)} \overline{D_l a_j^{(r)}} \, dz^k \wedge d\bar{z}^l\right). \end{aligned}$$

where, for simplicity, we have writen Ω_{ij*} for its image in B (cf. § 5.3). Now, put

$$\varphi_i^{(r)} = D_k a_i^{(r)} \, dz^k.$$

Then,

$$\begin{aligned} c_1 &= \frac{1}{(2\pi\sqrt{-1})^n} \det\left(\sum_{r=1}^{N} \varphi_i^{(r)} \wedge \overline{\varphi_j^{(r)}}\right) \\ &= \frac{1}{(2\pi\sqrt{-1})^n} \det({}^t\Phi \wedge \bar{\Phi}) \end{aligned}$$

where Φ is the matrix $(\varphi_i^{(r)})$ and ${}^t\Phi$ denotes its transpose.

The result follows after expressing Φ in terms of real analytic coordinates (x^i, y^i) with $z^i = x^i + \sqrt{-1}\, y^i$, since $dz^i \wedge d\bar{z}^i = -2\sqrt{-1}\, dx^i \wedge dy^i$.

Theorem 6.12.1. *The Euler characteristic of a compact complex manifold of complex dimension n on which there exists $N \geqq n$ closed holomorphic differentials $a_i^{(r)} dz^i$ such that rank$(a_i^{(r)}) = n$ satisfies the inequality*

$$(-1)^n \chi(M) \geqq 0.$$

Moreover, $\chi(M)$ vanishes, if and only if, the nth Chern class vanishes [8].

6.13. The effect of sufficiently many holomorphic differentials

It was shown in § 6.11 that the existence of sufficiently many independent holomorphic differentials which are, at the same time, d'-closed precludes the existence of holomorphic contravariant tensor fields of any order provided the Ricci curvature defined by the given differentials is negative. In fact, the condition that the differentials be d'-closed ensured the existence of a Kaehler metric relative to which the Ricci curvature was non-positive. By restricting the independence assumption on the holomorphic differentials we may drop the restriction on the curvature entirely, thereby obtaining interesting consequences from an algebraic point of view.

We consider a compact complex manifold M of complex dimension n. No assumption regarding a metric will be made, that is, in particular, M need not be a Kaehler manifold. Let α be a holomorphic form of bidegree (1,0) and X a holomorphic (contravariant) vector field on M. Then, since M is compact

$$i(X)\alpha = \text{const.},$$

for, $i(X)\alpha$ is a holomorphic function on M. If we assume that there are $N > n$ holomorphic 1-forms $\alpha^1, \cdots, \alpha^N$ defined on M, then

$$i(X)\alpha^r = c^r, \quad r = 1, \cdots, N \tag{6.13.1}$$

where the c^r, $r = 1, \cdots, N$ are constants. If, for any system of constants c^r (not all zero) the linear equations (6.13.1) are independent, that is, if the rank of the matrix

$$(a^{(r)}_i, c^r), \quad \alpha^r = a^{(r)}_i\, dz^i$$

is $n + 1$ at some point, the holomorphic vector field X must vanish.

Now, let t be a holomorphic contravariant tensor field of order p on M. Then, under the conditions, the same conclusion prevails, that is, t must vanish. Indeed, it is known for $p = 1$. Applying induction, assume the validity of the statement for holomorphic contravariant tensor fields of order $p - 1$ and consider the holomorphic tensor field

$$t = \xi^{i_1 \cdots i_p} \frac{\partial}{\partial z^{i_1}} \otimes \cdots \otimes \frac{\partial}{\partial z^{i_p}} \cdot$$

Then, the functions

$$a^{(r)}_{i_p}\, \xi^{i_1 \cdots i_p}, \quad r = 1, \cdots, N$$

are the components of N holomorphic contravariant tensor fields of order $p - 1$. By the inductive assumption they must vanish. But, we have assumed that at least n of the differentials α^r are independent. Thus, the coefficients of the $a^{(r)}_{i_p}$ in the system of linear equations

$$a^{(r)}_{i_p}\, \xi^{i_1 \cdots i_p} = 0$$

must vanish.

Theorem 6.13.1. *Let* $\alpha^r, r = 1, \cdots, N$ *be* $N > n$ *holomorphic differentials on the compact complex manifold* M *with the property: For any system of constants* $c^r, r = 1, \cdots, N$ *(not all zero) the rank of the matrix* $(a_i^{(r)}, c^r)$, $r = 1, \cdots, N$; $i = 1, \cdots, n$ *has its maximum value* $n + 1$ *at some point. Then, there do not exist (non-trivial) holomorphic contravariant tensor fields of any order on* M. *In particular, there are no holomorphic vector fields on* M *[9].*

This result is generalized in Chapter VII. In particular, it is shown that if $b_{n,0}(M) = 2$, M cannot admit a transitive Lie group of holomorphic homeomorphisms.

6.14. The vanishing theorems of Kodaira

A *complex line bundle* B over a Kaehler manifold M (of complex dimension n) is an analytic fibre bundle over M with fibre C—the complex numbers and structural group the multiplicative group of complex numbers acting on C. Let $\wedge^q(B)$ be the 'sheaf' (cf. § A.2 with $\Gamma = \wedge^q(B)$) over M of germs of holomorphic q-forms with coefficients in B (see below). Denote by $H^p(M, \wedge^q(B))$ the pth cohomology group of M with coefficients in $\wedge^q(B)$ (in the sense of § A.2). It is known that these groups are finite dimensional [47]. It is important in the applications of sheaf theory to complex manifolds to determine when the cohomology groups vanish. By employing the methods of § 3.2, Kodaira [47] was able to obtain sufficient conditions for the vanishing of the groups $H^p(M, \wedge^q(B))$. It is the purpose of this section to state these conditions in a form which indicates the connection with the results of § 3.2. The details have been omitted for technical reasons—the reader being referred to the appropriate references, principally [97].

In terms of a sufficiently fine locally finite covering $\mathcal{U} = \{U_\alpha\}$ of M (cf. Appendix A), the bundle B is determined by the system $\{f_{\alpha\beta}\}$ of holomorphic functions $f_{\alpha\beta}$ (the transition functions) defined in $U_\alpha \cap U_\beta$ for each α, β. In $U_\alpha \cap U_\beta \cap U_\gamma$, they satisfy $f_{\alpha\beta} f_{\beta\gamma} f_{\gamma\alpha} = 1$. Setting $a_{\alpha\beta} = |f_{\alpha\beta}|^2$, it is seen that the functions $\{a_{\alpha\beta}\}$ define a principal fibre bundle over M (cf. I.J) with structural group the multiplicative group of positive real numbers. This bundle is topologically a product. Hence, we can find a system of positive real functions $\{a_\alpha\}$ of class ∞ defined in $\{U_\alpha\}$ such that, for each pair α, β

$$|f_{\alpha\beta}|^2 = \frac{a_\alpha}{a_\beta} \quad \text{in } U_\alpha \cap U_\beta.$$

Since the functions $f_{\alpha\beta}$ are holomorphic in $U_\alpha \cap U_\beta$, it follows that

$$\frac{\partial^2 \log a_\alpha}{\partial z^i\, \partial \bar{z}^j} = \frac{\partial^2 \log a_\beta}{\partial z^i\, \partial \bar{z}^j} \quad \text{in } U_\alpha \cap U_\beta.$$

Thus, the 2-form

$$\gamma_{ij^*}\, dz^i \wedge d\bar{z}^j = \frac{\partial^2 \log a_\alpha}{\partial z^i\, \partial \bar{z}^j}\, dz^i \wedge d\bar{z}^j$$

is defined over the whole manifold M (cf. V.D).

A form ϕ (form of bidegree (p, q)) with coefficients in B is a system $\{\phi_\alpha\}$ of differential forms (forms of bidegree (p, q)) defined in $\{U_\alpha\}$ such that

$$\phi_\alpha = f_{\alpha\beta}\, \phi_\beta \quad \text{in} \quad U_\alpha \cap U_\beta.$$

Following § 5.4 we define complex analogs d', d'', δ' and δ'' of the operators d and δ for a form $\phi = \{\phi_\alpha\}$ with coefficients in B:

$$d'\phi = \{(d'\phi)_\alpha\}, \quad d''\phi = \{(d''\phi)_\alpha\}$$

and

$$\delta'\phi = \{(\delta'\phi)_\alpha\}, \quad \delta''\phi = \{(\delta''\phi)_\alpha\}$$

where

$$(d'\phi)_\alpha = d'\phi_\alpha, \quad (\delta'\phi)_\alpha = -\, *a_\alpha d''\left(\frac{1}{a_\alpha} *\phi_\alpha\right)$$

(α : not summed)—the star operator $*$ being defined as usual by the Kaehler metric of M. In terms of these operators it can be shown that

$$\Delta = 2(d'\delta' + \delta' d')$$

is the correct operator for the analogous Hodge theory — ϕ being called *harmonic* if it is a solution of $\Delta\phi = 0$.

If M is compact it is known that $H^p(M, \wedge^q(B)) \cong H^{q,p}(B)$—the vector space of all harmonic forms of bidegree (q, p) with coefficients in B [47]. It follows that dim $H^p(M, \wedge^q(B))$ is finite for all p and q.

Since $f_{\alpha\beta}\, f_{\beta\gamma}\, f_{\gamma\alpha} = 1$ in $U_\alpha \cap U_\beta \cap U_\gamma$

$$\log f_{\alpha\beta} + \log f_{\beta\gamma} + \log f_{\gamma\alpha} = 2\pi \sqrt{-1}\, c_{\alpha\beta\gamma}$$

is a constant in $U_\alpha \cap U_\beta \cap U_\gamma$ where $c_{\alpha\beta\gamma} \in Z$. The system $\{c_{\alpha\beta\gamma}\} \subset Z$ defines a 2-cocycle on the nerve $N(\mathscr{U})$ of the covering $\mathscr{U}$ (cf. Appendix A and [72]). It therefore determines a cohomology class $c_N \in H^2(N(\mathscr{U}), Z)$; indeed, by taking the direct limit

$$H^2(M,Z) = \lim_{\mathscr{U}} H^2(N(\mathscr{U}),Z)$$

we obtain an element $c = c(B) \in H^2(M, Z)$ called the *characteristic class* of the principal bundle associated with B.

Lemma 6.14.1 [47]. *The real closed 2-form*

$$\gamma = \frac{\sqrt{-1}}{2\pi}\frac{\partial^2 \log a_\alpha}{\partial z^i\, \partial \bar{z}^j}\, dz^i \wedge d\bar{z}^j$$

on M is a representative of the characteristic class $c(B)$. Conversely, if γ is a real closed form of bidegree (1,1) on M belonging to the characteristic class $c(B)$, there exists a system of positive functions a_α of class ∞ such that for each pair α, β

$$a_\alpha = |f_{\alpha\beta}|^2 a_\beta \quad \text{in} \quad U_\alpha \cap U_\beta$$

and

$$\gamma = \frac{\sqrt{-1}}{2\pi}\frac{\partial^2 \log a_\alpha}{\partial z^i\, \partial \bar{z}^j}\, dz^i \wedge d\bar{z}^j.$$

The 2-form γ is said to be *positive* ($\gamma > 0$) if the corresponding hermitian quadratic form is positive definite at each point of M. Let

$$\phi = \left\{\frac{1}{p!q!}\phi_{\alpha i_1 \ldots i_p j_1^* \ldots j_q^*}\, dz^{i_1} \wedge \ldots \wedge dz^{i_p} \wedge d\bar{z}^{j_1} \wedge \ldots \wedge d\bar{z}^{j_q}\right\}$$

be a differential form of bidegree (p, q) with coefficients in B and denote by $F^{p,q}(\gamma, v)$ the quadratic form (corresponding to $F(\alpha)$ in § 3.2—the operator Δ being given by $\Delta = 2(d'\delta' + \delta'd')$),

$$F^{p,q}(\gamma, v) = [\delta^r_s(\gamma^{i^*}{}_{j^*} + R^{i^*}{}_{j^*}) - p g^{ki^*} R^r{}_{skj^*}].$$

$$\cdot v_{rk_2 \ldots k_p i^* i_2^* \ldots i_q^*} \bar{v}^{sk_2 \ldots k_p j^* i_2^* \ldots i_q^*}$$

where $\gamma^{i^*}{}_{j^*} = g^{ki^*}\gamma_{kj^*}$.

We now state the vanishing theorems:

Theorem 6.14.1. *If the characteristic class $c(B)$ contains a real closed form*

$$\gamma = \frac{\sqrt{-1}}{2\pi}\gamma_{ij^*}\, dz^i \wedge d\bar{z}^j$$

with the property that the quadratic form $F^{p,q}(\gamma, v)$ is positive definite at each point of M, then

$$H^q(M, \wedge^p(B)) = \{0\}, \quad q = 1, \cdots, n.$$

Theorem 6.14.2. *If the form $\gamma > 0$,*

$$H^q(M, \wedge^n(B)) = \{0\}, \quad q = 1, \cdots, n.$$

For, then

$$F^{n,q}(\gamma, v) = n!\gamma^{i^*}{}_{j^*} v_{12\cdots ni^* i_2^* \cdots i_q^*} \bar{v}^{12\cdots nj^* i_2^* \cdots i_q^*}.$$

The proof of theorem 6.14.1 is an immediate consequence of the fact that $H^q(M, \wedge^p(B)) \cong H^{p,q}(B)$. For, by lemma 6.14.1, we may choose the system of functions $\{a_\alpha\}$ satisfying $a_\alpha = |f_{\alpha\beta}|^2 a_\beta$ in such a way that $(\sqrt{-1}/2\pi)\ (\partial^2 \log a_\alpha/\partial z^i \partial \bar{z}^j)\ dz^i \wedge d\bar{z}^j = \gamma$ (cf. VI. H. 2). Then, by the argument given below $F^{p,q}(\gamma, \phi_\alpha) = 0$, $q = 1, \cdots, n$ holds for any form $\phi = \{\phi_\alpha\} \in H^{p,q}(B)$. The result now follows since $F^{p,q}(\gamma, \phi_\alpha) > 0$ unless ϕ_α vanishes.

Let $-B$ denote the complex line bundle defined by the system $\{f_{\alpha\beta}^{-1}\}$. Then, the map $\phi \to \phi'$ defined by

$$\phi'_\alpha = \frac{1}{a_\alpha} *\bar{\phi}_\alpha$$

maps $H^{p,q}(B)$ isomorphically onto $H^{n-p,n-q}(-B)$. Hence,

$$H^q(M, \wedge^p(B)) \cong H^{n-q}(M, \wedge^{n-p}(-B)).$$

Corollary 6.14.1. *Under the hypothesis of theorem 6.14.1*

$$H^{n-q}(M, \wedge^{n-p}(-B)) = \{0\}, \qquad q = 1, \cdots, n.$$

Corollary 6.14.2. *If the form $\gamma > 0$,*

$$H^{n-q}(M, \wedge^0(-B)) = \{0\}, \qquad q = 1, \cdots, n.$$

By the *canonical bundle K* over M is meant the complex line bundle defined by the system of Jacobian matrices $k_{\alpha\beta} = \partial(z_\beta^1, \cdots, z_\beta^n)/\partial(z_\alpha^1, \cdots, z_\alpha^n)$, where the (z_α^i) are complex coordinates in U_α. It can be shown that the characteristic class $c(-K)$ of $-K$ is equal to the first Chern class of M.

The characteristic class $c(B)$ is said to be *positive definite* if it can be represented by a positive real closed form of bidegree (1,1). We are now in a position to state the following generalization of theorem 6.2.1.

Theorem 6.14.3. *There are no (non-trival) holomorphic p-forms $(0 < p \leq n)$ on a compact Kaehler manifold with positive definite first Chern class.*

This is almost an immediate consequence of theorem 6.14.2 (cf. [47]).

It is an open question whether there exists a compact Kaehler manifold with positive definite first Chern class whose Ricci curvature is not positive definite.

Proof of Theorem 6.14.1. Since M is compact, the requirement that $\phi \in H^{p,q}(B)$ is given by the equations $d''\phi_\alpha = \delta''\phi_\alpha = 0$ for each α. In the local complex coordinates (z^i), ϕ_α has the expression

$$\phi_\alpha = \frac{1}{p!q!} \phi_{\alpha k_1 \cdots k_p i_1^* \cdots i_q^*}\, dz^{k_1} \wedge \cdots \wedge dz^{k_p} \wedge d\bar{z}^{i_1} \wedge \cdots \wedge d\bar{z}^{i_q}.$$

Hence,

$$\sum_{t=0}^{q}(-1)^t D_{i_t^*}\phi_{\alpha k_1\ldots k_p i_0^*\ldots i_{t-1}^* i_{t+1}^*\ldots i_q^*}=0$$

and if I is the identity operator on forms

$$g^{lm^*}(D_l+\rho_{\alpha l}\cdot I)\,\phi_{\alpha k_1\ldots k_p m^* i_2^*\ldots i_q^*}=0, \qquad \rho_{\alpha l}=-\frac{\partial \log a_\alpha}{\partial z^l}\,.$$

Thus, for a harmonic form of bidegree (p, q) with coefficients in B

$$g^{lm^*}(D_l+\rho_{\alpha l}\cdot I)\,D_{m^*}\phi_{\alpha k_1\ldots k_p i_1^*\ldots i_q^*}$$

$$=\sum_{t=1}^{q}(\gamma^{m^*}{}_{i_t^*}+R^{m^*}{}_{i_t^*})\,\phi_{\alpha k_1\ldots k_p i_1^*\ldots i_{t-1}^* m^* i_{t+1}^*\ldots i_q^*} \tag{6.14.1}$$

$$-\sum_{j=1}^{p}\sum_{t=1}^{q}R^r{}_{k_j i_t^*}{}^{m^*}\phi_{\alpha k_1\ldots k_{j-1} r k_{j+1}\ldots k_p i_1^*\ldots i_{t-1}^* m^* i_{t+1}^*\ldots i_q^*}.$$

Consider the 1-form

$$\xi=\xi_{m^*}d\bar{z}^m$$

of bidegree (0,1) where

$$\xi_{m^*}=\frac{1}{a_\alpha}\bar{\phi}_\alpha{}^{k_1\ldots k_p i_1^*\ldots i_q^*}\,D_{m^*}\phi_{\alpha k_1\ldots k_p i_1^*\ldots i_q^*} \tag{6.14.2}$$

and

$$\bar{\phi}_\alpha{}^{k_1\ldots k_p i_1^*\ldots i_q^*}=g^{k_1 r_1^*}\ldots g^{k_p r_p^*}g^{s_1 i_1^*}\ldots g^{s_q i_q^*}\,\overline{\phi_{\alpha r_1\ldots r_p s_1^*\ldots s_q^*}}\,.$$

It is easily checked that it is a globally defined form on M. We compute its divergence:

$$-\delta\xi\equiv g^{lm^*}D_l\xi_{m^*}=G(\phi)+\lambda \tag{6.14.3}$$

where

$$G(\phi)=\frac{1}{a_\alpha}g^{lm^*}[(D_l+\rho_{\alpha l}\cdot I)\,D_{m^*}\phi_{\alpha k_1\ldots k_p i_1^*\ldots i_q^*}]\bar{\phi}_\alpha{}^{k_1\ldots k_p i_1^*\ldots i_q^*} \tag{6.14.4}$$

and

$$\lambda=\frac{1}{a_\alpha}g^{lm^*}D_{m^*}\phi_{\alpha k_1\ldots k_p i_1^*\ldots i_q^*}\cdot\overline{D_{l^*}\phi_\alpha{}^{k_1\ldots k_p i_1^*\ldots i_q^*}}\,.$$

Formula (6.14.3) should be compared with (6.5.3).

Note that equations (6.14.1)-(6.14.4) are vacuous unless $q \geqq 1$.

Now, by the Hodge-de Rham decomposition of a 1-form

$$\xi=df+\delta\eta+H[\xi]$$

where f is a real-valued function on M. Then,

$$\delta\xi = \delta df,$$

and so, from (6.14.3)

$$-\delta df = G(\phi) + \lambda.$$

Assume $G(\phi) \geqq 0$. Then, since $\lambda \geqq 0$, $\delta df \leqq 0$. Applying VI.F.3, we see that δdf vanishes identically. Thus $G(\phi) = -\lambda \leqq 0$. Consequently, $G(\phi) = 0$ and $\lambda = 0$. Finally, if $F^{p,q}(\gamma, v)$ is positive definite at each point of M, ϕ must vanish. For, by substituting (6.14.1) into (6.14.4), we derive

$$qF^{p,q}(\gamma,v) = G(v).$$

Remark: If the bundle B is the product of M and C, $B = M \times C$, the usual formulas are obtained.

EXERCISES

A. δ-pinched Kaehler manifolds [2]

1. Establish the following identities for the curvature tensor of a Kaehler manifold M with metric g (cf. I. I):

(a) $R(X,Y) = R(JX,JY)$,

(b) $K(X,Y) = K(JX,JY)$,

(c) $K(X,JY) = K(JX,Y)$,

and when X, Y, JX, JY are orthonormal

(d) $g(R(X,JX)Y,JY) = -K(X,Y) - K(JX,Y)$.

To prove (a), apply the interchange formula (1.7.21) to the tensor J defining the complex structure of M (see proof of lemma 7.3.2); to prove (b), (c), and (d) employ the symmetry properties (I.I. (a) - (d)) of the curvature tensor.

2. If the real dimension of M is $2n(n > 1)$, and M is δ-pinched, then $\delta \leqq \frac{1}{4}$.

To see this, let $\{X,JX,Y,JY\}$ be an orthonormal set of vectors in the tangent space T_P at $P \in M$. Then, from (3.2.23)

$$| g(R(X,JX)Y,JY) | \leqq \frac{2}{3}(1 - \delta).$$

Applying (1.(d)) we obtain

$$\delta \leqq K(X,Y) \leqq \frac{1}{3}(2 - 5\delta)$$

from which we conclude that $\delta \leqq \frac{1}{4}$.

3. The manifold M is said to be λ-*holomorphically pinched* if, for any holomorphic section π_H there exists a positive real number K_1 (depending on g) such that

$$\lambda K_1 \leqq R(P, \pi_H) \leqq K_1.$$

The metric g may be normalized so that $K_1 = 1$, in which case,

$$\lambda \leqq R(P, \pi_H) \leqq 1.$$

A δ-pinched Kaehler manifold is $\delta(8\delta + 1)/(1 - \delta)$ -holomorphically pinched.

To see this, apply the inequality

$$| R_{ijkl} | \leqq \tfrac{1}{3}[(PS)^{1/2} + (QR)^{1/2}] \tag{1}$$

valid for any orthonormal set of vectors $\{X_i, X_j, X_k, X_l\}$, $i,j,k,l = 1, \cdots, 2n$ where

$$P = 2K_{ij} - 2\delta, \quad Q = K_{ik} + K_{il} - 2\delta,$$
$$R = K_{jk} + K_{jl} - 2\delta, \quad S = 2K_{kl} - 2\delta.$$

This inequality is proved in a manner analogous to that of (3.2.21); indeed, set

$$L(a,i;b,k;c,j;d,l) = G(a,i;b,k;c,j;d,l) + G(a,i;b,l;c,j; -d,k)$$

and show that

$$L = Pa^2c^2 + Qa^2d^2 + Rb^2c^2 + Sb^2d^2 + 6R_{ijkl}\, abcd.$$

Put $X_{i^*} = JX_i (i = 1, \cdots, n)$ (cf. (5.2.6)) and apply (1) with $j = i^*$ and $l = k^*$. Hence, from 1.(b) - (d)

$$K_{ik} + K_{i^*k} \leqq \tfrac{1}{3}[2(K_{ii^*} - \delta)^{1/2}(K_{kk^*} - \delta)^{1/2} + K_{ik} + K_{ik^*} - 2\delta]$$

from which

$$[(K_{ii^*} - \delta)\,(K_{kk^*} - \delta)]^{1/2} \geqq K_{ik} + K_{ik^*} + \delta.$$

Since $K_{ik} \geqq \delta$, $K_{ik^*} \geqq \delta$ and $K_{kk^*} \leqq 1$, we conclude that

$$K_{ii^*} \geqq \frac{\delta(8\delta + 1)}{1 - \delta}.$$

(Note that a manifold of constant holomorphic curvature is $\frac{1}{4}$ - pinched.)

4. Prove that if M is λ-holomorphically pinched, then M is $3(7\lambda - 5)/8(4 - \lambda)$-pinched.

In the first place, for any orthonormal vectors X and Y, $g(aX + bY, aX + bY) = a^2 + b^2$. Applying 1.(b) and (c) as well as (I.I. 1(d)),

$$(a^2 + b^2)^2 K(aX + bY, J(aX + bY)) = a^4 K(X, JX) + b^4 K(Y, JY)$$
$$+ 2a^2b^2[K(X, Y) - 3g(R(Y, JX)Y, JX)]$$
$$+ 4ab^3 g(R(Y, JY)X, JY) + 4a^3 b g(R(X, JX)X, JY).$$

Put $g(Y,JX) = \sin\theta$; then,

$$g(R(Y,JX)Y,JX) = -K(Y,JX)\cos^2\theta.$$

Hence, since

$$\lambda \leqq K(aX + bY, J(aX + bY)) \leqq 1,$$

$$\lambda(a^2 + b^2)^2 \leqq a^4 K(X,JX) + 2a^2b^2[K(X,Y) + 3K(Y,JX)\cos^2\theta] + b^4 K(Y,JY) \leqq (a^2 + b^2)^2$$

for any $a, b \in R$, and so

$$2\lambda - 1 \leqq K(X,Y) + 3K(Y,JX)\cos^2\theta \leqq 2 - \lambda. \tag{2}$$

Similarly, from

$$\lambda \leqq K(aX + bJY, J(aX + bJY)) \leqq 1,$$

we deduce

$$2\lambda + 2\sin^2\theta - 1 \leqq 3K(X,Y) + K(JX,Y)\cos^2\theta \leqq 1 + 2\sin^2\theta. \tag{3}$$

Consequently,

$$\frac{3\lambda + 3\sin^2\theta - 1}{4} \leqq K(X,Y) \leqq \frac{3\sin^2\theta + 2 - \lambda}{4}$$

from which

$$K(X,Y) \geqq \tfrac{1}{4}(3\lambda - 2)$$

for any X and Y. In particular, $K(JX,Y) \geqq \frac{1}{4}(3\lambda - 2)$, and so from (3)

$$\tfrac{1}{4}(3\lambda - 2) \leqq K(X,Y) \leqq 1 - \tfrac{1}{4}\lambda\cos^2\theta \leqq 1.$$

5. Show that for every orthonormal set of vectors $\{X,Y,JX,JY\}$

$$K(X,Y) + K(JX,Y) \geqq \tfrac{1}{2}(2\lambda - 1).$$

B. Reduction of a real 2-form of bidegree (1,1)

1. At each point $P \in M$, show that there exists a basis of T_P of the form

$$\{X_i, X_{i^*}, X_{i+1}, X_{(i+1)^*}\} \cup \{X_k, X_{k^*}\}$$

($i = 1,3,\cdots,2p-1$; $k = 2p+1,\cdots,n$) such that only those components of a real 2-form α of bidegree (1,1) of the form a_{ii^*}, $a_{i+1,(i+1)^*}$, $a_{i,i+1} = a_{i^*,(i+1)^*}$, a_{kk^*} may be different from zero.

To see this, observe that T_P may be expressed as the direct sum of the 2-dimensional orthogonal eigenspaces of α. Since α is real and of bidegree (1,1), $\alpha(X,Y) = \alpha(JX,JY)$ for any two vectors X and Y (cf. V. C.6). Let V be such a subspace. Put $\tilde{V} = V + JV$. In general, $JV \neq V$; however, $J\tilde{V} = \tilde{V}$. T_P is a direct sum of subspaces of the type given by $\tilde{V}$. Only two cases are possible for $\tilde{V}$:

(a) $\tilde{V}$ is generated by X and JX. Then, $\alpha(X,Z) = \alpha(JX,Z) = 0$ for any $Z \in \{X,JX\}^{\perp}$—the orthogonal complement of the space generated by X and JX.

(b) $\tilde{V}$ is generated by X,JX,Y,JY where X and Y have the property that $\alpha(X,Z) = \alpha(Y,Z) = 0$ for any $Z \in \{X,Y\}^{\perp}$. Put $Y = aJX + bW$ where W is a vector defined by the condition that $\{X,JX,W,JW\}$ is an orthonormal set. The only non-vanishing components of α on $\tilde{V}$ are given by $\alpha(X,JX)$, $\alpha(W,JW)$, $\alpha(X,W) = \alpha(JX,JW)$. Therefore, when $Z \in \tilde{V}^{\perp}$, $\alpha(X,Z) = \alpha(JX,Z) = \alpha(W,Z) = \alpha(JW,Z) = 0$.

C. The Ricci curvature of a λ-holomorphically pinched Kaehler manifold

1. The Ricci curvature of a δ-pinched manifold is clearly positive. Show that the Ricci curvature of a λ-holomorphically pinched Kaehler manifold is positive for $\lambda \geqq \frac{1}{2}$.

In the notation of (1.10.10)

$$R_{ik}\, \xi^{i}_{(r)}\, \xi^{k}_{(r)} = \sum_{s=1}^{2n} K_{rs}\,.$$

Choose an orthonormal basis of the form $\{X, JX\} \cup \{X_i, JX_i\}$ $(i = 2, \cdots, n)$ and apply (A. 5).

D. The second betti number of a compact δ-pinched Kaehler manifold [2]

1. Prove that for a 4-dimensional compact Kaehler manifold M of strictly positive curvature, $b_2(M) = 1$.

In the first place, by theorem 6.2.1 a harmonic 2-form α is of bidegree (1,1). By cor. 5.7.3, $\alpha = r\Omega + \varphi$, $r \in R$ where Ω is the fundamental 2-form of M and φ is an effective form (of bidegree (1,1)). Since a basis may be chosen so that the only non-vanishing components of φ are of the form φ_{ii^*}, then, by (3.2.10),

$$F(\varphi) = \sum_{i} \sum_{j \neq i, i^*} (K_{ij} + K_{ij^*})\, (\varphi_{ii^*})^2 + 4 \sum_{i<j} R_{ii^*jj^*}\, \varphi_{ii^*}\, \varphi_{jj^*}.$$

Applying (A. 1(d)) we obtain

$$F(\varphi) = \sum_{i<j} (K_{ij} + K_{ij^*})\, (\varphi_{ii^*} - \varphi_{jj^*})^2\,.$$

Finally, since $K_{ij} + K_{ij^*} > 0$ and φ is effective, it must vanish.

2. If M is λ-holomorphically pinched with $\lambda > \frac{1}{2}$, then $b_2(M) = 1$.

Hint: Apply A.5.

3. Show that (D.2) gives the best possible result. (It has recently been shown that a 4-dimensional compact Kaehler manifold of strictly positive curvature is homeomorphic with P_2—the methods employed being essentially algebraic geometric, that is, a knowledge of the classification of surfaces being necessary.)

D.1 has been extended to all dimensions by R. L. Bishop and S. I. Goldberg [90].

E. Symmetric homogeneous spaces [26]

1. Let G be a Lie group and H a closed subgroup of G. The elements a, $b \in G$ are said to be congruent modulo H if $aH = bH$. This is an equivalence relation —the equivalence classes being left cosets modulo H. The quotient space G/H by this equivalence relation is called a *homogeneous space.*

Denote by $\pi : G \to G/H$ the *natural map* of G onto G/H (π assigns to $a \in G$ its coset modulo H). Since G and H are Lie groups G/H is a (real) analytic manifold and π is an analytic map. H acts on G by right translations: $(x,a) \to xa$, $x \in G$, $a \in H$. On the other hand, G acts on G/H canonically, since the left translations by G of G commute with the action of H on G. The group G is a Lie transformation group on G/H which is *transitive* and analytic, that is, for any two points on G/H, there is an element of G sending one into the other.

Let σ be a non-trivial involutory automorphism of $G : \sigma^2 = I$, $\sigma \neq I$. Denote by G_σ the subgroup consisting of all elements of G which are invariant under σ and let G_σ^0 denote the component of the identity in G_σ. If H is a closed subgroup of G with G_σ^0 as its component of the identity, G/H is called a *symmetric homogeneous space.*

Let G/H be a symmetric homogeneous space of the compact and connected Lie group G. Then, with respect to an invariant Riemannian metric on G/H an invariant form (by G) is harmonic, and conversely.

In the first place, since G is connected it can be shown by averaging over G that a differential form α on G/H invariant by G is closed. (Since G is transitive, an invariant differential form is uniquely determined by its value at any point of M). Let h be a Riemannian metric on G/H and denote by a^*h the transform of h by $a \in G$. Put

$$g = \int_G (a^*h) *1.$$

Then, g is a metric on G/H invariant by G. In terms of g, $*\alpha$ is also invariant and therefore closed. Thus, α is a harmonic form on G/H.

2. Show that

$$P_n = U(n+1)/\,U(n) \times U(1)$$

is a symmetric homogeneous space.

To see this, we define an involutory automorphism σ of $U(n+1)$ by

$$\sigma\begin{pmatrix} A & B \\ C & D \end{pmatrix} = \begin{pmatrix} A & -B \\ -C & D \end{pmatrix}, \quad A \in U(1), \quad D \in U(n).$$

Then,

$$G_\sigma = \begin{pmatrix} A & O \\ O & D \end{pmatrix} = U(1) \times U(n).$$

3. Prove that the curvature tensor (defined by the invariant metric g) of a symmetric space has vanishing covariant derivative.

Hint: Make essential use of the fact that an invariant form on a symmetric space is a closed form.

F. Bochner's lemma [*4*]

1. Let M be a differentiable manifold and U a coordinate neighborhood of M with the local coordinates $(u^1, \cdots, u^n)$. Consider the elliptic operator

$$L = g^{jk} \frac{\partial^2}{\partial u^j\, \partial u^k} + h^i \frac{\partial}{\partial u^i}$$

on $F(U)$—the algebra of differentiable functions of class 2 on U, where the coefficients g^{jk}, h^i are merely assumed to be continuous functions on U. (The condition that L is elliptic is equivalent to the condition that the matrix (g^{jk}) is positive definite). If for an element $f \in F$: (a) $Lf \geqq 0$ and (b) $f(u^1, \cdots, u^n) \leqq f(a^1, \cdots, a^n)$ for some point $P_0 \in U$ with coordinates $(a^1, \cdots, a^n)$, then $f(u^1, \cdots, u^n) = f(a^1, \cdots, a^n)$ everywhere in U.

This maximum principle is due to E. Hopf [*40*]. The corresponding minimum principle is given by reversing the inequalities. This result should be compared with (V. A. 2).

2. If M is compact and $f \in F(M)$ is a differentiable function (of class 2) for which $Lf \geqq 0$, then f is a constant function on M.

3. If M is a compact Riemannian manifold, then a function $f \in F(M)$ for which $\Delta f \geqq 0$ is a constant function on M.

This is the Bochner lemma [*4*].

Note that M need not be orientable. By applying the Hopf minimum principle the statements 2 and 3 are seen to be valid with the inequalities reversed.

G. Zero curvature

1. The results of § 6.7 may be described in the following manner:

Zero curvature is the integrability condition for the pfaffian system given by the connection forms on the bundle B of unitary frames over M. Hence, there exist integral manifolds; a maximal integral submanifold through a point will be a covering space of the manifold M. These manifolds are locally isometric since the mapping from the horizontal part of the tangent space of B to the tangent space of M is always an isometry (cf. the last paragraph of § 1.8 where in the description of an affine connection W_x is the horizontal part of T_x, by definition, and $(\pi^*(T_P^*))^*$ is the vertical part). Since B is parallelisable into horizontal and vertical fields, the horizontal parallelization yields a local parallelization on M which is covariant constant by the properties of the horizontal parallelization.

An integral manifold is called a *maximal integral manifold* if any integral manifold containing it coincides with it [27].

H. The vanishing theorems

1. Theorem 3.2.1 is a special case of Myers' theorem [62]: *The fundamental group of a compact Riemannian manifold M of positive definite Ricci curvature is finite.* The proof depends on his theorem on conjugate points which was established by means of the second variation of the length integral. It has recently been shown that *if M is Kaehlerian, it is simply connected* [81]. The proof depends on theorem 6.14.1 and the theorem of Riemann-Roch [80].

2. Given a real closed form γ of bidegree (1, 1) belonging to $c(B)$ there exists a system $\{a_\alpha\}$ of positive functions of class ∞ satisfying $a_\alpha = |f_{\alpha\beta}|^2 a_\beta$ in $U_\alpha \cap U_\beta$ such that $\sqrt{-1}\, d'd'' \log a_\alpha = 2\pi\gamma$.

To see this, let $\{a'_\alpha\}$ be a system of positive functions satisfying $a'_\alpha = |f_{\alpha\beta}|^2 a'_\beta$ and set $2\pi\gamma_0 = 2\pi\gamma - \sqrt{-1}\, d'd'' \log a'_\alpha$. Then, since $H[\gamma_0] = 0$, $\gamma_0 = 2\, d''\delta'' G\gamma_0$. Applying (5.6.1), show that $\gamma_0 = 2\sqrt{-1}\ \ d'd''\, \Lambda G\gamma_0$. Finally, put $a_\alpha = a'_\alpha \exp(-4\pi\Lambda G\gamma_0)$.

3. Show that the first betti number of a compact Kaehler manifold with positive definite first Chern class is zero.

I. Cohomology

1. For a compact Kaehler manifold, show that the cohomology groups defined by the differential operators d, d', and d'' are canonically isomorphic.

In the case of an arbitrary complex manifold, it can be shown that the de Rham isomorphism theorem is valid for d''-cohomology.

CHAPTER VII

GROUPS OF TRANSFORMATIONS OF KAEHLER AND ALMOST KAEHLER MANIFOLDS

In Chapter III the study of conformal transformations of Riemannian manifolds was initiated. Briefly, by a conformal map of a Riemannian manifold M is meant a differentiable homeomorphism preserving the metric up to a scalar factor. If the metric is preserved, the transformation is an isometry. The group of all the isometries of M onto itself is a Lie group (with respect to the natural topology). It was shown that the curvature properties of M affect the structure of its group of motions. More precisely, if M is compact, the existence or, rather, non-existence of 1-parameter groups of conformal maps is dependent upon the Ricci curvature of the manifold.

In § 3.8, an infinitesimal conformal transformation of a compact and orientable Riemannian manifold was characterized as a solution of a system of differential equations. This characterization is dependent upon the Ricci curvature, so that, if the curvature is suitably restricted there can be no non-trivial solutions of the system. In an analogous way, an infinitesimal holomorphic transformation X of a compact Kaehler manifold may be characterized as a solution of a differential system. Again, since this system of equations involves the Ricci curvature explicitly, conditions may be given in terms of this tensor under which X becomes an isometry. For example, if the 1st Chern class determined by the 2-form ψ (cf. (5.3.38)) is preserved ($\theta(X)\psi = 0$), X defines an isometry [*58*].

On the other hand, if the scalar curvature is a (positive) constant, the holomorphic vector field X may be expressed as a sum $Y + JZ$ where both Y and Z are Killing vector fields and J is the almost complex structure defining the complex structure of the manifold. If K denotes the subalgebra of Killing vector fields of the Lie algebra L_a of infinitesimal holomorphic transformations, then, under the conditions,

$L_a = K + JK$. In this way, it is seen that the Lie algebra of the group of holomorphic homeomorphisms of a compact Kaehler manifold with constant scalar curvature is reductive [*58*].

Moreover, for a compact Kaehler manifold M, with metric h, let $A_0(M)$ denote the largest connected group of holomorphic homeomorphisms of M and G a maximal compact subgroup. Suppose that the Lie algebra L_a of $A_0(M)$ is semi-simple. For every $a \in G$, let a^*h denote the transform of h by a. Then, since a is a holomorphic homeomorphism a^*h is again a Kaehlerian metric of M and $g = \int_G(a^*h)da$ is a Kaehlerian metric invariant by G. Since G is a maximal compact subgroup of $A_0(M)$, the subalgebra K of L_a corresponding to the subgroup G of $A_0(M)$ coincides with the Lie algebra generated by the Killing vector fields of the Kaehler manifold defined by M and g. Since L_a is a complex and semi-simple Lie algebra, and G is a maximal compact subgroup of $A_0(M)$, the complex subspace of L_a generated by K coincides with L_a, that is $L_a = K + JK$.

Let M be a compact complex manifold whose group of holomorphic homeomorphisms $A(M)$ is transitive. If the fundamental group of M is finite and its Euler characteristic is different from zero, $A(M)$ is semi-simple [*59*].

By an application of theorem 6.13.1, it is shown that a compact complex manifold for which $b_{n,0} = 2$ does not admit a complex Lie group of holomorphic homeomorphisms which is transitive [*9*].

Now, a conformal transformation of a Riemann surface is a holomorphic homeomorphism. For complex manifolds of higher dimension this is not necessarily the case. However, if M is a compact Kaehler manifold of complex dimension $n > 1$, an infinitesimal conformal transformation is holomorphic, if and only if, it is an infinitesimal isometry.

By an automorphism of a Kaehler manifold is meant a holomorphic homeomorphism preserving the symplectic structure. Hence, by theorem 3.7.1, the largest connected Lie group of conformal transformations of a compact Kaehler manifold coincides with the largest connected group of automorphisms of the Kaehlerian structure provided $n > 1$. For $n = 1$, it coincides with the largest connected group of holomorphic homeomorphisms [*58*, *36*].

The problem of determining the most general class of spaces for which an infinitesimal conformal transformation is an infinitesimal isometry is considered. To begin with, a (real analytic) manifold M of $2n$ real dimensions which admits a closed 2-form Ω of maximal rank everywhere is said to be symplectic. Let g be a Riemannian metric of M which commutes with Ω (cf. (5.2.8)). Such an inner product exists at each point of M. Assume that the operator $J : \xi^k \to (i(X)\Omega)^k$ acting in the tangent space at each point defines an almost complex structure on M and, together with g, an almost hermitian structure. If the manifold is symplectic with respect to Ω, the almost hermitian structure is called almost Kaehlerian. In this case, M is said to be an almost Kaehler manifold. Regarding conformal maps of such spaces, it is shown that the largest connected Lie group of conformal transformations coincides with the largest connected group of isometries of the manifold provided the space is compact and $n > 1$ [*36*]. More generally, if M is a compact Riemannian manifold admitting a harmonic form of constant length, then $C_0(M) = I_0(M)$ (cf. § 3.7 and [*78*]).

By considering infinitesimal transformations whose covariant forms are closed the above results may be partially extended to non-compact manifolds. Indeed, let X be a vector field on a Kaehler manifold whose image by J is an infinitesimal conformal map preserving the structure. The vector field X is then 'closed', that is its covariant form (by the duality defined by the metric) is closed. In general, a 'closed conformal map' is a homothetic transformation. In fact, a closed conformal map X of a complete Kaehler manifold (of complex dimension $n > 1$) which is not locally flat is an isometry [*45*]. In the locally flat case, if X is of bounded length, the same conclusion prevails [*42*].

7.1. Infinitesimal holomorphic transformations

In § 5.8, the concept of a holomorphic map is given. Indeed, a differentiable map $f: M \to M'$ of a complex manifold M into a complex mainfold M' is said to be holomorphic if the induced dual map $f^* : \wedge^{*c}(M') \to \wedge^{*c}(M)$ sends forms of bidegree (1,0) into forms of the same bidegree. It follows from this definition that f^* maps holomorphic forms into holomorphic forms. The connection with ordinary holomorphic functions was given in lemma 5.8.1: If $M' = C$, f is a holomorphic map, if and only if, it is a holomorphic function.

Let f be a holomorphic map of M (that is, a holomorphic map of M into itself) and denote by J the almost complex structure defining its complex structure. The structure defined by J is integrable, that is, in a coordinate neighborhood with the complex coordinates (z^i) operating with J is equivalent to sending $\partial/\partial z^i$ and $\partial/\partial \bar{z}^i$ into $\sqrt{-1}\ \partial/\partial z^i$ and $-\sqrt{-1}\ \partial/\partial \bar{z}^i$, respectively. Hence, J is a map sending vector fields of bidegree (1,0) into vector fields of bidegree (1,0), so that at each point $P \in M$

$$f_{*P} J_P = J_{f(P)} f_{*P} \tag{7.1.1}$$

where f_* denotes the induced map in the tangent space T_P at P and J_P is the linear endomorphism defined by J in T_P. Since two complex structures which induce the same almost complex structure coincide, the map f is holomorphic, if and only if, the relation (7.1.1) is satisfied. If the manifold is compact, it is known that the largest group of holomorphic transformations is a complex Lie group, itself admitting a natural complex structure [*13*].

Let G denote a connected Lie group of holomorphic transformations of the complex manifold M. To each element A of the Lie algebra of G is associated the 1-parameter subgroup a_t of G generated by A. The corresponding 1-parameter group of transformations R_{a_t} on $M(R_{a_t} P = P \cdot a_t, P \in M)$ induces a (right invariant) vector field X on M. From the action on the almost complex structure J, it follows that $\theta(X)J$ vanishes where $\theta(X)$ is the operator denoting Lie derivation with respect to the vector field X and J denotes the tensor field of type (1,1) defined by the linear endomorphism J. On the other hand, a vector field on M satisfying the equation

$$\theta(X)J = 0 \tag{7.1.2}$$

generates a local 1-parameter group of local holomorphic transformations of M.

An *infinitesimal holomorphic transformation* or *holomorphic vector field* X is an infinitesimal transformation defined by a vector field X satisfying (7.1.2).

In order that a connected Lie group G of transformations of M be a group of holomorphic transformations, it is necessary and sufficient that the vector fields on M induced by the 1-parameter subgroups of G define infinitesimal holomorphic transformations. If M is complete, an example due to E. Cartan [*19*] shows that not every infinitesimal holomorphic transformation generates a 1-parameter global group of holomorphic transformations of M.

Let L_a denote the set of all holomorphic vector fields on M. It is a subalgebra of the Lie algebra of all vector fields on M. If M is compact, L_a is finite dimensional and may be identified with the algebra of the group $A(M)$ of holomorphic transformations of M.

Lemma 7.1.1. *Let X be an infinitesimal holomorphic transformation of a Kaehler manifold. Then, the vector field X satisfies the system of differential equations*

$$F^A{}_C D_B \xi^C - F^C{}_B D_C \xi^A = 0 \tag{7.1.3}$$

where, in terms of a system of local coordinates (u^A), $A = 1, \cdots, 2n$, $X = \xi^A \partial/\partial u^A$, the $F^A{}_B$ denote the components of the tensor field defined *by J, and D_A indicates covariant differentiation with respect to the connection canonically defined by the Kaehler metric.*

We denote by the same symbol J the tensor field of type (1,1) defined by the linear endomorphism J:

$$J = F^A{}_B \frac{\partial}{\partial u^A} \otimes du^B.$$

(Note that we have written J in place of the tensor $\bar{J}$ of § 5.2.) Then,

$$\theta(X)J = (X F^A{}_B) \frac{\partial}{\partial u^A} \otimes du^B + F^A{}_B \left[X, \frac{\partial}{\partial u^A}\right] \otimes du^B + F^A{}_B \frac{\partial}{\partial u^A} \otimes dXu^B$$

$$= \xi^C \frac{\partial F^A{}_B}{\partial u^C} \frac{\partial}{\partial u^A} \otimes du^B + F^A{}_B \left[\xi^C \frac{\partial}{\partial u^C}, \frac{\partial}{\partial u^A}\right] \otimes du^B + F^A{}_B \frac{\partial \xi^B}{\partial u^C} \frac{\partial}{\partial u^A} \otimes du^C$$

$$= \left(\xi^C \frac{\partial F^A{}_B}{\partial u^C} - F^C{}_B \frac{\partial \xi^A}{\partial u^C} + F^A{}_C \frac{\partial \xi^C}{\partial u^B}\right) \frac{\partial}{\partial u^A} \otimes du^B$$

$$= \left(\xi^C D_C F^A{}_B + F^A{}_C D_B \, \xi^C - F^C{}_B D_C \, \xi^A \right) \frac{\partial}{\partial u^A} \otimes du^B.$$

Since the connection is canonically defined by the Kaehler metric, and $F_{ij^*} = \sqrt{-1}\, g_{ij^*}$ in terms of a J-basis (cf. (5.2.11)), $D_k F^i{}_j = 0$. Finally, since X is a holomorphic vector field, $\theta(X)J$ vanishes.

Corollary 1. *An infinitesimal holomorphic transformation X of a Kaehler manifold satisfies the relation*

$$\theta(X)JY = J\theta(X)Y$$

for any vector field Y.

Indeed, for any vector fields X and Y

$$(\theta(X)Y)^A = [X,Y]^A = \xi^C \frac{\partial \eta^A}{\partial u^C} - \eta^C \frac{\partial \xi^A}{\partial u^C}$$

$$= \xi^C \, D_C \, \eta^A - \eta^C \, D_C \, \xi^A.$$

Taking account of the fact that the covariant derivative of J vanishes the relation follows by a straightforward computation.

Corollary 2. *In terms of a system of local complex coordinates a holomorphic vector field satisfies the system of differential equations*

$$\frac{\partial \xi^i}{\partial \bar{z}^j} = 0, \quad i,j = 1, \cdots, n.$$

This follows from the fact that the coefficients of connection $\Gamma^i_{Aj^*}$ vanish.

It is easily checked that

$$\theta(JX)J = J\theta(X)J.$$

Therefore, if X is an infinitesimal holomorphic transformation, so is JX, and dim L_a is even.

In the sequel, we denote the covariant form of $\theta(X)J$ by $t(X)$, that is

$$t(X)_{AB} = g_{AC}(\theta(X)J)^C{}_B.$$

Lemma 7.1.2. *For any vector field X on a Kaehler manifold with metric g and fundamental 2-form Ω*

$$t(X) = J\theta(X)g + \theta(X)\Omega,$$

where by $\theta(X)\Omega$ we mean here the covariant tensor field defined by the 2-form $\theta(X)\Omega$.

For,

$$\begin{aligned} -t(X)_{AB} &= F^C{}_B(D_C\,\xi_A - D_A\,\xi_C + D_A\,\xi_C) + F^C{}_A\,D_B\,\xi_C \\ &= F^C{}_B(\theta(X)g)_{AC} + D_B(i(X)\Omega)_A - D_A(i(X)\Omega)_B \\ &= -(J\theta(X)g)_{AB} - (di(X)\Omega)_{AB} \end{aligned}$$

(cf. formula (5.2.10)).

Lemma 7.1.3. *A vector field X defines an infinitesimal holomorphic transformation of a Kaehler manifold, if and only if,*

$$J\theta(X)\Omega = \theta(X)g,$$

that is, when applied to the fundamental form the operators $\theta(X)$ and J commute or, when applied to the metric tensor, they commute.

This follows from the previous lemma, since $J\Omega = g$.

Let X be an infinitesimal holomorphic transformation of the Kaehler manifold M. Then, by the second corollary to lemma 7.1.1, $\partial\xi^i/\partial\bar{z}^j = 0$. But these equations have the equivalent formulation

$$D_{j^*}\,\xi^i = 0$$

since the coefficients of connection $\Gamma^i_{Aj^*}$ vanish. Hence, a necessary and sufficient condition that the vector field X be an infinitesimal holomorphic transformation is that it be a solution of the system of differential equations

$$D_j\,\xi_i = 0. \tag{7.1.4}$$

With this formulation (in local complex coordinates) of an infinitesimal holomorphic transformation we proceed to characterize these vector fields as the solutions of a system of second order differential equations.

To every real 1-form α, we associate a tensor field $a(\alpha)$ whose vanishing characterizes an infinitesimal holomorphic transformation (by means of the duality defined by the metric). Indeed, if $\alpha = \alpha_A dz^A$, we define $a(\alpha)$ by

$$a(\alpha)_{ij} = D_i \, \alpha_j, \quad a(\alpha)_{ij^*} = a(\alpha)_{j^* i} = 0, \quad a(\alpha)_{i^* j^*} = D_{i^*} \alpha_{j^*}.$$

Now, from

$$(\Delta\alpha)_A = - g^{BC} D_C D_B \alpha_A + R_{AB} \alpha^B$$

we obtain

$$(\Delta\alpha)_i = - g^{j^*k} D_k D_{j^*} \alpha_i - g^{jk^*} D_{k^*} D_j \alpha_i + R_{ij^*} \alpha^{j^*}. \tag{7.1.5}$$

Transvecting the Ricci identity

$$D_k D_{j^*} \alpha_i - D_{j^*} D_k \alpha_i = \alpha_l R^l{}_{ikj^*}$$

with g^{kj^*} we obtain

$$g^{j^*k} D_k D_{j^*} \alpha_i - g^{k^*j} D_{k^*} D_j \alpha_i = R_{ij^*} \alpha^{j^*}. \tag{7.1.6}$$

Hence, from (7.1.5) and (7.1.6)

$$(\Delta\alpha - 2Q\alpha)_i = - 2 g^{jk^*} D_{k^*} D_j \alpha_i.$$

From the definition of $a(\alpha)$, it follows that

$$(\Delta\alpha - 2Q\alpha)_i = - 2 g^{jk^*} D_{k^*} a(\alpha)_{ji}.$$

Hence, if $a(\alpha) = 0, \Delta\alpha = 2Q\alpha$. If M is compact, the converse is also true. To see this, define the auxiliary vector field $b(\alpha)$ by

$$b(\alpha)_j = \alpha^i a(\alpha)_{ji}.$$

Then, by means of a computation analogous to that of § 3.8

$$2\delta b(\alpha) = \langle \Delta\alpha - 2Q\alpha, \alpha \rangle - 4 \langle a(\alpha), a(\alpha) \rangle.$$

If we assume that M is compact, then, by integrating both sides of this relation and applying Stokes' formula, we obtain the integral formula

$$(\Delta\alpha - 2Q\alpha, \alpha) = 4(a(\alpha), a(\alpha)). \tag{7.1.7}$$

Theorem 7.1.1. *On a compact Kaehler manifold, a necessary and sufficient condition that a 1-form define an infinitesimal holomorphic transformation (by means of the duality defined by the metric) is that it be a solution of the equation*

$$\Delta \xi = 2Q\xi. \tag{7.1.8}$$

[76].

The fact that this equation involves the Ricci curvature (of the Kaehler metric) explicitly will be particularly useful in the study of the structure of the group of holomorphic transformations of Kaehler manifolds with specific curvature properties.

If a vector field X generates a 1-parameter group of motions of a compact Kaehler manifold, then, by theorem 3.8.2, cor.

$$\Delta \xi = 2Q\xi \text{ and } \delta\xi = 0. \tag{7.1.9}$$

Hence,

Corollary. *An infinitesimal isometry of a compact Kaehler manifold is a holomorphic transformation.*

In terms of the 2-form ψ defining the 1st Chern class of the compact Kaehler manifold M

$$Q\xi = -2\pi\, i(JX)\psi$$

for any vector field X on M. The equation (7.1.8) may then be written in the form

$$\Delta \xi = -4\pi i(JX)\psi. \tag{7.1.10}$$

Taking the exterior derivative of both sides of this relation we obtain, by virtue of the fact that ψ is a closed form,

$$\Delta d\xi = -4\pi\theta(JX)\psi. \tag{7.1.11}$$

Let $Y = JX$ be an infinitesimal holomorphic transformation preserving ψ. Then, since $X = -JY$, equation (7.1.11) yields

$$\Delta\theta(Y)\Omega = 4\pi\theta(Y)\psi = 0.$$

Hence, $\theta(Y)\Omega$ is a harmonic 2-form. But $\theta(Y)\Omega = di(Y)\Omega$. Thus, since a harmonic form which is exact must vanish, $i(Y)\Omega$ is a closed 1-form. Applying the Hodge-de Rham decomposition theorem

$$i(Y)\Omega = df + H[i(Y)\Omega] \tag{7.1.12}$$

for some real function f of class ∞.

Define the map $C : \wedge^1(M) \to \wedge^1(M)$ associated with J as follows

$$C\xi = i(\xi)\Omega.$$

Since $F^i{}_j F^j{}_k = -\delta^i_k$,

$$C^2 \equiv CC = -I.$$

The relation (7.1.12) may now be re-written as

$$C\eta = df + H[C\eta], \tag{7.1.13}$$

where η is the covariant form for Y. Applying the operator C to (7.1.13) we obtain

$$\eta = -Cdf + CH[C\eta].$$

Since df is a gradient field and $H[C\eta]$ is a harmonic 1-form, $\delta\eta$ vanishes (cf. lemma 7.3.2). We have proved

Theorem 7.1.2. *If an infinitesimal holomorphic transformation of a compact Kaehler manifold preserves the 1st Chern class it is an infinitesimal isometry* [*58*].

7.2. Groups of holomorphic transformations

The set L_a of all holomorphic vector fields on a compact complex manifold is a finite dimensional Lie algebra. As a vector space it may be given a complex structure in the following way: If $X, Y \in L_a$ so do JX and JY, and by lemma 7.1.1 (see remark in VII. A. 1),

$$J([X,Y]) = [X,JY] = [JX,Y];$$

the complex structure is defined by putting $\sqrt{-1}X = JX$ for every $X \in L_a$. Clearly, then, $J^2X = -X$ for all X, that is $J^2 = -I$ on L_a.

Let K denote the Lie algebra of Killing vector fields on the compact Kaehler manifold M. Since M is compact it follows from the corollary to theorem 7.1.1 that K is a subalgebra of L_a. We seek conditions on the Kaehlerian structure of M in order that the complex subspace of L_a generated by K coincides with L_a.

Let K be an arbitrary subalgebra of a Lie algebra L. The derivations $\theta(X)$, $X \in K$ define a linear representation of K with representation space $\wedge(L)$—the Grassman algebra over L. If this representation is completely reducible, K is said to be a *reductive subalgebra* of L or to be *reductive* in L. A Lie algebra L is said to be *reductive* if, considered as a subalgebra of itself, it is reductive in L [*48*].

Let K be a reductive subalgebra of L and H a subalgebra of L containing K. For every $X \in K$, the extension $\phi : \wedge(H) \to \wedge(L)$ of the identity map of H into L satisfies

$$\phi\theta(X) = \theta(\phi X)\phi.$$

Since ϕ is an isomorphism, it follows that the inverse image by ϕ of an irreducible subspace of $\wedge(L)$ invariant by K is an irreducible subspace of $\wedge(H)$ invariant by K. We conclude that K is reductive in H. In particular, a reductive subalgebra of L is reductive.

It can be shown, if L is the Lie algebra of a compact Lie group, that every subalgebra of L is a reductive subalgebra. In particular, L is then also reductive.

Now, let M be a compact Kaehler manifold and assume that its Lie algebra of holomorphic vector fields L_a is generated by the subalgebra K of Killing fields. More precisely, assume that

$$L_a = K + JK.$$

Then, the complex subspace of L_a generated by K coincides with L_a. Since M is compact, the largest group of isometries is compact. Hence, the Lie algebra K is reductive; in addition, its complexification K^c is also reductive. Since $L_a = K + JK$, there is a natural homomorphism of K^c on L_a and, therefore, L_a is a reductive Lie algebra. The last statement follows from the fact that the homomorphic image of a reductive Lie algebra is a reductive Lie algebra.

Lemma 7.2.1. *If the Lie algebra L_a of holomorphic vector fields on a compact Kaehler manifold can be represented in the form*

$$L_a = K + JK$$

where K is the Lie algebra of Killing vector fields, then L_a is a reductive Lie algebra.

As a consequence, if the manifold is a Kaehler-Einstein manifold, we may prove

Theorem 7.2.1. *The Lie algebra of the group of holomorphic transformations of a compact Kaehler-Einstein manifold is reductive* [59].

For an element $X \in L_a$, $\Delta\xi = 2Q\xi = c\xi$ for some constant c since the manifold is an Einstein space. By the Hodge decomposition of a 1-form $\xi = df + \delta\alpha + H[\xi]$ for some function f of class ∞ and 2-form α. Applying Δ to both sides of this relation, we obtain $\Delta\xi = d\Delta f + \delta\Delta\alpha$

and, since $\Delta\xi = c\xi$, $d\Delta f + \delta\Delta\alpha = dcf + \delta c\alpha + cH[\xi]$. Thus, $d(\Delta f - cf) + \delta(\Delta\alpha - c\alpha) - H[c\xi] = 0$; again, by the decomposition theorem, $d(\Delta f - cf) = 0$ and $\delta(\Delta\alpha - c\alpha) = 0$, that is $\Delta df = cdf$, $\Delta\delta\alpha = c\delta\alpha$. Consequently, the (contravariant) vector fields defined by df and $\delta\alpha$ (due to the duality defined by the metric) are holomorphic. But $\delta\delta\alpha = 0$, and so by the corollary to theorem 3.8.2, $\eta = \delta\alpha$ defines a Killing vector field. Since df is a gradient field, the 1-form $-\ \zeta = Cdf$ has zero divergence. Thus,

$$\xi = \eta + C\zeta + H[\xi]$$

where η and ζ define Killing fields.

If $c > 0$, $H[\xi]$ vanishes by theorem 3.2.1. If $c = 0$, the Ricci curvature vanishes, and therefore $\Delta\xi = 2Q\xi = 0$. ξ is thus harmonic, and so $\delta\xi = 0$, that is, ξ defines a Killing field. If $c < 0$, $\eta = \zeta = 0$ by theorem 3.8.1, that is ξ is harmonic, and consequently defines a Killing field. In all cases, ξ is of the form $\eta + C\zeta$.

Conversely, if η and ζ define Killing fields, $\xi = \eta + C\zeta$ defines a holomorphic vector field.

Lemma 7.2.2. *A necessary and sufficient condition that a Lie algebra L over R be reductive is that it be the direct sum of a semi-simple Lie algebra and an abelian Lie algebra* [*48*].

If L is reductive, the endomorphisms $ad(X)$ which are the restrictions of $\theta(X)$ to $\wedge^1(L)$ define a completely reducible linear representation of L. The L-invariant subspaces of $\wedge^1(L)$ are therefore the ideals of L. Moreover, L is the direct sum of the derived algebra L' of L and an ideal C (supplementary to L') belonging to the center of L. Let K be the radical of L'. Since K is an ideal of L, there exists an ideal of L supplementary to K. Therefore, the derived algebra K' of K is the intersection of K with L'. Hence, $K' = K$ and thus $K = \{0\}$. We conclude that L' is semi-simple and C the center of L.

Conversely, let L be the direct sum of a semi-simple Lie algebra and an abelian Lie algebra. Then, the endomorphisms $\theta(X)$ define a linear representation of the semi-simple part since $\theta(X)$ vanishes on the abelian summand. Since this representation is completely reducible, L is reductive.

We have seen that the Lie algebra of the group of holomorphic transformations of a compact Kaehler-Einstein manifold is reductive. It is now shown that the group $A(M)$ of holomorphic transformations of a compact complex manifold M, with no restriction on the metric, but with the topology of the manifold suitably restricted, is a semi-simple Lie group, and hence the Lie algebra of $A(M)$ is reductive.

Theorem 7.2.2. *If the group of holomorphic transformations $A(M)$ of a compact complex manifold M with finite fundamental group and non-vanishing Euler characteristic is transitive, it is a semi-simple Lie group* [*59*].

Since M is a connected manifold and $A(M)$ is transitive, the component of the identity $A_0(M)$ of $A(M)$ is transitive on M. Let G be a maximal compact subgroup of $A_0(M)$. Then, since M is compact and has a finite fundamental group, G is also transitive on M [*61*]. Let B be the isotropy subgroup of G at a point P of M. Since the Euler characteristic of M is different from zero, B is a subgroup of G of maximal rank [*41*]. Since G is effective on M it must be semi-simple; for, otherwise B contains the center of G. Applying a theorem due to Koszul [*49*], M admits, as a result, a Kaehler-Einstein metric invariant by G. It follows from the proof of theorem 7.2.1 that $L_a = K + JK$ where L_a is the Lie algebra of $A_0(M)$ and K the Lie algebra of G. Finally, since K is semi-simple, L_a is also semi-simple.

7.3. Kaehler manifolds with constant Ricci scalar curvature

The main results of the previous section are now extended to manifolds with metric not necessarily a Kaehler-Einstein metric.

To begin with let $\tau(X)$ denote the 2-form corresponding to the skew-symmetric part of $t(X)$ (cf. § 7.1). Then, by a straightforward application of lemma 7.1.2 and equation (3.7.11) we obtain

Lemma 7.3.1. *For any vector field X on a Kaehler manifold*

$$\bar{\theta}(X)\Omega - \theta(X)\Omega = \delta\xi \cdot \Omega - 2\tau(X).$$

We shall require the following

Lemma 7.3.2. *On a Kaehler manifold*

$$\Delta C = C\Delta \quad \text{and} \quad QC = CQ.$$

The first relation follows from the fact that the covariant derivative of J vanishes, and the second is a consequence of the relation

$$R_{ij^*}F^i{}_k F^{j^*}{}_{l^*} = R_{kl^*}.$$

which may be established as follows. In the first place,

$$D_D D_C F^A{}_B - D_C D_D F^A{}_B = F^N{}_B R^A{}_{NCD} - F^A{}_N R^N{}_{BCD}.$$

Hence,

$$F^N{}_B R^A{}_{NCD} = F^A{}_N R^N{}_{BCD},$$

that is,

$$F^N{}_B R_{NACD} = F^N{}_A R_{NBCD}$$

or,

$$R_{ABCD} = F^K{}_A F^L{}_B R_{KLCD}.$$

Thus, in terms of a J-basis

$$R_{kl^*ij^*} = F^a{}_k F^{b^*}{}_{l^*} R_{ab^*ij^*}.$$

The desired result is obtained by transvecting with g^{ij^*}.

This may also be seen as follows: Since the affine connection preserves the almost complex structure J, and the curvature tensor (which, as we have seen is an endomorphism of the tangent space) is an element of the holonomy algebra [*63*], it becomes clear that J and $R(X, Y)$ commute (cf. VI.A.1).

As an immediate consequence, we obtain a previous result:

Corollary 1. *If X is a holomorphic vector field so is JX.*

Corollary 2. *On a compact Kaehler manifold the operators C and H commute.*

This follows from the fact that $\xi = \Delta G\xi + H[\xi]$ for any p-form ξ. For, then, $C\xi = \Delta CG\xi + CH[\xi]$. But $C\xi = \Delta GC\xi + H[C\xi]$. Hence, $\Delta(GC\xi - CG\xi) = CH[\xi] - H[C\xi]$, and so, by § 2.10, the right-hand side is orthogonal to $\wedge^p_H(T^{c*})$ and therefore must vanish.

Let $X \in L_a$—the Lie algebra of holomorphic vector fields on the compact Kaehler manifold M. Then, as in the proof of theorem 7.2.1, decompose the 1-form ξ:

$$\xi = \eta + \zeta \qquad (7.3.1)$$

where η is co-closed and ζ exact, that is $\eta = \delta\alpha + H[\xi]$ and $\zeta = df$. We show that $\theta(\eta)\Omega$ vanishes. Indeed, by lemma 7.3.1

$$\bar{\theta}(X)\Omega - \theta(X)\Omega = \delta\xi\cdot\Omega.$$

Applying δ to both sides of this relation, we derive

$$\delta\theta(X)\Omega = Cd\delta\xi$$

(see proof of theorem 7.5.1). Hence, from (7.3.1), $\delta\theta(\eta)\Omega + \delta\theta(\zeta)\Omega = Cd\delta\zeta$. Taking the global scalar product of this relation with $C\eta$,

$$\|\, \theta(\eta)\Omega \,\|^2 + (\theta(\zeta)\Omega,\, \theta(\eta)\Omega) = 0$$

where we have employed the notation $\|\, \alpha \,\|^2 = \int_M \alpha \wedge *\alpha$. But $(\theta(\zeta)\Omega,\, \theta(\eta)\Omega) = (\delta dC\zeta,\, C\eta) = (\delta dCdf,\, C\eta) = (\Delta Cdf,\, C\eta) = (C\Delta df,\, C\eta) = (\Delta df,\, \eta) = (\Delta df,\, \delta\alpha + H[\xi]) = (\Delta df,\, \delta\alpha) = (d\Delta df,\, \alpha) = (\Delta ddf,\, \alpha) = 0$.

Since $\theta(\eta)\Omega = dC\eta$, it follows that

$$D_{j^*}\eta_i + D_i\eta_{j^*} = 0.$$

Consequently, since

$$2\delta Q\eta = -\, 2D_A(R^{AB}\eta_B) = -\, \eta^B \frac{\partial R}{\partial u^B} - R^{AB}(D_B\eta_A + D_A\eta_B),$$

we deduce from the previous statement that

$$2\delta Q\eta = -\, \langle \eta, dR \rangle.$$

Hence, assuming $R =$ const.,

$$\delta Q\eta = 0.$$

Thus, since $\Delta\eta$ is co-closed, so is

$$\Delta\zeta - 2Q\zeta = -\, (\Delta\eta - 2Q\eta).$$

Applying formula (7.1.7) to the 1-form ζ, we obtain

$$0 = (\Delta\zeta - 2Q\zeta, \zeta) = 4(a(\zeta), a(\zeta))$$

since ζ is exact. Hence, ζ defines a holomorphic vector field, and consequently so does η. In fact, η defines a Killing vector field.

We show that $H[\xi]$ has vanishing covariant derivative. In the first place, since $dC\eta = 0$ and $C\eta = C\delta\alpha + H[C\xi]$, $C\delta\alpha$ is closed. Thus, $C\delta\alpha = \zeta' + H[C\delta\alpha]$ where ζ' is exact. It follows, as above, that $H[C\xi + C\delta\alpha]$ defines a holomorphic vector field. Hence, $H[\xi + \delta\alpha]$ defines a holomorphic vector field. But $H[\xi + \delta\alpha] = H[\xi]$. Applying (7.1.4), the result follows.

Summarizing, we have the following generalization of theorem 7.2.1:

Theorem 7.3.1. *The Lie algebra of the group of holomorphic transformations of a compact Kaehler manifold with (positive) constant scalar curvature is reductive. Moreover, the harmonic part of a 1-form defining an infinitesimal holomorphic transformation has zero covariant differential* [58].

Corollary. *If M is a homogeneous Kaehlerian space of a compact Lie group G of holomorphic transformations of M, the Lie algebra of G is reductive.*

This follows from the fact that the manifold M with the invariant Kaehlerian metric (by G) constructed from the (original) metric of M has constant scalar curvature (cf. VI.E.1 and proof of theorem 3.7.5).

In particular, if the group of holomorphic transformations $A(M)$ is transitive and the fundamental group of M is finite, M is a homogeneous Kaehlerian space of a compact Lie group G [58]. For, then, a maximal compact subgroup G of the component of the identity of the group $A(M)$ operates transitively on M.

7.4. A theorem on transitive groups of holomorphic transformations

In this section, it is shown if the dimension of the vector space of holomorphic n-forms of a compact complex manifold M of complex dimension n is suitably restricted, M cannot admit a transitive group of holomorphic transformations.

To begin with, we state the special case of theorem 6.13.1:

Let $\alpha^r = a_i^{(r)} dz^i$, $r = 1, \cdots, N$ be $N > n$ holomorphic differentials on the compact complex manifold M with the property: 'For any system of constants c^r (not all zero), the rank of the matrix $(a_i^{(r)}, c^r)_{r=1,\cdots,N; i=1,\cdots,n}$ has its maximum value $n + 1$ at some point.' Then, there are no (non-trivial) holomorphic vector fields on M.

We generalize this statement in the following manner:

Let t and t' be holomorphic tensor fields of type (s, r) and (r, s), respectively. They each have n^{r+s} components which we denote by ξ_α and η^α, respectively, $\alpha = 1, \cdots, n^{r+s}$, in a fixed ordering, that is, by ξ_α we mean the component $\xi_{\alpha(1)\ldots\alpha(r)}{}^{\alpha(r+1)\ldots\alpha(r+s)}$ and by η^α the component $\eta^{\alpha(1)\ldots\alpha(r)}{}_{\alpha(r+1)\ldots\alpha(r+s)}$. Now, since t and t' are holomorphic, the product $\xi_\alpha \eta^\alpha$ is a constant. Thus,

Theorem 7.4.1. *Let t^m, $m = 1, \cdots, N$ be $N > n^{r+s}$ holomorphic tensor fields of type (r, s) on the compact complex manifold M with the property:*

'For any system of constants c^m (not all zero) the rank of the matrix $(\xi^{(m)}_{\alpha}, c^m)_{m=1,\ldots,N;\alpha=1,\ldots,n^{r+s}}$ *is* $n^{r+s}+1$ *at some point.' Then, there are no (non-trivial) holomorphic tensor fields of type* (s, r) *on* M [9].

If the tensor fields have symmetries, the integer N can be reduced. In particular, if ϕ^m, $m = 1,2$ are two holomorphic n-forms, the number of components of the coefficients of each is essentially one, and we have

Corollary 1. *A compact complex manifold for which* $b_{n,0}(M) = 2$ *cannot carry a (non-trivial) skew-symmetric holomorphic contravariant tensor field of order* n.

Corollary 2. *A compact complex manifold for which* $b_{n,0}(M) = 2$ *does not admit a transitive Lie group of holomorphic transformations.*

For, by the previous corollary, M does not admit n independent holomorphic vector fields (locally).

7.5. Infinitesimal conformal transformations. Automorphisms

Conformal transformations of Riemannian manifolds were studied in Chapter III. The problem of determining when an infinitesimal conformal transformation is an infinitesimal isometry was omitted. In this, as well as the following section, this problem is studied for compact manifolds. Indeed, it is shown that for a rather large class of Riemannian manifolds, an infinitesimal conformal transformation is an infinitesimal isometry. This class includes the so-called almost Kaehler manifolds which, as the name signifies, are more general than Kaehler manifolds.

Consider a $2n$-dimensional real analytic manifold M admitting a 2-form Ω of rank $2n$ everywhere. If Ω is closed, the manifold is said to be *symplectic*. Assume that M admits a metric g such that

$$g(JX,JY) = g(X,Y),$$

that is, assume g defines an hermitian structure on M admitting Ω as *fundamental 2-form*—the 'almost complex structure' J being determined by g and Ω: $g(X,Y) = \Omega(X,JY)$ (cf. VII.B). The manifold M with metric g and almost complex structure J is called an *almost hermitian manifold* (Ω need not be closed). If the manifold is symplectic with respect to Ω, the almost hermitian structure is said to be *almost Kaehlerian*. In this case, M is called an *almost Kaehler manifold.*

Lemma 7.5.1. *In an almost Kaehler manifold with metric g the fundamental form Ω is both closed and co-closed.*

In the first place, the Riemannian connection of g is defined by the (self adjoint) forms $\theta^A{}_B$:

$$\theta^{i*}_{j*} = \overline{\theta^i{}_j}, \quad \theta^{i*}_j = \overline{\theta^i{}_{j*}},$$

$$\theta^i{}_j + \overline{\theta^j{}_i} = 0$$

in the bundle of unitary frames (cf. § 5.3). Since this connection is torsion free

$$d\theta^A = \theta^C \wedge \theta^A{}_C;$$

consequently, in terms of the complex coframes (θ^i, θ^{i*}), $i = 1, \cdots, n$

$$d\theta^i = \theta^k \wedge \theta^i{}_k + \theta^{k*} \wedge \theta^i{}_{k*}$$

and

$$d\theta^{i*} = \theta^k \wedge \theta^{i*}_k + \theta^{k*} \wedge \theta^{i*}_{k*}.$$

We put

$$\theta^A{}_B = \Gamma^A_{BC}\theta^C.$$

Then, since

$$F_{ij*} = \sqrt{-1}\, g_{ij*}$$

(where the g_{ij*} are the components of g with respect to the coframes (θ^i, θ^{i*})), and the connection is a metrical connection

$$D_k F_{ij*} = \sqrt{-1}\, D_k g_{ij*} = 0$$

where D_k denotes covariant differentiation with respect to the Riemannian connection. Moreover, it can be shown that $D_k F^i{}_{j*} = 2\sqrt{-1}\,\Gamma^i_{j*k}$ and $D_{k*}F^i{}_{j*} = 2\sqrt{-1}\,\Gamma^i_{j*k*}$. Hence, since Ω is closed, it follows from (2.12.2) that

$$D_k F_{ij*} + D_i F_{j*k} + D_{j*} F_{ki} = 0.$$

Thus, since $D_k F_{ij*} = 0$,

$$D_{j*} F_{ki} = 0.$$

In conclusion, then,

$$-(\delta\Omega)_i = g^{jk^*}D_{k^*}F_{ji} + g^{j^*k}D_kF_{j^*i} = 0.$$

If J defines a completely integrable almost complex structure, M is Kaehlerian (cf. § 5.2). A Kaehler manifold is therefore an hermitian manifold which is symplectic for the fundamental 2-form of the hermitian structure.

We have seen that on a compact and orientable Riemannian manifold M the Lie derivative of a harmonic form with respect to a Killing vector field X vanishes. If M is Kaehlerian, the 1-parameter group of isometries φ_t generated by X preserves the fundamental 2-form Ω, that is

$$\varphi_t^*\Omega = \Omega. \tag{7.5.1}$$

Moreover, from theorem 7.1.1, cor., φ_t is a holomorphic transformation for each t, and so from (7.1.1) $\varphi_t^* J\Omega = J\varphi_t^*\Omega$. This may also be seen in the following way

$$\varphi_t^* J\Omega = \varphi_t^* g = g = J\Omega = J\varphi_t^*\Omega$$

by (7.5.1).

A holomorphic transformation f preserving the symplectic structure (that is, for which $f^*\Omega = \Omega$) will be called an *automorphism of the Kaehlerian structure*. A holomorphic vector field satisfying the equation $\theta(X)\Omega = 0$ will be called an *infinitesimal automorphism of the Kaehlerian structure*.

Now, an infinitesimal isometry is an infinitesimal conformal transformation. The converse, however, is not necessarily true. For, a conformal map X of a Riemann surface S with the conformally invariant metric (see p. 158) need not be an infinitesimal isometry. In any case, the vector field X defines an infinitesimal holomorphic transformation of S. For higher dimensional compact manifolds however, we prove

Theorem 7.5.1. *An infinitesimal conformal transformation of a compact Kaehler manifold of complex dimension $n > 1$ is an infinitesimal isometry* [*57, 35*].

This statement is also an immediate consequence of theorem 3.7.4. From equation (3.7.12)

$$\theta(X)\Omega + \bar{\theta}(X)\Omega = \left(1 - \frac{2}{n}\right) \delta\xi\cdot\Omega.$$

Applying the operator δ to both sides of this relation we derive since $\bar{\theta}(X)$ and δ commute and Ω is co-closed

$$\begin{aligned}\delta\theta(X)\Omega &= \left(1 - \frac{2}{n}\right) \delta(\delta\xi \cdot \Omega)\\ &= -\left(1 - \frac{2}{n}\right) D_B(\delta\xi \cdot F^B{}_A)\, du^A\\ &= -\left(1 - \frac{2}{n}\right) C d\delta\xi.\end{aligned}$$

Taking the global scalar product with $C\xi$, we have

$$(\delta\theta(X)\Omega, C\xi) = (\theta(X)\Omega, dC\xi) = \|\, \theta(X)\Omega \,\|^2$$

and

$$(Cd\delta\xi, C\xi) = (d\delta\xi, \xi) = \|\, \delta\xi \,\|^2.$$

Hence,

$$\|\, \theta(X)\Omega \,\|^2 = -\left(1 - \frac{2}{n}\right) \|\, \delta\xi \,\|^2.$$

Thus, for $n > 1$, since one side is non-positive and the other non-negative, we conclude that $\theta(X)\Omega$ vanishes. For $n > 2$, it is immediate that $\delta\xi = 0$, that is, X is an infinitesimal isometry, whereas for $n = 2$, a previous argument gives the same result.

Corollary. *The largest connected Lie group of conformal transformations of a compact Kaehler manifold of complex dimension $n > 1$ coincides with the largest connected group of automorphisms of the Kaehlerian structure. For $n = 1$, it coincides with the largest connected group of holomorphic transformations. Moreover, in this case, in terms of the norm $\|\ \|$ defined by the Kaehler metric,*

$$\|\, \theta(X)\Omega \,\| = \|\, \delta\xi \,\|.$$

This is an immediate consequence of lemma 7.1.3; for, an infinitesimal automorphism of a Kaehler manifold is an infinitesimal isometry.

We give a proof of theorem 7.5.1 which, although valid only for the dimensions $4k$ is instructive since it involves the hermitian structure in an essential way [35]. In the first place, by lemma 5.6.8, Ω^k is a harmonic $2k$-form. Applying theorem 3.7.3, it follows that $\theta(X)\Omega^k = 0$. Now, since $\theta(X)$ is a derivation, $\theta(X)\Omega^k = k\theta(X)\Omega \wedge \Omega^{k-1}$, and so by corollary 5.7.2, $L^{k-2}\theta(X)\Omega$ vanishes. It follows by induction that $\theta(X)\Omega$ vanishes, that is X defines an infinitesimal isometry of the manifold.

The operators L and Δ do not commute, in general, for almost Kaehlerian manifolds. However, it can be shown that Ω^k is harmonic in this case as well.

Theorem 7.5.1 may be extended to the almost Kaehler manifolds without restriction. For, the proof of this theorem does not involve the complex structure of the manifold, but rather, its almost complex structure. In fact, insofar as the fundamental form is concerned only the facts that it is closed and co-closed are utilized. That the covariant differential of Ω vanishes has no bearing on the result. Hence,

Theorem 7.5.2. *An infinitesimal conformal transformation of a compact almost Kaehler manifold of dimension 2n, $n > 1$ is an infinitesimal isometry [36, 68].*

Note that theorem 7.5.2 follows directly from theorem 3.7.4. For, Ω is harmonic and $\langle\Omega, \Omega\rangle$ is a constant.

Corollary. *The largest connected Lie group of conformal transformations of a compact almost Kaehler manifold of dimension 2n, $n > 1$ coincides with the largest connected group of isometries of the manifold.*

Remarks: For almost Kaehlerian manifolds, the conditions $\theta(X)\Omega = 0$ and $\theta(X)J = 0$ (X is an *infinitesimal automorphism*) are sufficient in order to conclude that $\theta(X)g = 0$. Conversely, if X is an infinitesimal isometry, it does not follow that $\theta(X)J = 0$. For, the first term on the right in

$$\theta(X)J = (\xi^C D_C F^A{}_B + F^A{}_C D_B\, \xi^C - F^C{}_B D_C\, \xi^A)\, \frac{\partial}{\partial u^A} \otimes du^B$$

does not vanish. Moreover, one cannot conclude that $\theta(X)\Omega$ vanishes. In fact, the best that can be said is that $\langle\theta(X)\Omega, \Omega\rangle$ vanishes.

7.6. Conformal maps of manifolds with constant scalar curvature

With respect to the left invariant metric g, we have seen that the Ricci scalar curvature of a compact semi-simple Lie group is a positive constant. Moreover, with respect to g, an infinitesimal conformal transformation is an infinitesimal isometry. The same statements are valid for complex projective space $P_n (n > 1)$ with the Fubini metric. However, the n-sphere may be given a metric of positive (constant) scalar curvature relative to which there exist infinitesimal non-isometric conformal transformations. On the other hand, for compact manifolds of constant non-positive scalar curvature we show, with no further restriction, that the only infinitesimal conformal maps are infinitesimal isometries [58].

To begin with, an infinitesimal conformal transformation must satisfy equation (3.8.4):

$$\varDelta\alpha + \left(1 - \frac{2}{m}\right) d\delta\alpha = 2Q\alpha.$$

Hence, since $d\delta\alpha + \delta d\alpha = \varDelta\alpha$,

$$\left(2 - \frac{2}{m}\right) \varDelta\alpha - \left(1 - \frac{2}{m}\right) \delta d\alpha = 2Q\alpha.$$

Taking the divergence of both sides of this relation, we obtain

$$\left(2 - \frac{2}{m}\right) \varDelta\,\delta\alpha = 2\delta Q\alpha.$$

Therefore, since

$$\begin{aligned} -\delta Q\alpha &= D_i(R^i{}_j\alpha^j) \\ &= D_iR^i{}_j\alpha^j + R^i{}_jD_i\alpha^j \\ &= \frac{1}{2}\left[\frac{\partial R}{\partial u^j}\,\alpha^j + R^{ij}(D_j\alpha_i + D_i\alpha_j)\right] \\ &= \frac{1}{2}\,[\langle dR,\alpha\rangle + R^{ij}(\theta(\alpha)g)_{ij}] \\ &= \frac{1}{2}\,\langle dR,\alpha\rangle - \frac{1}{m}\,R\cdot\delta\alpha, \end{aligned}$$

it follows that

$$\left(1 - \frac{1}{m}\right) \varDelta\,\delta\alpha = \frac{1}{m}\,R\cdot\delta\alpha - \frac{1}{2}\,\langle dR,\alpha\rangle.$$

But $R =$ const., and so

$$\left(1 - \frac{1}{m}\right) \varDelta\,\delta\alpha = \frac{1}{m}\,R\cdot\delta\alpha.$$

Hence, since this constant is non-positive, by taking the global scalar product of the last relation with $\delta\alpha$, we obtain the desired conclusion.

Theorem 7.6.1. *If M is a compact Riemannian manifold of constant non-positive scalar curvature, then $C_0(M) = I_0(M)$.*

Let M be a compact Riemannian manifold of positive constant scalar curvature. If M admits a non-isometric infinitesimal conformal transformation it is not known whether M is isometric with a sphere. In fact, it is not even known whether M is a rational homology sphere (cf. theorem 3.7.5).

7.7. Infinitesimal transformations of non-compact manifolds

Let X be a vector field on a Kaehler manifold whose image by the almost complex structure operator J (inducing the complex structure of the manifold) is an infinitesimal transformation preserving the Kaehlerian structure. The vector field X is then 'closed', that is its covariant form (by the duality defined by the metric) is closed. We show that a *closed conformal map* X (that is, an infinitesimal conformal transformation whose covariant form ξ is closed) is a homothetic transformation.

Indeed, since ξ is closed

$$\begin{aligned} -t(X)_{AB} &= F^C{}_B(D_C\,\xi_A - D_A\,\xi_C + D_A\,\xi_C) + F^C{}_A D_B\,\xi_C \\ &= F^C{}_B D_A\,\xi_C + F^C{}_A D_B\,\xi_C \\ &= (\theta(C\xi)g)_{AB}, \end{aligned}$$

that is $t(X)$ is a symmetric tensor field. On the other hand, since $\theta(X)g = -1/n\,\delta\xi\cdot g$, it follows from lemma 7.1.2 that

$$t(X) = \frac{1}{n}\delta\xi\cdot\Omega + \theta(X)\Omega.$$

Hence, $t(X)$ is also skew-symmetric and must therefore vanish. Therefore,

$$-\frac{1}{n}\,d\delta\xi\wedge\Omega = d\theta(X)\Omega = \theta(X)d\Omega = 0.$$

Thus, for $n > 1$, we may conclude that $d\delta\xi$ vanishes, that is, the vector field X defines a homothetic transformation.

Moreover, we have proved that a closed conformal map is an infinitesimal holomorphic transformation. However, it need not be an infinitesimal isometry, as in the compact case. For, by lemma 7.3.1

$$\bar\theta(X)\Omega - \theta(X)\Omega = \delta\xi\cdot\Omega.$$

Applying δ to both sides of this relation, we obtain

$$\delta\theta(X)\Omega = Cd\delta\xi = 0.$$

Consequently, $\theta(X)\Omega$ is both closed and co-closed, that is harmonic. But, although it is exact, it need not vanish; for, the decomposition theorem is valid for compact manifolds and, in the case of open manifolds further restrictions are necessary [*31*]. Conversely, an infinitesimal isometry need not be a holomorphic transformation. Thus, an infinitesimal isometry of a Kaehler manifold need not be an automorphism of the Kaehlerian structure. The best that can be said in this context is given by

Theorem 7.7.1. *A closed conformal map of a Kaehler manifold is a holomorphic homothetic trans, rmation* [*36*].

Conditions may be given in order to ensure that a closed conformal map X be an infinitesimal isometry. Indeed, if the manifold is complete but not locally flat this situation prevails [*45*]. In the locally flat case, if X is of bounded length, the same conclusion may be drawn [*42*].

A Riemannian manifold M can be shown to be complete if every geodesic may be extended for infinitely large values of the arc length parameter. By a well-known theorem in topology this assertion can be shown to be equivalent to the statement: "Every infinite bounded set (with respect to d, cf. I.K.1) of M has a limit point." For the relationship with complete vector fields, the reader is referred to [*63*].

EXERCISES

A. Groups of holomorphic transformations

1. For any vector fields X, Y and Z on a Kaehler manifold M show

(a) $(\theta(X)\Omega)(Y,Z) = \theta(X)(g(JY,Z)) - g(J[X,Y],Z) - g(JY,[X,Z])$

and

(b) $(\theta(X)g)(JY,Z) = \theta(X)(g(JY,Z)) - g([X,JY],Z) - g(JY,[X,Z])$.

Hence, if M is compact, prove that a Killing vector field is holomorphic.

Hint: Express $J[X, Y]$ and $[X, JY]$ in local complex coordinates. Incidentally, one may then show that cor. 1, lemma 7.1.1 and its converse hold for complex manifolds, in general.

2. If $b_1(M) = 0$ prove that $L_a = K + JK$, if and only if,

$$L_a^* = L_a^* \cap \delta \wedge^2(T^{c*}) + L_a^* \cap d \wedge^0(T^{c*})$$

where L_a^* is the dual space of L_a.

3. If M has constant scalar curvature show that

$$\dim L_a = 2 \dim K - \dim K_c$$

where $K_c \subset K$ is the ideal determined by the elements of K^* which are closed [*58*]. Indeed,

$$K_c^* = \{\alpha \in \wedge^1(T^{c*}) \mid D_X \alpha = 0, X \in T\}.$$

It can be shown that

(i) $\dim K \leqq n^2 + 2n, 2n = \dim M$; hence, the maximum dimension attained by L_a is $2(n^2 + 2n)$.

(ii) The largest connected group of isometries of P_n is $SU(n+1)$; hence, for P_n (since $\dim SU(n+1) = n^2 + 2n$)

$$\dim L_a = 2(n^2 + 2n).$$

4. Prove that there are no holomorphic vector fields on a compact Kaehler manifold with negative definite Ricci curvature.

S. Nakano has shown that the hypothesis of negative definite Ricci curvature can be replaced by negative definite 1st Chern class [cf. § 6.14 and K. Kodaira-D. C. Spencer, On deformations of complex analytic structures I, *Ann. Math.* **67**, 328-401 (1958)]. Moreover, *the group of holomorphic transformations of a compact Kaehler manifold with negative definite* 1st *Chern class is finite* [S. Kobayashi, On the automorphism group of a certain class of algebraic manifolds, *Tôhoku Math. J.* **11**, 184-190 (1959)].

B. Almost hermitian metric

1. Let Ω be an element of $\wedge^2(T_P^*)$ of maximal rank $2n$ ($\dim T = 2n$) and h an inner product in T_P. Construct an inner product g which is hermitian relative to Ω, that is

$$g(JX, JY) = g(X, Y)$$

for any $X, Y \in T_P$ where J is the tensor of type (1,1) defined by Ω and h [*56*].

(As usual J denotes the linear transformation defined by the tensor J with components $F^A{}_B = h^{AC}F_{CB}$ relative to a given base of T_P—the F_{CB} being the coefficients of Ω).

Proceed as follows: Define the inner product k in terms of h by

$$k(X, Y) = h(JX, JY).$$

Next, compute the eigenvalues and eigenvectors of the matrix $k = (k_{AB})$. Let X be an eigenvector corresponding to the eigenvalue $\lambda^2 (\lambda > 0)$:

$$kX = \lambda^2 X,$$

that is

$$k^A{}_B X^B = \lambda^2 X^A \quad (k^A{}_B = h^{AC}k_{CB}).$$

Then, JX is also an eigenvector of λ^2 and

$$J^2 X = -\lambda^2 X.$$

The linear operator $(1/\lambda)J$ therefore defines a complex structure on the eigenspace of λ^2. Denote by λ_ρ^2, S_ρ ($\rho = 1, \cdots, r$) the eigenvalues and corresponding eigenspaces of k of the kind prescribed. The vector space T_P then has the decomposition

$$T_P = \sum_{\rho=1}^{r} S_\rho$$

—the S_ρ being invariant by J and orthogonal in pairs. Hence, for $\rho \neq \sigma$

$$F^{A_\rho}{}_{B_\sigma} = 0, \quad h_{A_\rho B_\sigma} = 0$$

in terms of a basis of T_P defined by this decomposition. Moreover,

$$k_{A_\rho B_\sigma} = 0$$

and

$$k_{A_\rho B_\rho} = \lambda_\rho^2 h_{A_\rho B_\rho}, \quad \rho = 1, \cdots, r.$$

The required inner product g is given by

$$g_{A_\rho B_\sigma} = 0 (\rho \neq \sigma), \quad g_{A_\rho B_\rho} = \lambda_\rho h_{A_\rho B_\rho}, \quad \lambda_\rho > 0, \rho = 1, \cdots, r.$$

C. Automorphisms

1. For any infinitesimal automorphisms X and Y of an almost Kaehler manifold, $[X,Y]$ is also an infinitesimal automorphism.

2. Denote the covariant forms of X, Y and $Z = [X,Y]$ by ξ, η and ζ, respectively. Hence, show if the Lie algebra of infinitesimal automorphisms is abelian

$$i(\xi \wedge \eta)\Omega = \text{const.}$$

Hint:

$$C\zeta = di(\xi \wedge \eta)\Omega.$$

3. Show that an infinitesimal automorphism of an almost Kaehler manifold is not, in general, an infinitesimal isometry.

D. A non-Kaehlerian hermitian manifold

1. Consider the shell between the spheres (cf. example 6, § 5.1).

$$\Sigma \mid z^i \mid^2 = 1, \quad \Sigma \mid z^i \mid^2 = 2$$

in C_n and denote by M the manifold obtained by identifying points on the spheres lying on the same radial lines. Let G denote the properly discontinuous group of automorphisms of $C_n - 0$ consisting of the homothetic transformations

$$(z^1, \cdots, z^n) \to (2^k z^1, \cdots, 2^k z^n)$$

for each integer k. The compact manifold M is a *fundamental domain* for this group. Since the quotient space $(C_n - 0)/G$ has a complex structure, M can be endowed with a natural real analytic structure. By showing that $b_2(M) = 0$, $(n > 1)$ prove that M is not Kaehlerian for $n > 1$. In fact,

$$b_0 = b_1 = 1,$$

$$b_p = 0, \quad 2 \leqq p < 2n - 1,$$

$$b_{2n-1} = b_{2n} = 1.$$

Note that b_1 is odd whereas in a Kaehler manifold all odd dimensional betti numbers are even (cf. theorem 5.6.2).

For a differential geometric characterization of a Hopf manifold see [98].

APPENDIX A

DE RHAM'S THEOREMS

The idea of the proof of the existence theorems of de Rham given below is due to A. Weil [*71*]. The method employed is due to Leray, namely, his theory of sheaves Without developing the general theory, a proof adapted to the object under consideration, namely, the de Rham sheaf, is given.

A.1. The 1-dimensional case

The existence theorems of de Rham are concerned with the periods of a closed differential form over the singular cycles of a compact differentiable manifold M. The periods are definite integrals. Let α be a closed 1-form and Γ a singular 1-cycle. We proceed to show how the period

$$\int_\Gamma \alpha$$

is related to an indefinite integral.

To this end, let $\mathscr{U} = \{U_i\}$ be a (countable open) covering of M by coordinate neighborhoods such that each U_i corresponds to an open ball in R^n. (We make a slight change in notation at this point so as to avoid confusion. In Chapters I and V Greek letters were generally employed as subscripts). Now, subdivide Γ until each 1-simplex is contained in some U_i. We may then represent Γ as a sum

$$\Gamma = \sum_i \Gamma_i$$

where each Γ_i is a chain contained in some U_j. Moreover, each boundary $\partial\Gamma_i$ is a 0-chain which may also be subdivided into parts each of which belongs to a U_k. It is important that each 0-simplex is assigned to a U_k

independently of the boundaries $\partial \Gamma_i$ containing it. For example, let Γ be a closed curve and consider the diagram

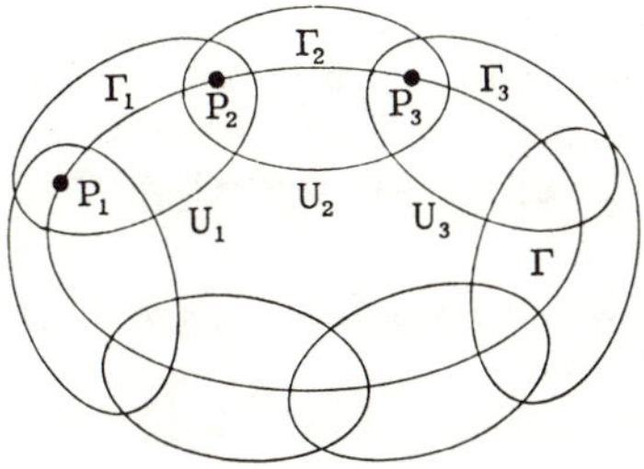

Then, it is easily seen that α has an integral in each U_i. By the Poincaré lemma (cf. § A.6) $\alpha = df_i$ in each U_i for some function f_i depending on α and U_i, and so

$$\int_\Gamma \alpha = \sum_i [f_i(P_{i+1}) - f_i(P_i)] = \sum_i (f_{i-1} - f_i)(P_i)$$

since the first sum is cyclic. More precisely, since there may be more than one P_j in a given U_i

$$\int_\Gamma \alpha = \sum_i [f_{k_i}(P_{i+1}) - f_{k_i}(P_i)]$$
$$= \sum_i (f_{k_{i-1}} - f_{k_i})(P_i)$$

where k_i is the index chosen such that U_{k_i} is the neighborhood for P_i. Since $df_{k_{i-1}} = df_{k_i}$ in $U_{k_i} \cap U_{k_{i-1}}$, $f_{k_i} - f_{k_{i-1}}$ is constant on the intersection. In this way, the integration has been reduced to the trivial case of integrating closed 0-forms (constants) over 0-chains (points).

The same general idea prevails in higher dimensions, although the situation there is more involved.

A.2. Cohomology

The above considerations motivate the theory to be developed below. Indeed, we shall consider (local) forms and chains defined only in U_i or $U_i \cap U_j$ where again $\mathscr{U} = \{U_i\}$ is any countable open covering of M.

The *nerve* of $\mathscr{U}$, denoted by $N(\mathscr{U})$ is the simplicial complex whose *vertices* (0-simplexes) are the elements of $\mathscr{U}$ and where any finite number of vertices span a simplex of $N(\mathscr{U})$, if and only if, they have a non-empty intersection. By a *p-simplex* $\sigma = \Delta(i_0, \cdots i_p)$ we mean an ordered finite set $(i_0, \cdots, i_p)$ of indices such that $U_{i_0} \cap \cdots \cap U_{i_p} \neq \square$. If $U_0, \cdots, U_p$ are the vertices of a p-simplex σ, their intersection $U_0 \cap \cdots \cap U_p$ will occasionally be denoted by $\cap\sigma$. By hypothesis $\cap\sigma \neq \square$.

For any open sets U and V, $U \supset V$, let ρ_{UV} denote the *restriction map* on differential forms

$$\rho_{UV}: \wedge^q(U) \to \wedge^q(V), \quad q = 0, 1, \cdots, n$$

defined by

$$\rho_{UV}(\alpha) = \alpha \mid V, \quad \alpha \in \wedge^q(U).$$

These maps have the following property: if $U \supset V \supset W$, then $\rho_{UW} = \rho_{VW}\rho_{UV}$.

A *p-cochain* of $N(\mathscr{U})$ is a function f which assigns to each p-simplex σ an element of an abelian group or vector space $\Gamma(\cap\sigma)$. In the sequel $\Gamma(U)$ will be one of the following:

(i) R: the real numbers,

(ii) $\wedge^q = \wedge^q(U)$: the space of q-forms over U,

(iii) $\wedge^q_c = \wedge^q_c(U)$: the space of closed q-forms over U.

It is important that Γ is allowed to depend on the simplex. This generalizes the usual definition in which to each simplex an element of a fixed module or abelian group is assigned (cf. § 2.1). More precisely, (a) for every open set U there is a vector space $\Gamma(U)$ and (b) if $U \supset V$, then $\rho_{UV}: \Gamma(U) \to \Gamma(V)$. (The map $\Gamma(U) \to \Gamma(V)$ need not be a monomorphism, that is an isomorphism into $\Gamma(V)$). The value $f(i_0, \cdots, i_p) \equiv f(\Delta(i_0, \cdots, i_p))$ of a p-cochain is an element of $\Gamma(U_{i_0} \cap \cdots \cap U_{i_p})$.

If $\sigma = \Delta(i_0, \cdots, i_p)$, let the faces of σ be the simplexes $\sigma^j = \Delta(i_0, \cdots, i_{j-1}, i_{j+1}, \cdots, i_p)$, $j = 0, \cdots, p$. Then, $\cap\sigma^j \supset \cap\sigma$ and there is a homomorphism

$$\rho_{\sigma^j,\sigma}: \Gamma(\cap\,\sigma^j) \to \Gamma(\cap\,\sigma)$$

defined by the restriction map, that is $\rho_{\sigma^j,\sigma} f(\sigma^j) = f(\sigma^j) \mid \cap\sigma$ is an element of the vector space $\Gamma(\cap\sigma)$. (In case $f(\sigma^j)$ is a real number consider $f(\sigma^j)$ as a constant function).

If f and g are p-cochains of $N(\mathscr{U})$ with values in the same abelian group $\Gamma(\cap\sigma)$, then cochains $f + g$ and $r \cdot f$, $r \in R$ are defined by

$$(f + g)(\sigma) = f(\sigma) + g(\sigma), \quad (r \cdot f)(\sigma) = rf(\sigma),$$

for each simplex $\sigma \in N(\mathscr{U})$. In this way, the p-cochains form a vector space over the reals (cf. § 2.1) which we denote by $C^p(N(\mathscr{U}), \Gamma)$. (No confusion should arise between the Γ employed here and the one in § 2.1 denoting a cycle.) The *coboundary operator* δ, (not to be confused with the operator δ employed previously) assigning a cochain δf to each p-cochain f is defined by

$$(\delta f)(\sigma) = \sum_{j=0}^{p} (-1)^j \rho_{\sigma^j,\sigma} f(\sigma^j), \quad \sigma = \Delta(i_0, \cdots i_{p+1}).$$

Thus $\delta: C^p(N(\mathscr{U}), \Gamma) \to C^{p+1}(N(\mathscr{U}), \Gamma)$; in fact, δf can be different from zero only on the $(p+1)$-simplexes of $N(\mathscr{U})$. It is easily checked that $\delta\delta f = 0$. In the usual manner one may therefore define the p-dimensional cohomology group $H^p(N(\mathscr{U}), \Gamma)$ as the quotient of $Z^p(N(\mathscr{U}), \Gamma)$—the p-cocycles by $B^p(N(\mathscr{U}), \Gamma)$—the p-coboundaries:

$$H^p(N(\mathscr{U}),\Gamma) = Z^p(N(\mathscr{U}),\Gamma)/B^p(N(\mathscr{U}),\Gamma).$$

In particular, if M is connected

$$H^0(N(\mathscr{U}),\Gamma) = \Gamma(M)$$

For, a 0-cochain f assigns to each $U \in \mathscr{U}$ an element α_U of $\Gamma(U)$. The condition $\delta f = 0$ requires that if $f(V) = \alpha_V \in \Gamma(V)$, $V \in \mathscr{U}$, and $U \cap V \neq \square$, then

$$\rho_{V,U \cap V}\alpha_V = \rho_{U,U \cap V}\alpha_U.$$

Conversely, for any globally defined α $(\in \wedge^q(M))$, a 0-cochain satisfying $\delta f = 0$ is given by defining $f(U) = \rho_{MU}\alpha$, $U \in \mathscr{U}$ (and $f(\sigma) = 0$ for all other $\sigma \in N(\mathscr{U})$). That the map $\Gamma(M) \to H^0(N(\mathscr{U}), \Gamma)$ is a monomorphism is left as an exercise.

A 1-cochain is defined by $f(U, V) = \alpha_{UV} \in \Gamma(U \cap V)$. It is a cocycle if $\rho_{U \cap V, U \cap V \cap W}\, \alpha_{UV} - \rho_{W \cap V, U \cap V \cap W}\, \alpha_{WV} + \rho_{W \cap U, U \cap V \cap W}\, \alpha_{WU} = 0$, $\alpha_{UV}, \alpha_{WV}, \alpha_{WU} \in \Gamma(U \cap V \cap W)$. If $U = V = W$, we conclude that $\alpha_{UU} = 0$ from which it follows that $\alpha_{UV} = -\alpha_{VU}$. The cocycle α_{UV} is a coboundary, if it can be expressed as $\alpha_V - \alpha_U$.

In the sequel, we shall write $(\delta f)(\sigma) = \Sigma(-1)^j f(\sigma^j)$ for simplicity.

A covering $\mathscr{V} = \{V\}$ of M is called a *refinement* of $\mathscr{U}$ if there is a map

$$\phi: \mathscr{V} \to \mathscr{U}$$

defined by associating with each $V \in \mathscr{V}$ a set $U \in \mathscr{U}$ such that $V \subset U$. If $\sigma = (V_0, \cdots, V_p) \in N(\mathscr{V})$, let $\phi\sigma = (\phi V_0, \cdots, \phi V_p)$. Then, $\cap\phi\sigma \supset \cap\sigma \neq \square$ and $\phi\sigma$ is an element of $N(\mathscr{U})$. Hence, there is (simplicial) map

$$\phi: N(\mathscr{V}) \to N(\mathscr{U}).$$

This map in turn induces a map $\tilde{\phi}$ sending each cochain $f \in C^p(N(\mathscr{U}), \Gamma)$ to a cochain $\tilde{\phi} f \in C^p(N(\mathscr{V}), \Gamma)$ where for each $\sigma \in N(\mathscr{V})$

$$\tilde{\phi} f(\sigma) = \rho_{\phi\sigma,\sigma} f(\phi\sigma).$$

The map ϕ is not unique. However, all such maps are contiguous and therefore induce the same homomorphisms (see below)

$$\phi^* : H^p(N(\mathscr{U}),\Gamma) \to H^p(N(\mathscr{V}),\Gamma).$$

Moreover, if $\mathscr{W} = \{W\}$ is a refinement of $\mathscr{V}$, the combined homomorphism

$$H^p(N(\mathscr{U}),\Gamma) \to H^p(N(\mathscr{V}),\Gamma) \to H^p(N(\mathscr{W}),\Gamma)$$

is equal to the direct homomorphism

$$H^p(N(\mathscr{U}),\Gamma) \to H^p(N(\mathscr{W}),\Gamma)$$

since a map $N(\mathscr{W}) \to N(\mathscr{V}) \to N(\mathscr{U})$ is contiguous to any direct map $N(\mathscr{W}) \to N(\mathscr{U})$.

To show that ϕ^* depends only on the pair $\mathscr{U}$, $\mathscr{V}$ we proceed as follows: Let ϕ' be another choice for ϕ. For $p = 0$, the assertion is clear. For $p \geqq 1$ let λf be the $(p-1)$-cochain on $\mathscr{V}$ defined by

$$(\lambda f)(V_0, \cdots, V_{p-1}) = \sum_{i=0}^{p-1} (-1)^i f(\phi' V_0, \cdots, \phi' V_i, \phi V_i, \cdots, \phi V_{p-1}).$$

Then,

$$\begin{aligned}(\lambda \delta f)(V_0, \cdots, V_p) &= \sum_{i=0}^{p} (-1)^i (\delta f)(\phi' V_0, \cdots, \phi' V_i, \phi V_i, \cdots \phi V_p)\\ &= \sum_{0 \leq j \leq i \leq p} (-1)^{i+j} f(\phi' V_0, \cdots, \phi' V_{j-1}, \phi' V_{j+1}, \cdots, \phi' V_i, \phi V_i, \cdots, \phi V_p)\\ &\quad + \sum_{0 \leq i \leq j \leq p} (-1)^{i+j+1} f(\phi' V_0, \cdots, \phi' V_i, \phi V_i, \cdots, \phi V_{j-1}, \phi V_{j+1}, \cdots, \phi V_p)\end{aligned}$$

and

$$\begin{aligned}(\delta \lambda f)(V_0, \cdots, V_p) &= \sum_{i=0}^{p} (-1)^i (\lambda f)(V_0, \cdots, V_{i-1}, V_{i+1}, \cdots, V_p)\\ &= \sum_{0 \leq j < i \leq p} (-1)^{i+j} f(\phi' V_0, \cdots, \phi' V_j, \phi V_j, \cdots, \phi V_{i-1}, \phi V_{i+1}, \cdots, \phi V_p)\\ &\quad + \sum_{0 \leq i < j \leq p} (-1)^{i+j+1} f(\phi' V_0, \cdots, \phi' V_{i-1}, \phi' V_{i+1}, \cdots, \phi' V_j, \phi V_j, \cdots, \phi V_p).\end{aligned}$$

It follows that

$$(\lambda\delta f + \delta\lambda f)(V_0, \cdots, V_p) = \sum_{0 \leq j \leq p} [f(\phi' V_0, \cdots, \phi' V_{j-1}, \phi V_j, \cdots, \phi V_p)$$

$$- f(\phi' V_0, \cdots, \phi' V_j, \phi V_{j+1}, \cdots, \phi V_p)]$$

$$= f(\phi V_0, \cdots, \phi V_p) - f(\phi' V_0, \cdots, \phi' V_p).$$

Hence, if f is a cocycle, $\tilde{\phi} f - \tilde{\phi}' f$ is a coboundary, that is $\phi^* = \phi'^*$.

In the sequel we denote the homomorphism ϕ^* by $\phi_{\mathscr{U}\mathscr{V}}$.

The set of all coverings of M is partially ordered by inclusion where $\mathscr{V}$ is *contained* in $\mathscr{U}$, if and only if, $\mathscr{V}$ is a refinement of $\mathscr{U}$. If $\mathscr{V}$ is a refinement of $\mathscr{U}$ we shall write $\mathscr{V} < \mathscr{U}$. It is not difficult to show that any two coverings have a common refinement.

If $\mathscr{W} < \mathscr{V} < \mathscr{U}$, it is readily shown that

$$\phi_{\mathscr{U}\mathscr{W}} = \phi_{\mathscr{V}\mathscr{W}}\, \phi_{\mathscr{U}\mathscr{V}}.$$

The *direct limits*

$$H^p(M,\Gamma) = \lim_{\mathscr{U}} H^p(N(\mathscr{U}),\Gamma)$$

of the groups $H^p(N(\mathscr{U}), \Gamma)$, $p = 0,1,\cdots$ are defined in the following way: Two elements $h_i \in H^p(N(\mathscr{U}_i), \Gamma)$, $i = 1,2$ are said to be equivalent if there exists an element $h_3 \in H^p(N(\mathscr{U}_3), \Gamma)$ with $\mathscr{U}_3 < \mathscr{U}_i$, $i = 1,2$ such that $h_3 = \phi_{\mathscr{U}_i \mathscr{U}_3} h_i$, $i = 1,2$; the direct limit is the set of these equivalence classes.

The sum of two cohomology classes of $H^p(M, \Gamma)$ is defined as follows: If $h_i \in H^p(N(\mathscr{U}_i), \Gamma)$, $i = 1,2$ are the elements to be added, we first find a common refinement $\mathscr{U}_3$ of $\mathscr{U}_1$ and $\mathscr{U}_2$ and then form the element $\phi_{\mathscr{U}_1 \mathscr{U}_3} h_1 + \phi_{\mathscr{U}_2 \mathscr{U}_3} h_2$. Multiplication by elements of R is clear. An element $h \in H^p(N(\mathscr{U}), \Gamma)$ represents the zero cohomology class, if and only if, there is a $\mathscr{V} < \mathscr{U}$ such that $\phi_{\mathscr{U}\mathscr{V}} h = 0$. We may therefore conclude that $H^p(M, \Gamma)$ is a vector space for each $p = 0,1, \cdots$.

Finally, a cochain f will be called a *finite cochain* if there exists a compact set S such that $f(i_0, \cdots, i_p) = 0$ whenever $U_{i_0} \cap \cdots \cap U_{i_p} \cap S = \square$. One may construct a cohomology theory in terms of finite cochains.

A.3. Homology

In this section we develop a theory dual to that of § A.2. Indeed, we associate as in the previous section with every open set $U \in \mathscr{U}$ a vector space which is again denoted by $\Gamma(U)$ (see (i)-(iii) below). Our first distinction now arises, namely, if $V \subset U$, then ρ_{VU}: $\Gamma(V) \to \Gamma(U)$, that is $\Gamma(V)$ is identified with a subspace of $\Gamma(U)$. (As before, the map $\Gamma(V) \to \Gamma(U)$ need not be a monomorphism).

By a *p-chain g* is meant a formal sum

$$g = \sum_{(i)} g(i_0, \cdots, i_p)\, \Delta(i_0, \cdots, i_p), \quad g(i_0, \cdots, i_p) \in \Gamma(U_{i_0} \cap \cdots \cap U_{i_p})$$

where $\Delta(i_0, \cdots, i_p)$ is a p-simplex on $N(\mathscr{U})$ and (i) implies summation on $(i_0, \cdots, i_p)$. Whereas the values of a p-cochain are in $\Gamma(U_{i_0} \cap \cdots \cap U_{i_p})$, the coefficients of a p-chain lie in $\Gamma(U_{i_0} \cap \cdots \cap U_{i_p})$. In the applications Γ will be either

(i) R : the real numbers,

(ii) $S_q(U)$: the space of finite singular chains (cf. § 2.2) with support in U, or

(iii) $S_q^c(U)$: the subspace of finite singular cycles.

A *boundary operator* ∂ mapping p-chains into $(p - 1)$-chains is defined on p-simplexes as follows:

$$\partial[\Delta(i_0, \cdots, i_p)] = \sum_{k=0}^{p} (-1)^k \Delta(i_0, \cdots, i_{k-1}, i_{k+1}, \cdots, i_p)$$

and on p-chains by linear extension, that is

$$\partial g = \Sigma g(i_0, \cdots, i_p)\, \partial[\Delta(i_0, \cdots, i_p)].$$

(In order to simplify notation we have written $g(i_0, \cdots, i_p)$ for the corresponding images $\rho_{..}\ g(i_0, \cdots, i_p)$). Denoting the coefficients of ∂g by $(\partial g)(j_0, \cdots, j_{p-1})$ we obtain

$$(\partial g)\,(j_0, \cdots, j_{p-1}) = \sum_{k=0}^{p} \sum_{i} (-1)^k g(j_0, \cdots, j_{k-1}, i, j_k, \cdots, j_{p-1})$$

where i runs over all indices for which the corresponding intersection is not empty. In order that this sum be finite it is assumed that the covering $\mathscr{U}$ of M is *locally finite*, that is every point of M has a neighborhood meeting only a finite number of $U_i \in \mathscr{U}$ (cf. §§ A.10-11).

It is easily checked that $\partial \partial g = 0$. One may then define the *p-dimensional homology group* $H_p(N(\mathscr{U}), \Gamma)$ as the quotient of $Z_p(N(\mathscr{U}), \Gamma)$—the *p-cycles* by $B_p(N(\mathscr{U}), \Gamma)$—the *p-boundaries*:

$$H_p(N(\mathscr{U}),\Gamma) = Z_p(N(\mathscr{U}),\Gamma)/B_p(N(\mathscr{U})\Gamma).$$

Let $\mathscr{V} = \{V\}$ be a refinement of $\mathscr{U}$. Then, as in the previous section there is a map ϕ: $\mathscr{V} \to \mathscr{U}$ defined by associating with each $V \in \mathscr{V}$ a set $U \in \mathscr{U}$ such that $V \subset U$. To a p-chain g on $\mathscr{V}$ one may then assign a chain $\tilde{\phi} g$ on $\mathscr{U}$ as follows:

$$\tilde{\phi}\colon \Sigma_{(i)}\, g(i_0, \cdots, i_p)\Delta(i_0, \cdots, i_p) \to \Sigma g(i_0, \cdots, i_p)\Delta(\phi(i_0), \cdots, \phi(i_p)), \phi(i_r) = \phi(V_r).$$

Evidently, cycles are mapped into cycles and boundaries into boundaries. Hence, $\tilde{\phi}$ induces a homomorphism

$$\phi_* : H_p(N(\mathscr{V}),\Gamma) \to H_p(N(\mathscr{U}),\Gamma).$$

As before, this homomorphism does not depend on ϕ but rather on the pair $\mathscr{V}$, $\mathscr{U}$ and so, we denote ϕ_* by $\phi_{\mathscr{V}\mathscr{U}}$. Moreover, if $\mathscr{W} < \mathscr{V} < \mathscr{U}$, it is easily checked that $\phi_{\mathscr{W}\mathscr{U}} = \phi_{\mathscr{V}\mathscr{U}} \cdot \phi_{\mathscr{W}\mathscr{V}}$. The *inverse limits*

$$H_p(M,\Gamma) = \lim_{\mathscr{U}} H_p(N(\mathscr{U}),\Gamma)$$

of the groups $H_p(N(\mathscr{U}), \Gamma)$, $p = 0,1,\cdots$ are defined as follows: Two elements $h_i \in H_p(N(\mathscr{U}_i), \Gamma)$, $i = 1,2$ are equivalent if there exists an element $h_3 \in H_p(N(\mathscr{U}_3), \Gamma)$ with $\mathscr{U}_3 < \mathscr{U}_i$, $i = 1,2$ such that $h_i = \phi_{\mathscr{U}_3\mathscr{U}_i} h_3$, $i = 1,2$; the inverse limit is the set of these equivalence classes. With the obvious definitions of addition and scalar multiplication $H_p(M, \Gamma)$ is a vector space for each $p = 0,1, \cdots$.

A.4. The groups $H^p(M, \wedge^q)$

It is now shown that in the cases $\Gamma = \wedge^q$, $q = 0,1, \cdots$, the cohomology groups $H^p(M, \wedge^q)$ vanish for all $p > 0$ provided M is compact (see remarks at end of § A.10 as well as at the end of this appendix). By the definition of the direct limit, it is sufficient to show that every covering $\mathscr{U}$ has a refinement $\mathscr{V}$ such that $H^p(N(\mathscr{V}), \wedge^q) = \{0\}$ for all q, and $p > 0$.

A refinement $\mathscr{V}$ of $\mathscr{U}$ is called a *strong refinement* if each $\bar{V}$ (the closure of V) is compact and contained in some U. In this case, we write $\mathscr{V} \ll \mathscr{U}$, and for a pair V, $U(\phi\colon V \to U)$ we write $V \Subset U$.

Lemma A.4.1. *For a compact differentiable manifold M,*

$$H^p(M, \wedge^q) = \{0\}$$

for all $p > 0$ and $q = 0,1, \cdots$.

Let $\mathscr{V}$ be a locally finite strong refinement of the open covering $\mathscr{U}$ of M and $\{e_j\}$ a partition of unity subordinated to $\mathscr{V}$ (cf. Appendix D). For an element $f \in C^p(N(\mathscr{V}), \wedge^q)$ let $f_j = e_j f$. Then, $\delta f_j = (\delta f)_j$, and so if f is a cocycle, so is $e_j f$.

Let f be a p-cocycle, $p > 0$. By definition, $f = \Sigma f_j$ is a locally finite sum. We shall prove that each cocycle f_j is a coboundary, that is $f_j = \delta g_j$ where $g_j(V_0, \cdots, V_{p-1}) = 0$ if $V_0 \cap \cdots \cap V_{p-1}$ does not intersect V_j. This being the case, $g = \Sigma g_j$ is well-defined and $f = \Sigma f_j = \Sigma\, \delta g_j = \delta g$.

To this end, consider a fixed j and put

$$g_j(V_0, \cdots, V_{p-1}) = f_j(V_j, V_0, \cdots, V_{p-1})$$

if $V_j \cap V_0 \cap \cdots \cap V_{p-1} \neq \square$ and $g_j = 0$, otherwise. In the first case,

$$(\delta g_j)(V_0, \cdots, V_p) = \Sigma(-1)^i f_j(V_j, V_0, \cdots, V_{i-1}, V_{i+1}, \cdots, V_p).$$

Since f_j is a cocycle,

$$0 = (\delta f_j)(V_j, V_0, \cdots, V_p) = f_j(V_0, \cdots, V_p) - \Sigma(-1)^i f_j(V_j, V_0, \cdots, V_{i-1}, V_{i+1}, \cdots, V_p).$$

Hence, $f_j = \delta g_j$.

In the second case, since $V_j \cap V_0 \cap \cdots \cap V_p = \square$, $f_j(V_0, \cdots, V_p) = 0$. But δg_j also vanishes; for, in

$$(\delta g_j)(V_0, \cdots, V_p) = \Sigma(-1)^i g_j(V_0, \cdots, V_{i-1}, V_{i+1}, \cdots, V_p)$$

each term on the right is either zero, by the definition of g_j, or else it is the restriction of $f_j(V_j, V_0, \cdots, V_{i-1}, V_{i+1}, \cdots, V_p)$ to the set $V_0 \cap \cdots \cap V_p$ and, since e_j vanishes outside of V_j, the value is again zero.

We conclude that $f_j = \delta g_j$ in all cases, and so by the above remark, the proof is complete.

A.5. The groups $H_p(M, S_q)$

Since the groups $H_p(M, S_q)$ are in a certain sense dual to the groups $H^p(M, \wedge^q)$, it is to be expected by the result of the previous section that they also vanish for $p > 0$. It is the purpose of this section to show that this is actually the case. To this end, it is obviously sufficient to show that for any open covering $\mathscr{U}$ of M and $g \in Z_p(\mathscr{U}, S_q)$, g is a boundary.

Lemma A.5.1. *For a differentiable manifold M,*

$$H_p(M,S_q) = \{0\}$$

for all $p > 0$ and $q = 0,1, \ldots$. Moreover, in order that a 0-chain be a boundary, it is necessary and sufficient that the sum of its coefficients be zero.

Consider all singular q-simplexes. Divide these simplexes into classes so that all those simplexes in the jth class are contained in U_j. (This can, of course, be done in many ways). For each cycle g construct a singular chain g_j by deleting those singular simplexes not in the jth class. That g_j is a cycle follows from the fact that $\partial(g_j) = (\partial g)_j$ (the cancellations occurring in $\partial g (= 0)$ occur amongst those simplexes in the same class). Since $g = \Sigma g_j$ it suffices to show that each g_j is a boundary. For simplicity, we take $j = 0$. Define a $(p + 1)$-chain h as follows:

$$h(i_0, \cdots, i_{p+1}) = \begin{cases} g_0(i_1, \cdots, i_{p+1}) \text{ if } i_0 = 0 \\ 0 \text{ if } i_0 \neq 0, \end{cases}$$

that is,

$$h = \sum_{(i)} g_0(i_1, \cdots, i_{p+1}) \, \Delta(0, i_1, \cdots i_{p+1}).$$

Now, since

$$\partial h = \sum_{(i)} g_0(i_1, \cdots, i_{p+1}) \, \Delta(i_1, \cdots, i_{p+1})$$

$$- \sum_{(i)} \sum_{k=1}^{p+1} (-1)^k \, g_0(i_1, \cdots, i_{p+1}) \, \Delta(0, i_1, \cdots, i_{k-1}, i_{k+1} \cdots, i_{p+1})$$

and

$$0 = \partial g_0 = \sum_{(i)} \sum_{k=1}^{p+1} (-1)^k \, g_0(i_1, \cdots, i_{p+1}) \, \Delta(i_1, \cdots, i_{k-1}, i_{k+1}, \cdots, i_{p+1})$$

where $g_0 = \Sigma_{(i)} \, g_0(i_1, \cdots, i_{p+1}) \, \Delta(i_1, \cdots, i_{p+1})$, $g_0 = \partial h$, For by comparing the expression for ∂g_0 with the last sum in ∂h, we see that (except for notation) they are identical. We conclude that each g_j is a boundary, and so g is a boundary.

For $p = 0$,

$$h = \Sigma g_0(i) \, \Delta(0, i),$$

and thus

$$\partial h = \Sigma g_0(i) \, \Delta(i) - \Sigma g_0(i) \, \Delta(0).$$

The condition $\partial g_0 = 0$ gives no information in this case. Hence, in order that g_0 be a boundary, it is necessary that $\Sigma g_0(j)$ vanish. On the other hand, if $\Sigma g(i)$ vanishes so does $\Sigma g_0(i)$. Therefore, a 0-chain is a boundary, if and only if, the sum of its coefficients is zero.

The above argument is based on the so-called cone construction.

A.6. Poincaré's lemma

It is not true, in general, that a closed form is exact. However, an exact form is closed. A partial converse is true. For a p-form, $p > 0$ this is the

Poincaré lemma. *On a starshaped region (open ball) Δ in R^n every closed p-form ($p > 0$) is exact.*

To establish this result we define a homotopy operator

$$I : \wedge^p(\Delta) \to \wedge^{p-1}(\Delta), \quad p > 0$$

with the property that

$$dI\alpha + Id\alpha = \alpha$$

for any p-form α defined in a neighborhood of Δ. Hence, if α is closed in Δ, then $Id\alpha = 0$ and $\alpha = d\beta$, where $\beta = I\alpha$.

Let $u^1, \cdots, u^n$ be a coordinate system at $P \in \Delta$ (where P is assumed to be at the origin). Denote by tu the vector with components $(tu^1, \cdots, tu^n)$, $0 \leqq t \leqq 1$. Then, for $\alpha = a_{(i_1 \ldots i_p)}(u) du^{i_1} \wedge \cdots \wedge du^{i_p}$, put

$$I\alpha = \sum_{k=1}^{p} (-1)^{k-1} \int_0^1 t^{p-1} a_{(i_1 \ldots i_p)}(tu) dt \cdot$$

$$\cdot u^{i_k} du^{i_1} \wedge \cdots \wedge du^{i_{k-1}} \wedge du^{i_{k+1}} \wedge \ldots \wedge du^{i_p}.$$

Thus,

$$dI\alpha = p \int_0^1 t^{p-1} a_{(i_1 \ldots i_p)}(tu) dt \cdot du^{i_1} \wedge \cdots \wedge du^{i_p}$$

$$+ \sum_{k=1}^{p} \sum_{j=1}^{n} (-1)^{k-1} \int_0^1 t^p \frac{\partial a_{(i_1 \ldots i_p)}}{\partial u^j}(tu) dt \cdot u^{i_k} du^j \wedge du^{i_1} \wedge \cdots \wedge \widehat{du^{i_k}} \wedge \cdots \wedge du^{i_p}.$$

On the other hand,

$$Id\alpha = \sum_{j=1}^{n} \int_0^1 t^p \frac{\partial a_{(i_1 \ldots i_p)}}{\partial u^j}(tu) dt \cdot u^j du^{i_1} \wedge \cdots \wedge du^{i_p}$$

$$- \sum_{j=1}^{n} \sum_{k=1}^{p} (-1)^{k-1} \int_0^1 t^p \frac{\partial a_{(i_1 \ldots i_p)}}{\partial u^j}(tu) dt \cdot u^{i_k} du^j \wedge du^{i_1} \wedge \cdots \wedge \widehat{du^{i_k}} \wedge \cdots \wedge du^{i_p}.$$

Hence,

$$dI\alpha + Id\alpha = [p\int_0^1 t^{p-1}a_{(i_1\dots i_p)}(tu)dt$$

$$+ u^j \sum_{j=1}^{n} \int_0^1 t^p \frac{\partial a_{(i_1\dots i_p)}}{\partial u^j}(tu)dt]du^{i_1} \wedge \cdots \wedge du^{i_p}$$

$$= \int_0^1 \frac{\partial}{\partial t}[t^p a_{(i_1\dots i_p)}(tu)]dt \cdot du^{i_1} \wedge \cdots \wedge du^{i_p}$$

$$= a_{(i_1\dots i_p)}(u)du^{i_1} \wedge \cdots \wedge du^{i_p}$$

$$= \alpha$$

provided that $p > 0$.

A.7. Singular homology of a starshaped region in R^n

In analogy with the previous section it is shown next that the singular homology groups $H_p(\Delta)$, $p > 0$ of a starshaped region in R^n are trivial.

Let us recall that by a singular p-simplex $s^p = [f : P_0, \cdots, P_p]$ we mean an Euclidean simplex $(P_0, \cdots, P_p)$ together with a map f of class 1 defined on $\Delta(P_0, \cdots, P_p)$—the convex hull of $(P_0, \cdots, P_p)$. Now, f can be extended to the Euclidean $(p + 1)$-simplex $(O, P_0, \cdots, P_p)$ by setting

$$\bar{f}(r_0P_0 + \cdots + r_pP_p) = (r_0 + \cdots + r_p)\, f\left(\frac{r_0P_0 + \cdots + r_pP_p}{r_0 + \cdots + r_p}\right)$$

and

$$\bar{f}(O) = 0.$$

Analogous to the map I of § A.6 we define the map $\mathbf{P}$ by

$$\mathbf{P}s^p = \sum_{i=0}^{p} (-1)^i [\bar{f} : \underbrace{O, \cdots, O}_{i+1}, P_i, \cdots, P_p].$$

Then,

$$\partial\mathbf{P}s^p = \sum_{i=0}^{p}\sum_{j=0}^{i} (-1)^{i+j} [\bar{f} : \underbrace{O, \cdots, O}_{i}, P_i, \cdots, P_p]$$

$$+ \sum_{i=0}^{p}\sum_{j=i}^{p} (-1)^{i+j+1} [\bar{f} : \underbrace{O, \cdots, O}_{i+1}, P_i, \cdots, P_{j-1}, P_{j+1}, \cdots, P_p].$$

On the other hand, from

$$\partial s^p = \sum_{j=0}^{p} (-1)^j [f : P_0, \cdots, P_{j-1}, P_{j+1}, \cdots, P_p]$$

we obtain

$$\mathbf{P}\partial s^p = \sum_{j=1}^{p}\sum_{i=0}^{j-1} (-1)^{i+j} [f : \underbrace{O, \cdots, O}_{i+1}, P_i, \cdots, P_{j-1}, P_{j+1}, \cdots, P_p]$$

$$+ \sum_{j=0}^{p} \sum_{i=j+1}^{p} (-1)^{i+j+1} [f : \underbrace{O, \cdots, O}_{i}, P_i, \cdots, P_p].$$

Hence,

$$\mathbf{P}\partial s^p + \partial\mathbf{P}s^p = \sum_{i=0}^{p} [f : \underbrace{O, \cdots, O}_{i}, P_i, \cdots, P_p] - \sum_{i=0}^{p} [f : \underbrace{O, \cdots, O}_{i+1}, P_{i+1}, \cdots, P_p]$$

$$= [f : P_0, \cdots, P_p] - [f : \underbrace{O, \cdots, O}_{p+1}]$$

$$= s^p - [f : \underbrace{O, \cdots, O}_{p+1}].$$

Now, put

$$\mathbf{P}_0 s^p = [f : \underbrace{O, \cdots, O}_{p+2}].$$

Then,

$$\partial\mathbf{P}_0 s^p = \epsilon_p [f : \underbrace{O, \cdots, O}_{p+1}], \quad \epsilon_p = \begin{cases} 0, p \text{ even} \\ 1, p \text{ odd} \end{cases}$$

and

$$\mathbf{P}_0 \partial s^p = \epsilon_{p+1} [f : \underbrace{O, \cdots, O}_{p+1}].$$

Hence, since $\epsilon_p + \epsilon_{p+1} = 1$

$$\partial\mathbf{P}_0 s^p + \mathbf{P}_0 \partial s^p = [f : \underbrace{O, \cdots, O}_{p+1}],$$

from which

$$\partial\bar{\mathbf{P}} s^p + \bar{\mathbf{P}}\partial s^p = s^p$$

where we have put $\bar{\mathbf{P}} = \mathbf{P} + \mathbf{P}_0$.

That any cycle is a boundary now follows by linearity.

Again, the above argument is based on the cone construction.

A.8. Inner products

The results of §§ A.2 and A.3 are now combined by defining an inner product of a cochain $f \in C^p(N(\mathscr{U}),\wedge^q)$ and a chain $g \in C_p(N(\mathscr{U}),S_q)$ as the integral of f over g. More precisely, the values $f(i_0, \cdots, i_p)$ are q-forms over $U_0 \cap \cdots \cap U_p$ whereas the values of g are singular q-chains in $U_0 \cap \cdots \cap U_p$. We define

$$(f(i_0, \cdots, i_p), g(i_0, \cdots, i_p)) = \int_g f \tag{A.8.1}$$

and

$$(f,g) = \sum_{(i)} (f(i_0, \cdots, i_p), g(i_0, \cdots, i_p)) \tag{A.8.2}$$

where the sum is extended over all p-simplexes on $N(\mathscr{U})$.

The notation $\int_g f$ is an abbreviation. To be more precise the form $f(i_0, \cdots, i_p)$ and chain $g(i_0, \cdots, i_p)$ should be written rather than the variables f and g.

Either f or g is assumed to be finite, In this case, the sum is finite. The elements $f \in C^p(N(\mathscr{U}), \wedge^q)$ and $g \in C_p(N(\mathscr{U}), S_q)$ are said to be of *type* (p, q).

Lemma A.8.1. *For elements* $f \in C^p(N(\mathscr{U}), \wedge^q)$ *and* $g \in C_{p+1}(N(\mathscr{U}), S_q)$

$$(\delta f,g) = (f,\partial g).$$

To begin with, since the bracket is linear in each variable we may assume that $g = g(0, \cdots, p + 1)\Delta(0, \cdots, p + 1)$. Then,

$$(\delta f,g) = \sum_i (-1)^i \int_{g(0,\ldots,p+1)} f(0, \cdots, i-1, i+1, \cdots, p+1)$$

$$= (f, \partial g)$$

since $(\partial g)(0, \cdots, i-1, i+1, \cdots, p+1) = (-1)^i g(0, \cdots, p+1)$.

We denote once more by d the operator on the cochain groups $C^p(N(\mathscr{U}), \wedge^q)$ defined as follows:

$$d : C^p(N(\mathscr{U}), \wedge^q) \rightarrow C^p(N(\mathscr{U}), \wedge^{q+1})$$

where to an element $f \in C^p(N(\mathscr{U}), \wedge^q)$ we associate the element df whose values are obtained by applying the differential operator d to the forms $f(i_0, \cdots, i_p) \in \wedge^q(U_{i_0} \cap \cdots \cap U_{i_p})$. Evidently, $dd = 0$.

An operator

$$D : C_p(N(\mathscr{U}),S_q) \rightarrow C_p(N(\mathscr{U}),S_{q-1})$$

is defined in analogy as follows: D is the operator replacing each coefficient of an element $g \in C_p(N(\mathscr{U}), S_q)$ by its boundary. Clearly, $DD = 0$.

Lemma A.8.2. *For elements* $f \in C^p(N(\mathscr{U}), \wedge^q)$ *and* $g \in C_p(N(\mathscr{U}), S_{q+1})$

$$(f,Dg) = (df,g).$$

This is essentially another form of Stokes' theorem.

The following commutativity relations are clear:

Lemma A.8.3. $\delta d = d\delta$ *and* $\partial D = D\partial$.

In § A.1 the problem of computing the period of a closed 1-form over a singular 1-cycle was considered—the resulting computation being reduced to the 'trivial' problem of integrating a closed 0-form over a 0-chain. The problem of computing the period of a closed q-form α (with compact carrier) over a singular q-cycle Γ is now considered.

In the first place, as in § A.1, by passing to a barycentric subdivision we may write $\Gamma = \Sigma\, \Gamma_i$ with Γ_i contained in U_i. If α_i denotes the restriction of α to U_i and f_0 the 0-cochain whose values are α_i, that is, $f_0(U_i) = \alpha_i$, then, if we denote by g_0 the chain whose coefficients are Γ_i,

$$\int_\Gamma \alpha = (f_0,g_0)$$

—the independence of the subdivision being left as an exercise. (Since more than one Γ_j may be contained in a single U_i choose one U_{i_j} to contain each Γ_j. Then, $g_0(U_k) = \Sigma_{i_j=k}\, \Gamma_j$).

A.9. De Rham's isomorphism theorem for simple coverings

Before establishing this result in its most general form we first prove it for a rather restricted type of covering. Indeed, a covering $\mathscr{U}$ of M is said to be *simple* if, (a) it is strongly locally finite (cf. § A.4) and (b) every non-empty intersection $U_0 \cap \cdots \cap U_p$ of open sets of the covering is homeomorphic with a starshaped region in R^n. It can be shown that such coverings exist. For, every point of M has a convex Riemannian normal coordinate neighborhood U, that is, for every P, $Q \in U$ there is a unique geodesic segment in U connecting P and Q [*23*]. Clearly, the intersection of such neighborhoods is starlike with respect to the Riemannian normal coordinate system at any point of the intersection. The neighborhoods U may also be taken with compact closure. Now, for every $V \in \mathscr{U}$ we can take a finite covering of $\bar{V}$ by such convex U, say $U_{V,1}, \cdots, U_{V,p_V}$. Then, the collection of all $\{U_{V,i} \mid V \in \mathscr{U},\ i = 1, \cdots, p_V\}$ is a simple covering of M provided (a) $\mathscr{U}$ is a strong refinement of a strong locally finite covering $\hat{\mathscr{U}}$ such that the $U_{V,i}$ refine $\hat{\mathscr{U}}$ strongly, that is, for $\bar{V} \subset \hat{V} \in \hat{\mathscr{U}}$, the neighborhoods $U_{V,1}, \cdots, U_{V,p_V}$ are all contained in $\hat{V}$ and (b) only finitely many $V \in \mathscr{U}$ are contained in a given $\hat{V} \in \hat{\mathscr{U}}$. From the above conditions it follows that $\{U_{V,i}\}$ is a locally finite covering.

Now, let $f_0 \in Z^0(N(\mathscr{U}), \wedge_c^q)$, $g_0 \in C_0(N(\mathscr{U}), S_q)$ and consider the systems of equations

$$\begin{aligned} f_0 &= df_1 & Dg_0 &= \partial g_1 \\ \delta f_1 &= df_2 & Dg_1 &= \partial g_2 \\ \delta f_2 &= df_3 & Dg_2 &= \partial g_3 \\ &\ \ \vdots & &\ \ \vdots \\ \delta f_{q-1} &= df_q & Dg_{q-1} &= \partial g_q. \end{aligned} \tag{A.9.1}$$

Clearly, f_i, $i = 1, 2, \cdots$ is of type $(i-1, q-i)$ and g_i is of type $(i, q-i)$. In the event there exist cochains f_i and chains g_i satisfying these relations it follows that

$$\begin{aligned} &(f_0,g_0) = (df_1,g_0) = (f_1,Dg_0) = (f_1,\partial g_1) \\ &= (\delta f_1,g_1) = (df_2,g_1) = (f_2,Dg_1) = (f_2,\partial g_2) \\ &= \quad \cdot \quad \cdot \quad \cdot \quad \cdot \quad \cdot \\ &= (\delta f_{q-1},g_{q-1}) = (df_q,g_{q-1}) = (f_q,Dg_{q-1}) = (f_q, \partial g_q) \\ &= (\delta f_q,g_q). \end{aligned}$$

Whereas f_0 and g_0 are of type $(0, q)$, δf_q and g_q are of type $(q, 0)$. Since $d\delta f_q = \delta d f_q = \delta\delta f_{q-1} = 0$, the coefficients of δf_q are constants. It follows that δf_q may be identified with a cocycle z^q with constant coefficients.

For a chain of type $(p, 0)$ let D_0 be the operator denoting addition of the coefficients in each singular 0-chain. Evidently, $\partial D_0 = D_0 \partial$ and $D_0 D = 0$. Thus, since $\partial g_q = Dg_{q-1}$, $\partial D_0 g_q$ vanishes, that is $D_0 g_q$ is a cycle ζ_q. We conclude that

$$(\delta f_q,g_q) = (z^q, \zeta_q)$$

(cf. formula A.8.2), that is

$$\int_\Gamma \alpha = (z^q, \zeta_q)$$

(cf. § A.1). The problem of computing the period of a closed q-form over a q-cycle has once again been reduced to that of integrating a closed 0-form over a 0-chain.

That the f_i exist follows from Poincarés lemma and the fact that from the equations (A.9.1), $d\delta f_i = \delta df_i = \delta\delta f_{i-1} = 0$, $i = 1, \cdots, q$. For, since f_0 is closed, there exists a $(q-1)$-form f_1 such that $f_0 = df_1$; since δf_1 is closed, there exists a $(q-2)$-form f_2 such that $\delta f_1 = df_2$, etc. To be precise, suppose that $f_1, \cdots, f_i$ exist satisfying $\delta df_k = 0$, $k = 1, \cdots, i$. Then, since $d\delta f_k = \delta df_k = 0$, the equation $\delta f_i = df_{i+1}$ has a solution satisfying $\delta df_{i+1} = 0$.

The dual argument shows that chains $g_i \in C_i(N(\mathscr{U}), S_{q-i})$ exist satisfying the system (A.9.1). That this argument works follows from property (b) of a simple covering and the fact that the homology of a ball (starshaped region in R^n) is trivial, as well as the equations $\partial Dg_i = 0$.

Now, suppose that a cocycle z^q (of type $(q, 0)$ with constant coefficients) and a cycle ζ_q (of type $(q, 0)$) are given. Since $\mathscr{U}$ is strongly locally finite, it is known from § A.4 that $H^q(N(\mathscr{U}), \wedge^0)$ vanishes. This being the case, there is an f_q such that $z^q = \delta f_q$. Hence, since z^q has constant coefficients $d\delta f_q$ must vanish. Since the operators d and δ commute (cf. lemma A.8.3) and the cohomology groups are trivial, the existence of an f_{q-1} with $df_q = \delta f_{q-1}$ is assured. In this manner, $f_{q-2}, \cdots, f_1$ are defined—the condition $d\delta f_1 = 0$ implying that $f_0 = df_1$ is a cocycle. Hence, f_0 determines a closed q-form. In a similar manner a $g_0 \in C_0(N(\mathscr{U}), S_q^c)$ can be constructed from ζ_q.

We have shown that cochains f_i of type $(i-1, q-i)$ exist satisfying the system of equations (A.9.1). Now, set

$$A_i = \{f_i \mid d\delta f_i = 0\},$$

$$X_i = \{f_i \mid df_i = 0\},$$

and

$$Y_i = \{f_i \mid \delta f_i = 0\}.$$

The values of f_i on the nerve of $\mathscr{U}$ are $(q-i)$-forms. The set X_i consists of all such closed $(q-i)$-forms.

The operator d maps the spaces A_i, Y_i and X_i homomorphically onto $Z^{i-1}(N(\mathscr{U}), \wedge_c^{q-i+1})$, $B^{i-1}(N(\mathscr{U}), \wedge_c^{q-i+1})$ and $\{0\}$, respectively, $2 \leqq i \leqq q$. For A_i, this follows from the Poincaré lemma since $q - i + 1 > 0$. Now, for an element $f_i \in Y_i$, $\delta f_i = 0$. Hence, since the cohomology is trivial for $i > 1$, there exists an f' such that $f_i = \delta f'$ from which $df_i = d\delta f' = \delta df'$, that is $df_i \in B^{i-1}(N(\mathscr{U}), \wedge_c^{q-i+1})$. To show that d is onto, let f' be an element of $B^{i-1}(N(\mathscr{U}), \wedge_c^{q-i+1})$. Then, $f' = \delta f_i$ for some $f_i \in C^{i-1}(N(\mathscr{U}), \wedge_c^{q-i})$ from which, since $q - i + 1 > 0$, by the Poincaré lemma, $f_i = df''$. It follows that $f' = \delta df'' = d\delta f''$, and so since $\delta f'' \in Y_i$, d is onto.

The following isomorphisms are a consequence of the previous paragraph:

$$A_i/X_i \cong Z^{i-1}(N(\mathscr{U}), \wedge_c^{q-i+1}),$$

$$(X_i + Y_i)/X_i \cong Y_i/X_i \cap Y_i \cong B^{i-1}(N(\mathscr{U}), \wedge_c^{q-i+1}).$$

We therefore conclude that

$$A_i/X_i + Y_i \cong H^{i-1}(N(\mathscr{U}), \wedge_c^{q-i+1}). \tag{A.9.2}$$

A similar discussion shows that the operator δ maps A_i, X_i, and Y_i onto $Z^i(N(\mathscr{U}), \wedge_c^{q-i})$, $B^i(N(\mathscr{U}), \wedge_c^{q-i})$ and $\{0\}$, respectively for $1 \leqq i \leqq q-1$ from which, as before, we conclude that

$$A_i/X_i + Y_i \cong H^i(N(\mathscr{U}), \wedge_c^{q-i}).$$

Consider now the following diagram:

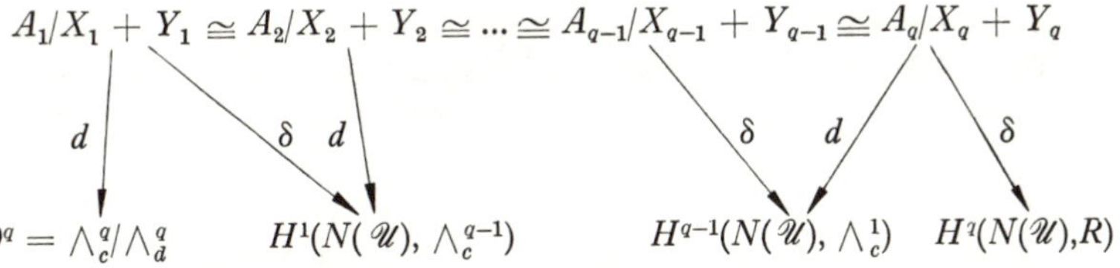

We show that $d: A_1/X_1+Y_1 \to D^q$ and $\delta: A_q/X_q+Y_q \to H^q(N(\mathscr{U}), R)$ are isomorphisms onto. Indeed, d sends $f_1 \in A_1$ into $df_1 \in Z^0(N(\mathscr{U}), \wedge_c^q)$, and so may be identified with a closed q-form α. Since the elements of X_1 are mapped into 0 we need only consider the effect of d on Y_1. Let y be an element of Y_1. Then, since $\delta y = 0$, dy represents an exact form. On the other hand, a closed q-form may be represented as df_1 and an exact form as dy with $\delta y = 0$. This establishes the first isomorphism. To prove the second isomorphism, let f_q be an element of A_q. Then, since $d\delta f_q = 0$, δf_q has constant coefficients and must therefore belong to $Z^q(N(\mathscr{U}), R)$. Since Y_q is annihilated by δ we need only consider the effect of δ on X_q. But an element $x \in X_q$ has constant coefficients, and so $\delta x \in B^q(N(\mathscr{U}), R)$.

From the complete sequence of isomorphisms, it follows that

$$D^q \cong H^q(N(\mathscr{U}),R).$$

It is now shown by means of the dual argument that the singular homology groups are dual to the groups $H^q(N(\mathscr{U}), R)$.

We have shown that chains g_i of type $(i, q-i)$ exist satisfying the system of equations (A.9.1). Now, set

$$A'_i = \{g_i \mid D\partial g_i = 0\},$$

$$X'_i = \{g_i \mid Dg_i = 0\}$$

and

$$Y'_i = \{g_i \mid \partial g_i = 0\}.$$

The values of g_i on $N(\mathscr{U})$ are $(q-i)$-singular chains. The set X'_i consists of all such $(q-i)$-singular cycles.

The operator D maps the spaces A'_i, X'_i and Y'_i homomorphically onto $Z_i(N(\mathscr{U}), S^c_{q-i-1})$, $\{0\}$ and $B_i(N(\mathscr{U}), S^c_{q-i-1})$, respectively whereas ∂ is a homomorphism onto $Z_{i-1}(N(\mathscr{U}), S^c_{q-i})$, $B_{i-1}(N(\mathscr{U}), S^c_{q-i})$ and $\{0\}$, respectively provided that the indices never vanish. This leads to the diagram

$$\begin{array}{ccccccccc}
A'_1/X'_1 + Y'_1 & \cong & A'_2/X'_2 + Y'_2 & \cong \cdots \cong & A'_{q-2}/X'_{q-2} + Y'_{q-2} & \cong & A'_{q-1}/X'_{q-1} + Y'_{q-1} \\
 & {\scriptstyle D}\searrow \quad \swarrow{\scriptstyle \partial} & & & & {\scriptstyle D}\searrow \quad \swarrow{\scriptstyle \partial} & \\
 & H_1(N(\mathscr{U}), S^c_{q-2}) & & & & H_{q-2}(N(\mathscr{U}), S^c_1). &
\end{array}$$

Let ∂_0 be the operator denoting addition of the coefficients of each chain in $C_0(N(\mathscr{U}), S_q)$; denote by Z_{00} the space annihilated by ∂_0 and put $H_{00} = Z_{00}/B_0$. Then, the diagram can be completed on the left in the following way

$$\begin{array}{ccc}
A'_0/X'_0 + Y'_0 & \cong & A'_1/X'_1 + Y'_1 \\
\downarrow{\scriptstyle \partial_0} & {\scriptstyle D}\searrow & \downarrow{\scriptstyle \partial} \\
S^c_q/S^b_q & & H_{00}(N(\mathscr{U}), S^c_{q-1})
\end{array}$$

For, ∂_0 maps the spaces A'_0, X'_0 and Y'_0 homomorphically onto S^c_q, S^b_q and $\{0\}$, respectively.

On the right, we have the diagram

$$A'_{q-1}/X'_{q-1} + Y'_{q-1} \cong A'_q/X'_q + Y'_q$$

$$D \qquad \partial \qquad D_0$$

$$H_{q-1}(N(\mathcal{U}),S_0^c) \qquad H_q(N(\mathcal{U}),R)$$

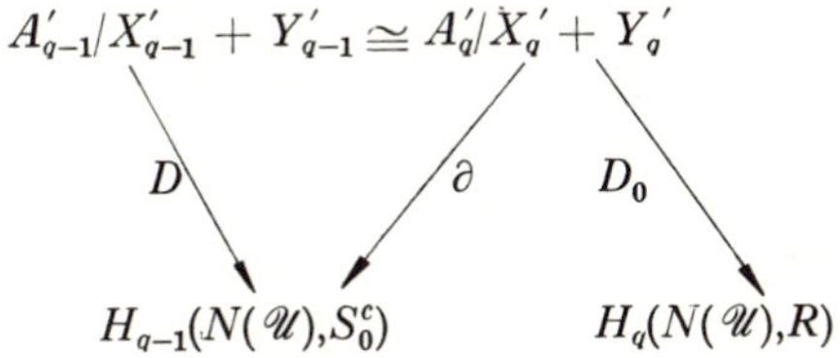

Recall that D_0 is the operator denoting addition of the coefficients in each singular chain. It maps the spaces A'_q, Y'_q and X'_q homomorphically onto $Z^q(N(\mathcal{U}), R)$, $B^q(N(\mathcal{U}), R)$ and $\{0\}$, respectively.

From the complete sequence of isomorphisms we are therefore able to conclude that

$$S_q^c/S_q^b \cong H_q(N(\mathcal{U}),R).$$

A.10. De Rham's isomorphism theorem

The results of the previous section hold for simple coverings. That they hold for any covering is a consequence of the following

Lemma A.10.1. *For any covering $\mathcal{U} = \{U_i\}$ of a differentiable manifold M there exists a covering $\mathcal{W} = \{W_i\}$ by means of coordinate neighborhoods with the properties (a) $\mathcal{W} < \mathcal{U}$ and (b) there exists a map $\phi: W_i \to U_i$ such that $W_{i_0} \cap \cdots \cap W_{i_p} \neq \square$ implies $W_{i_0} \cup \cdots \cup W_{i_p} \subset U_{i_0} \cap \cdots \cap U_{i_p}$.*
    To begin with, there exist locally finite coverings $\mathcal{V}$ and $\mathcal{U}'$ such that $\mathcal{V} \ll \mathcal{U}' < \mathcal{U}$. Hence, for any point $P \in M$, there is a ball $W(P)$ around P such that

(i) $P \in U'$ implies $W(P) \subset U'$,

(ii) $P \in V$ implies $W(P) \subset V$,

(iii) $P \notin \bar{V}$ implies $W(P) \cap \bar{V} = \square$.

For, since P belongs to only a finite number of U' and V, (i) and (ii) are satisfied. That (iii) is satisfied is seen as follows: Let $P \in V_0 \in \mathcal{V}$. Then, either $\bar{V} \cap V_0 = \square$ or $\bar{V} \cap V_0 \neq \square$. In the first case, (iii) is obviously fulfilled. As for the latter case, since $\mathcal{V}$ is locally finite there are only a finite number of sets V meeting $\bar{V}_0$, and so by choosing $W(P)$ sufficiently small (iii) may be satisfied.
    Let $W_i = W(P_i)$ be a covering of M by coordinate neighborhoods.

Then, there is an open set V_i with $P_i \in V_i$ and, by (ii) $W_i \subset V_i \Subset U_i'$ $\subset U_i$. Hence, property (a) is satisfied. That property (b) is fulfilled is seen as follows: Suppose that $W_i \cap W_j \neq \square$; then, $W_i \cap \bar{V}_j \neq \square$. Hence, by (iii) $P_i \in \bar{V}_j \subset U_j'$, and so by (i) $W_i \subset U_j' \subset U_j$. By symmetry we conclude that $W_i \cup W_j \subset U_i \cap U_j$ and (b) follows.

We are now in a position to complete the proof of de Rham's isomorphism theorem. To this end, let $\bar{A}_i$ be the direct limit of the $A_i = A_i(\mathscr{U})$ and $\bar{X}_i$, $\bar{Y}_i$ the corresponding direct limits. The proof is completed by showing that

$$\bar{A}_i/\bar{X}_i + \bar{Y}_i \cong H^{i-1}(N(\mathscr{U}), \wedge_c^{q-i+1})$$

for any open covering $\mathscr{U}$ of M thereby proving that the isomorphisms (A.9.2) are independent of the given covering. The above isomorphism follows directly from two lemmas which we now establish.

Lemma A.10.2. *The maps $\bar{d}$ and δ induce homomorphisms*

$$\bar{d} : \bar{A}_i \to H^{i-1}(N(\mathscr{U}), \wedge_c^{q-i+1}),$$

$$\delta : \bar{A}_i \to H^{i}(N(\mathscr{U}), \wedge_c^{q-i}).$$

Moreover, these maps are epimorphisms (homomorphisms onto).

Indeed, for any $f_i \in A_i(\mathscr{U})$, $\bar{d}f_i$ and $\bar{\delta}f_i$ are defined as the cohomology classes of df_i and δf_i, respectively. That they are well-defined is clear from the notion of direct limit. We must show that both $\bar{d}$ and $\bar{\delta}$ are onto. For $\bar{d}$, let z be an element of $Z^{i-1}(N(\mathscr{U}), \wedge_c^{q-i+1})$ and $\mathscr{W}$ be a refinement of $\mathscr{U}$ as in lemma A.10.1: ϕ: $W_j \to U_j$; then, the values of $\phi^* z$ are defined on $W_0 \cap \cdots \cap W_{i-1} \subset U_0$ and may be extended to W_0. By the Poincaré lemma, $\phi^* z$ is exact, that is there is a $y \in C^{i-1}(N(\mathscr{W}), \wedge^{q-i})$ for which $\phi^* z = dy$ on W_0 and consequently in $W_0 \cap \cdots \cap W_{i-1}$. That $\bar{\delta}$ is onto is clear. For, since the cohomology is trivial, any $z \in Z^i(N(\mathscr{U}), \wedge_c^{q-i})$ is of the form δy, $y \in C^{i-1}(N(\mathscr{U}), \wedge^{q-i})$. The element y represents an element of $\bar{A}_i$.

Lemma A.10.3.

$$\textit{kernel}\, \bar{d} = \textit{kernel}\, \bar{\delta} = \bar{X}_i + \bar{Y}_i.$$

The images of $x_i(U) + y_i(U)$ under d and δ are the cohomology classes of $dy_i(U)$ and $\delta x_i(U)$, respectively (cf. § A.9). The lemma is therefore trivial for $\bar{d}$. Now, as in the proof of the previous lemma, there is a refinement $\mathscr{W}$ of $\mathscr{U}$ such that $\tilde{\phi} x_i(\mathscr{U}) = dz(\mathscr{W})$. Hence $\delta x_i(U)$

is equivalent to $\delta dz(\mathscr{W}) = d\delta z(\mathscr{W})$, that is to an element which is cohomologous to zero.

We show finally that the kernels are precisely $\bar{X}_i + \bar{Y}_i$. To this end, let $dz(\mathscr{U})$ represent $\{0\}$. Then, for a suitable refinement ψ, $\tilde{\psi}dz = \delta u$ where $du = 0$. For a further refinement ϕ, $\tilde{\phi}u = dv$ by the Poincaré lemma. Hence, $d(\tilde{\phi}\tilde{\psi}z - \delta v) = \tilde{\phi}\tilde{\psi}dz - d\delta v = \tilde{\phi}\delta u - d\delta v = \delta\tilde{\phi}u - \delta dv = 0$, and so, since $\tilde{\phi}\tilde{\psi}z = (\tilde{\phi}\tilde{\psi}z - \delta v) + \delta v$, z is an element of $\bar{X}_i + \bar{Y}_i$.

Analogous reasoning applies to the map δ.

Remarks: 1. De Rham's isomorphism theorem has been established for compact spaces. That it holds for *paracompact manifolds*, that is, a manifold for which every open covering has a locally finite open refinement, is left as an exercise. Indeed, it can be shown that every covering of a paracompact space has a locally finite strong refinement.

2. The isomorphism theorem extends to the cohomology rings (cf. Appendix B).

A.11. De Rham's existence theorems

We recall these statements referred to as (R_1) and (R_2) in § 2.11.

(R_1) Let $\{\zeta_q^i\}$ $(i = 1, \cdots, b_q(M))$ be a basis for the singular q-cycles of a compact differentiable manifold M and $\omega_q^i(i = 1, \cdots, b_q(M))$, b_q arbitrary real constants. Then there exists a closed q-form α on M having the ω_q^i as periods.

(R_2) A closed form with zero periods is exact.

Proof of (R_1). Due to the isomorphism theorem, (R_1) need only be established for the cycles and cocycles (with real coefficients) on the nerve of a given covering $\mathscr{U}$.

Let L be a linear functional on $Z_q(N(\mathscr{U}), R)$ (the singular q-cycles) which vanishes on $B_q(N(\mathscr{U}), R)$ (the singular boundaries). L may be extended to $C_q(N(\mathscr{U}), R)$ in the following way: Let ξ_i be a basis of the vector space $C_q(N(\mathscr{U}), R)/Z_q(N(\mathscr{U}),R)$. Then, every $\xi \in C_q(N(\mathscr{U}), R)$ has a unique representation in the form

$$\xi = \Sigma r_i \xi_i + \zeta, \quad \zeta \in Z_q(N(\mathscr{U}),R), r_i \in R;$$

We extend L to $C_q(N(\mathscr{U}), R)$ by putting $L(\xi) = L(\zeta)$.

Now, there is a (unique) cochain $x \in C^q(N(\mathscr{U}), R)$ such that $(x, \xi) = L(\xi)$, namely, the cochain whose values are $L(\Delta(i_0, \cdots, i_q))$. It remains to be shown that x is a cocycle. Indeed,

$$(\delta x, \xi) = (x, \partial\xi) = L(\partial\xi) = 0$$

since L vanishes on the boundaries. Thus, since ξ is an arbitrary chain δx vanishes.

Proof of (R_2). Suppose that $(x, \partial\xi) = 0$ for all $\xi \in C_{q+1}(N(\mathscr{U}), R)$. We wish to show that x is a coboundary. To this end, let L be the linear functional on $B_{q-1}(N(\mathscr{U}), R)$ defined by

$$L(\partial\eta) = (x, \eta). \tag{A.11.1}$$

Since $\partial\eta = \partial\eta'$ implies $(x, \eta) = (x, \eta')$, L is well defined. Now, extend L to $C_{q-1}(N(\mathscr{U}), R)$ and determine y by the condition

$$(y, \xi) = L(\xi). \tag{A.11.2}$$

Then,

$$\begin{aligned}(x - \delta y, \eta) &= (x, \eta) - (\delta y, \eta)\\ &= (x, \eta) - (y, \partial\eta) = 0\end{aligned}$$

by (A.11.1) and (A.11.2). Since this holds for all η, $x - \delta y$ vanishes and x is a coboundary.

Remarks: 1. The cohomology theory defined in § A.2 is a straightforward generalization of the classical Čech definition of cohomology. The idea of cohomology with 'coefficients' in a sheaf Γ is due to Leray and is a generalization of Steenrod's cohomology with 'local coefficients'.

2. It can be shown that a topological manifold is paracompact. In fact, there exists a locally finite strong refinement of every covering (cf. Appendix D). Hence, by the remark at the end of § A.10, de Rham's isomorphism theorem is valid for differentiable manifolds. The existence theorems, however, require compactness.

3. There are at least two distinct cohomology theories on a manifold. The de Rham cohomology is defined on the graded algebra $A[M]$ of *all* differential forms of class 1 on M. On the other hand, cohomology theories may be defined on $A_k[M]$—the graded algebra consisting of those forms of class k (> 1), and on $A_c[M]$—the graded algebra of forms with compact carriers. If $M = R^n$, Poincaré's lemma for forms with compact carriers asserts that a closed p-form (with compact carrier) is the differential of a $(p - 1)$-form with compact carrier if $p \leqq n - 1$, and an n-form α is the differential of an $(n - 1)$-form with compact carrier if, and only if, $(\alpha, 1) = 0$. Hence, $b_p(A_c[R^n]) = 0$, $p \leqq n - 1$, and $b_n(A_c[R^n]) \neq 0$. But, $b_0(A[R^n]) = 1$ and, from §A.6, for $p > 0$, $b_p(A[R^n]) = 0$.

De Rham's theorem states that there are precisely two cohomology theories, namely, those on $A[M]$ and $A_c[M]$. Moreover, if M is compact, there is only one.

APPENDIX B

THE CUP PRODUCT

For a compact manifold M, we have seen that each element of the singular homology group SH_p acts as a linear functional on the de Rham cohomology group $D^p(M)$, and that each element of $D^p(M)$ may be considered as a linear functional on SH_p. In fact, the correspondences

$$SH_p \to (D^p(M))^*$$

and

$$D^p(M) \to (SH_p)^* = H^p(M)$$

(where ( )* denotes the dual space of ( )) are isomorphisms. In this appendix we should like to show how the second map may be extended to the cohomology ring structures. To this end, a product is defined in $(SH_p)^*$.

B.1. The cup product

Let α and β be closed p- and q-forms, corresponding to the cohomology classes z_α and z_β, respectively. Let f_α and f_β be representative p- and q-cocycles. We shall show that $\alpha \wedge \beta$ corresponds to the cohomology class $z_{\alpha\wedge\beta}$ defined by the $(p+q)$-cocycle $f_{\alpha\wedge\beta}$ where

$$f_{\alpha\wedge\beta}(U_0, \cdots, U_{p+q}) = f_\alpha(U_0, \cdots, U_p) f_\beta(U_p, \cdots, U_{p+q}). \qquad \text{(B.1.1)}$$

The product so defined will henceforth be denoted by $f_\alpha \cup f_\beta$ and called the *cup product* of f_α and f_β.

Lemma B.1.1. *The operator δ is an anti-derivation:*

$$\delta(f_\alpha \cup f_\beta) = \delta f_\alpha \cup f_\beta + (-1)^p f_\alpha \cup \delta f_\beta.$$

Indeed, for a given covering $\mathscr{U} = \{U_i\}$

$$(\delta f_{\alpha\wedge\beta})(U_0, \cdots U_{p+q+1}) = \sum_r (-1)^r f_{\alpha\wedge\beta}(U_0, \cdots, U_{r-1}, U_{r+1}, \cdots, U_{p+q+1})$$

$$= \sum_{r=0}^{p+1} (-1)^r f_\alpha(U_0, \cdots, U_{r-1}, U_{r+1}, \cdots, U_{p+1}) f_\beta(U_{p+1}, \cdots, U_{p+q+1})$$

$$+ \sum_{r=p}^{p+q+1} (-1)^r f_\alpha(U_0, \cdots, U_p) f_\beta(U_p, \cdots, U_{r-1}, U_{r+1}, \cdots, U_{p+q+1})$$

$$= (\delta f_\alpha)(U_0, \cdots, U_{p+1}) f_\beta(U_{p+1}, \cdots, U_{p+q+1})$$

$$+ (-1)^p f_\alpha(U_0, \cdots, U_p)(\delta f_\beta)(U_p, \cdots, U_{p+q+1})$$

$$= (\delta f_\alpha \cup f_\beta)(U_0, \cdots, U_{p+q+1}) + (-1)^p (f_\alpha \cup \delta f_\beta)(U_0, \cdots, U_{p+q+1}).$$

Corollary.

cocycle $\cup$ cocycle = cocycle,
cocycle $\cup$ coboundary = coboundary,
coboundary $\cup$ cocycle = coboundary.

The cup product is thus defined for cohomology classes and gives a pairing of the cohomology groups $H^p(N(\mathscr{U}), R)$ and $H^q(N(\mathscr{U}), R)$ to the cohomology group $H^{p+q}(N(\mathscr{U}), R)$.

Lemma B.1.2. *The cup product has the anti-commutativity property*

$$f \cup g = (-1)^{pq} g \cup f, \quad f \in H^p(N(\mathscr{U}), R), g \in H^q(N(\mathscr{U}), R).$$

This is clear from the formula (B.1.1).

B.2. The ring isomorphism

As in § A.9, let $f_0 \in Z^0(N(\mathscr{U}), \wedge_c^p)$, $f_0' \in Z^0(N(\mathscr{U}), \wedge_c^q)$ and consider the relations

$$\begin{aligned} f_0 &= df_1 & f_0' &= df_1' \\ \delta f_1 &= df_2 & \delta f_1' &= df_2' \\ &\vdots & &\vdots \\ \delta f_{p-1} &= df_p & \delta f_{q-1}' &= df_q' \end{aligned}$$

Assume that f_0, f_0' have the values α and β, respectively, and put $f_0''(U_0, \cdots, U_{p+q}) = \alpha \wedge \beta$. Moreover, let

$$f_i''(U_0, \cdots, U_{i-1}) = f_i(U_0, \cdots, U_{i-1}) \wedge \beta, \quad 1 \leqq i \leqq p \tag{B.2.1}$$

and

$$f_{p+j}''(U_0, \cdots, U_{p+j-1}) = f_\alpha(U_0, \cdots, U_p) f_j'(U_p, \cdots, U_{p+j-1}), \quad 1 \leqq j \leqq q. \tag{B.2.2}$$

For $i \neq p$, $\delta f_i'' = df_{i+1}''$. Now, for $i = p$ we have from (B.2.1)

$$(\delta f_p'')(U_0, \cdots, U_p) = (\delta f_p)(U_0, \cdots, U_p) \wedge \beta$$

and, from (B.2.2)

$$(df_{p+1}'')(U_0, \cdots, U_p) = f_\alpha(U_0, \cdots, U_p) \wedge (df_1')(U_p).$$

Hence, since $\delta f_p = f_\alpha$ and $(df_1')(U_p) = \beta$, $\delta f_p'' = df_{p+1}''$. In this way, we see that

$$\begin{aligned}
&(\delta f_{p+q}'')(U_0, \cdots, U_{p+q}) \\
&= (\delta f_\alpha)(U_0, \cdots, U_{p+1}) f_q'(U_{p+1}, \cdots, U_{p+q}) + f_\alpha(U_0, \cdots, U_p)(\delta f_q')(U_p, \cdots, U_{p+q}) \\
&= f_\alpha(U_0, \cdots, U_p) f_\beta(U_p, \cdots, U_{p+q}) \\
&= (f_\alpha \cup f_\beta)(U_0, \cdots, U_{p+q})
\end{aligned}$$

since $\delta f_\alpha = 0$ and $\delta f_q' = f_\beta$. We conclude that $\alpha \wedge \beta$ determines $f_\alpha \cup f_\beta$.

Summarizing, we have shown that the direct sum $D(M)$ of the vector spaces (cohomology groups) $D^p(M)$ has a ring structure, and that the de Rham isomorphism between the cohomology groups extends to a ring isomorphism.

Remark: Many of the methods of sheaf theory have apparently resulted from the developments of Appendix A. In fact, perhaps the most important applications of the theory are in proving isomorphism theorems as, for example, those in § 6.14.

APPENDIX C

THE HODGE EXISTENCE THEOREM

Let M be a compact and orientable Riemannian manifold with metric tensor g of class $k \geqq 5$. We have tacitly assumed that M is of class $k + 1$. Denote by $\wedge^p$ the Hilbert space of all measurable p-forms α on M such that (α, α) is finite. (The notation follows closely that of Chapter II). The norm in $\wedge^p$ is defined by the global scalar product. We assume some familiarity with Hilbert space methods. The properties of the Laplace-Beltrami operator Δ are to be developed from this point of view. The idea of the proof of the existence theorem is to show that $\bar{\Delta}^{-1}$—the inverse of the closure of Δ is a completely continuous operator with domain $(\wedge^p_H)^\perp$—the orthogonal complement of $\wedge^p_H$ [*31*]. The Green's operator G (cf. II.B) defined by

$$G = \begin{cases} \bar{\Delta}^{-1} & \text{on } (\wedge^p_H)^\perp \\ 0 & \text{on } \wedge^p_H \end{cases}$$

is therefore completely continuous.

Since $R(\bar{\Delta})$—the range of $\bar{\Delta}$ is all of $(\wedge^p_H)^\perp$, we obtain the

Decomposition theorem

A regular form α of degree $p(0 < p < \dim M)$ has the unique decomposition

$$\alpha = d\delta\gamma + \delta d\gamma + H[\alpha]$$

where γ is of class 2 and $H[\alpha]$ is of class $k - 4$ (cf. § 2.10). (If $k = 5$, $H[\alpha]$ is of class 2).

For, since $\alpha - H[\alpha] \in (\wedge^p_H)^\perp$, it belongs to $R(\bar{\Delta})$. Hence, there is a p-form γ such that $\bar{\Delta}\gamma = \alpha - H[\alpha]$. However, $\alpha - H[\alpha]$ is of class 1. Consequently, by lemma C. 1 below, γ is of class 2 from which we conclude that it belongs to the domain of Δ.

The complete continuity of the operator $(\bar{\Delta} + I)^{-1}$ is used, on the other hand, for the proof that dim $\wedge_H^p$ (where $\wedge_H^p$ is the null space of $\bar{\Delta}$) is finite.

The following lemma given without proof is of fundamental importance [46]:

Lemma C.1. *Let $\alpha \in \wedge^p$ and $\beta = \gamma + r\alpha$ where γ is a p-form of class $l(1 \leqq l \leqq k-5)$ and $r \in R$. (When $k = 5$, take $l = 1$). If $(\Delta\theta, \alpha) = (\theta, \beta)$ for every p-form θ of class $k-2$, then α is a form of class $l+1$ (almost everywhere) and $\Delta\alpha = \beta$.*

For forms α of class 2, this is clear. In this case $(\Delta\theta, \alpha) = (\theta, \Delta\alpha)$. Consequently, $(\theta, \Delta\alpha - \beta)$ vanishes for all p-forms θ of class 2. Hence, $\Delta\alpha - \beta = 0$ almost everywhere on M.

We begin by showing that $\bar{\Delta}$ is self-adjoint, (or, self-dual) that is, $\bar{\Delta}$ is its maximal adjoint operator. (The closure of an operator on $\wedge^p$ is the closure of its graph in $\wedge^p \times \wedge^p$). Let $\Delta_1 = \Delta + I$ ($I \equiv$ identity). Since $\overline{\Delta + I} = \bar{\Delta} + I$, $\bar{\Delta}$ is self-adjoint, if and only if, $\bar{\Delta}_1$ is self-adjoint. We show that $\bar{\Delta}_1$ is self-adjoint. In the first place, Δ_1 is (1-1). For, since

$$(\Delta_1\alpha, \alpha) = (d\alpha, d\alpha) + (\delta\alpha, \delta\alpha) + (\alpha, \alpha) \geqq (\alpha, \alpha),$$

the condition $\Delta_1\alpha = 0$ implies $\alpha = 0$. Again, since

$$\| \alpha \| \leqq \| \Delta_1\alpha \|$$

(cf. § 7.3 for notation), the inverse mapping $(\Delta_1)^{-1}$ is bounded. Thus $R(\bar{\Delta}_1)$ is closed in $\wedge^p$. That $R(\bar{\Delta}_1)$ is all of $\wedge^p$ may be seen in the following way: Let $\alpha \neq 0$ be a p-form with the property $(\Delta_1\beta, \alpha) = 0$ for all β. Applying lemma C. 1 with $r = -1$ this implies that α is of class 2 and $\Delta_1\alpha = 0$. Hence, since Δ_1 is (1-1), $\alpha = 0$, and so $R(\bar{\Delta}_1) = \wedge^p$.

We have shown that $(\bar{\Delta}_1)^{-1}$ is a bounded, symmetric operator on $\wedge^p$. It is therefore self-adjoint, and hence its inverse is self-adjoint. Thus,

Lemma C.2. *The closure of Δ is a self-adjoint operator on $\wedge^p$.*

We require the following lemma in order to establish the complete continuity of the operator $(\bar{\Delta}_1)^{-1}$:

Lemma C.3. *There exists a coordinate neighborhood U of every point $P \in M$ such that for all forms α of class 2 vanishing outside U*

$$D(\alpha) \leqq C(\Delta_1\alpha, \alpha) = C[(\Delta\alpha, \alpha) + (\alpha, \alpha)]$$

where C is a constant depending on U and

$$D(\alpha) = \int_M \sum_k \left(\frac{\partial\alpha_{(i_1\cdots i_p)}}{\partial u^k}\right)^2 *1$$

is the Dirichlet integral.

Remark: If, in E^n we integrate the right hand side of

$$(\Delta\alpha,\alpha) = -\int_M \sum_k \frac{\partial^2 \alpha_{(i_1\cdots i_p)}}{\partial(u^k)^2} \alpha^{(i_1\dots i_p)} *1$$

by parts, we obtain by virtue of the computation following (3.2.8)

$$D(\alpha) = (\Delta\alpha,\alpha).$$

(The lemma is therefore clear in E^n.)

Let g_{ij} denote the components of the metric tensor g relative to a geodesic coordinate system at P: $g_{ij}(P) = \delta_{ij}$. Denote by U' the neighborhood of P in which this coordinate system is valid. Define a new metric g' in U' by $g'_{ij} = \delta_{ij}$. (The existence of such a metric in U' is clear). Then, by the above remark

$$D(\alpha) = (\Delta'\alpha,\alpha)' = \|\, d\alpha \,\|'^2 + \|\, \delta'\alpha \,\|'^2$$

where the prime indicates that the corresponding quantity has been computed with respect to the metric g', and α is a form vanishing outside U'. Since

$$\|\, \beta \,\|'^2 \leqq C_1 \|\, \beta \,\|^2$$

for some constant C_1 and any form β,

$$\begin{aligned} D(\alpha) &\leqq C_1(\|\, d\alpha \,\|^2 + \|\, \delta'\alpha \,\|^2) \\ &= C_1(\|\, d\alpha \,\|^2 + \|\, \delta\alpha \,\|^2 + \|\, \delta'\alpha \,\|^2 - \|\, \delta\alpha \|^2) \\ &\leqq 2C_1(\|\, d\alpha \,\|^2 + \|\, \delta\alpha \,\|^2 + \|\, \delta'\alpha - \delta\alpha \,\|^2) \\ &= 2C_1[(\Delta\alpha,\alpha) + \|\, \delta'\alpha - \delta\alpha \,\|^2] \end{aligned}$$

—the second inequality following from the parallelogram law. The following estimate is left as an exercise:

$$\|\, \delta'\alpha - \delta\alpha \,\|^2 \leqq C_2 \|\, \alpha \,\|^2 + \epsilon(U)D(\alpha)$$

where $\epsilon(U) \to 0$ as U shrinks to P. The proof is straightforward. We conclude that

$$\begin{aligned} D(\alpha) &\leqq 2C_1[(\Delta\alpha,\alpha) + C_2 \|\, \alpha \,\|^2 + \epsilon(U)D(\alpha)] \\ &\leqq 2C_1[(\Delta\alpha,\alpha) + \|\, \alpha \,\|^2] + \tfrac{1}{2}D(\alpha) \end{aligned}$$

by taking U small enough so that $2C_1\epsilon(U) \leqq \frac{1}{2}$ and $C_2 \leqq 1$.

Lemma C.4. *The operator $(\bar{\Delta}_1)^{-1}$ is completely continuous, that is, it sends bounded sets into relatively compact sets.*

We employ the following well-known fact regarding operators on a Hilbert space. Let $\{\alpha_i\}$ be a sequence of forms and assume that the sequence $\{\bar{\Delta}_1 \alpha_i\}$ is defined and bounded. If from the former sequence, a norm convergent subsequence can be selected, $(\bar{\Delta}_1)^{-1}$ is completely continuous. We need only consider those forms in the domain of Δ_1. In the first place, since

$$|| \Delta_1 \alpha ||^2 = || \Delta\alpha ||^2 + 2 \, || d\alpha ||^2 + 2 \, || \delta\alpha ||^2 + || \alpha ||^2,$$

the sequences $\{\alpha_i\}$, $\{d\alpha_i\}$ and $\{\delta\alpha_i\}$ are also bounded (in norm). If we take a partition of unity $\{g_\beta\}$ (cf. § 1.6), the corresponding sequences $\{g_\beta \alpha_i\}$, $\{dg_\beta \alpha_i\}$ and $\{\delta g_\beta \alpha_i\}$ are also bounded. Since the terms of the sequence $\{g_\beta \alpha_i\}$ are bounded (in norm), the same is true of the first partial derivatives of their coefficients by virtue of lemma C.3, provided we choose a sufficiently fine partition of unity. Lemma C.4 is now an immediate consequence of the Rellich selection theorem, namely, "if a sequence of functions together with their first derivatives is bounded in norm, then a convergent subsequence can be selected".

Proposition (Hodge-de Rham). *The number of linearly independent harmonic forms on a compact and orientable Riemannian manifold is finite.*

Since the operator $(\bar{\Delta}_1)^{-1}$ is (1-1), self-adjoint and completely continuous, its spectrum has infinitely many eigenvalues (each of finite multiplicity) which are bounded and with zero as their only limit. However, 0 is not an eigenvalue. The eigenvalues of $\bar{\Delta}_1$ are the reciprocals of those of $(\bar{\Delta}_1)^{-1}$—the multiplicities being preserved; moreover, the spectrum of $\bar{\Delta}$ has no limit points. Since $\bar{\Delta}_1 = \bar{\Delta} + I$, the spectrum of $\bar{\Delta}$ is obtained from that of $\bar{\Delta}_1$ by means of a translation. Thus, the spectrum of $\bar{\Delta}$ has no (finite) limit points; in addition, the eigenvalues of $\bar{\Delta}$ have finite multiplicities. In particular, *if zero is an eigenvalue, the number of linearly independent harmonic forms is finite* since each eigenspace has finite dimension. (In the original proof due to Hodge, this was a consequence of the Fredholm theory of integral equations).

Finally, we show that $\bar{\Delta}^{-1}$ is a completely continuous operator on $(\wedge^p_H)^\perp$. In the first place, $\bar{\Delta}$ is (1-1) on $(\wedge^p_H)^\perp$. Thus, if we restrict $\bar{\Delta}$ to $(\wedge^p_H)^\perp$, it has an inverse. (It is this inverse which we denote by $\bar{\Delta}^{-1}$). By lemma C.2, $\bar{\Delta}^{-1}$ is self-adjoint. Consequently, its domain is dense in $(\wedge^p_H)^\perp$; for, an element orthogonal to the range of a self-adjoint operator is in its null space. Moreover, $\bar{\Delta}^{-1}$ has a bounded spectrum with zero as the only limit point. This follows from the fact that the eigenvalues of $\bar{\Delta}$ on $(\wedge_H^p)^\perp$ have no limit points.

Summarizing, $\bar{\Delta}^{-1}$ has the properties:

(a) it is self-adjoint with domain $(\wedge_H^p)^\perp$,

(b) its spectrum is bounded with the zero element as its only limit point, and

(c) each of its eigenspaces is finite dimensional.

This allows us to conclude that $\bar{\Delta}^{-1}$ is completely continuous. That its domain is $(\wedge_H^p)^\perp$ follows from the fact that a completely continuous operator is bounded. The remaining portions of the proof of the existence theorem appear in § 2.11.

Remark: Lemma C.2 is not essential to the argument. For, the complete continuity of $\bar{\Delta}^{-1}$ can be shown directly from that of $(\bar{\Delta}_1)^{-1}$, which is defined on the whole space $\wedge^p$, since $\wedge_H^p$ is an invariant subspace of this operator.

APPENDIX D

PARTITION OF UNITY

To show that to a locally finite open covering $\mathscr{U} = \{U_i\}$ of a differentiable manifold M there is associated a partition of unity (cf. §1.6) we shall make use of the following facts: (a) M is normal (since a topological manifold is regular), that is, to every pair of disjoint closed sets, there exist disjoint open sets containing them. (b) Since M is normal, there exist locally finite open coverings $\mathscr{V} = \{V_i\}$, $\mathscr{W}^0 = \{W_i^0\}$, $\mathscr{W} = \{W_i\}$ and $\mathscr{W}^1 = \{W_i^1\}$ such that

$$\bar{W}_i^1 \subset W_i \subset \bar{W}_i \subset W_i^0 \subset \bar{W}_i^0 \subset V_i \subset \bar{V}_i \subset U_i$$

for each i.

In the construction given below, it will be assumed (with no loss in generality) that each U_i is contained in a coordinate neighborhood and has compact closure.

In constructing a partition of unity, it is convenient to employ a smoothing function in E^n, that is a function $g_\epsilon \geqq 0$ of class k corresponding to an arbitrary $\epsilon > 0$ such that

(i) $\text{carr}(g_\epsilon) \subset \{r \leqq \epsilon\}$ where r denotes the distance from the origin;

(ii) $g_\epsilon > 0$ for $r < \epsilon$;

(iii) $\int_{E^n} g_\epsilon(u^1, \cdots, u^n) du^1 \cdots du^n = 1$.

An example of a smoothing function is given by

$$g_\epsilon(u) = \begin{cases} 0, & r \geqq \epsilon \\ \dfrac{c}{\epsilon^n} \exp \dfrac{-\epsilon^2}{\epsilon^2 - r^2}, & r < \epsilon \end{cases}$$

where c is chosen so that

$$\int_{E^n} g_\epsilon(u) du = c \int_{r \leqq 1} \exp\left(\frac{-1}{1 - r^2}\right) du = 1.$$

For each U_i, let f_i be the continuous function

$$f_i(P) = \begin{cases} 1, & P \in W_i^1 \\ 0, & P \in \text{the complement of } W_i, \end{cases}$$

$$0 \leqq f_i(P) \leqq 1, \quad P \in W_i - \bar{W}_i^1.$$

Let $(u^1, \ldots, u^n)$ be a local coordinate system in U_i and define "distance" between points of U_i to be the ordinary Euclidean distance between the corresponding points of B_i where B_i is the ball in E^n homeomorphic with U_i. Let ϵ_i be chosen so small that a sphere of radius ϵ_i with center P is contained in U_i for all $P \in V_i$ and does not meet W_i for $P \in V_i - \bar{W}_i^0$. Consider the function

$$h_i(P) \equiv h_i(u) = \int f_i(v) g_{\epsilon_i}(u - v)\,dv, \quad P \in V_i.$$

It has the following properties.

(i) h_i is of class k;

(ii) $h_i \geqq 0$, $h_i(P) > 0$, $P \in W_i^1$; $h_i(P) = 0$, $P \in V_i - \bar{W}_i^0$.

Thus, if we define h_i to be 0 in the complement of V_i, it is a function of class k on M.

(iii) $W_i^1 \subset \operatorname{carr}(h_i) \subset \bar{W}_i^0 \subset U_i$.

(iv) $h(P) = \Sigma_i h_i(P)$ is defined for each $P \in M$ (since $\mathscr{U}$ is a locally finite covering); $h(P)$ is of class k and is never 0 since $\mathscr{W}^1$ is a covering of M.

We may therefore conclude that the functions

$$g_i(P) = \frac{h_i(P)}{h(P)}$$

form a partition of unity subordinated to the covering $\mathscr{U}$.

Remarks: 1. The above theorem shows that there are many non-trivial differential forms of class k on M.

2. A topological space is said to be *regular* if to each closed set S and point $P \notin S$, there exist disjoint open sets containing S and P. Since M is a topological manifold, it is locally homeomorphic with R^n. Hence, it is locally compact. That M is regular is a consequence of the fact that it is locally compact and Hausdorff. That it is normal follows from regularity and the existence of a countable basis. Finally, from these properties, it can be shown that M is paracompact.

REFERENCES

1. M. Berger, Sur quelques variétés Riemanniennes suffisamment pincées, *Bull. soc. math. France* **88**, 57-71 (1960).
2. M. Berger, Pincement Riemannien et pincement holomorphe, *Ann. scuola norm. super. Pisa III*, **14**, 151-159 (1960).
3. G. Bidal and G. de Rham, Les formes différentielles harmoniques, *Comm. Math. Helv.* **19**, 1-49 (1946).
4. S. Bochner, Vector fields and Ricci curvature, *Bull. Am. Math. Soc.* **52**, 776-797 (1946).
5. S. Bochner, Curvature in Hermitian metric, *Bull. Am. Math. Soc.* **53**, 179-195 (1947).
6. S. Bochner, Curvature and Betti numbers, *Ann. Math.* **49**, 379-390 (1948).
7. S. Bochner, Curvature and Betti numbers II, *Ann. Math.* **50**, 77-93 (1949).
8. S. Bochner, Euler-Poincaré characteristic for locally homogeneous and complex spaces, *Ann. Math.* **51**, 241-261 (1950).
9. S. Bochner, Vector fields on complex and real manifolds, *Ann. Math.* **52**, 642-649 (1950).
10. S. Bochner, Complex spaces with transitive commutative groups of transformations, *Proc. Natl. Acad. Sci. U.S.* **37**, 356-359 (1951).
11. S. Bochner, Tensor fields and Ricci curvature in Hermitian metric, *Proc. Natl. Acad. Sci. U. S.* **37**, 704-706 (1951).
12. S. Bochner, Curvature and Betti numbers in real and complex vector bundles, *Rend. seminar. mat. Torino* **15**, 225-253 (1955-1956).
13. S. Bochner and D. Montgomery, Groups of differentiable and real or complex analytic transformations, *Ann. Math.* **46**, 685-694 (1945).
14. S. Bochner and K. Yano, Tensor-fields in non-symmetric connections, *Ann. Math.* **56**, 504-519 (1952).
15. W. Boothby, Some fundamental formulas for Hermitian manifolds with non-vanishing torsion, *Am. J. Math.* **76**, 509-534 (1954).
16. W. Boothby, Hermitian manifolds with zero curvature, *Mich. Math. J.* **5**, 229-233 (1958).
17. S. S. Cairns, Transformation of regular loci, *Ann. Math.* **35**, 579-587 (1934).
18. E. Calabi, The space of Kähler metrics, *Proc. Intern. Congr. Mathematicians Amsterdam* **2**, 206 (1954).
19. E. Cartan, Sur les domaines bornés homogènes de l'espace de *n* variables complexes, "Œuvres complètes," Part I, Vol. 2, p. 1262.
20. H. Cartan, Variétés riemanniennes, variétés analytique complexes, variétés Kählériennes Semin. École Normale Paris, Mimeographed Notes (1951-1952).
21. S.-S. Chern, Characteristic classes of hermitian manifolds, *Ann. Math.* **47**, 85-121 (1946).
22. S.-S. Chern, Topics of differential geometry, Inst. for Adv. Study, Mimeographed Notes (1951).
23. S.-S. Chern, Differentiable manifolds, University of Chicago, Mimeographed Notes (1953).

24. S.-S. Chern, Relations between Riemannian and Hermitian geometries, *Duke Math. J.* **20**, 575-587 (1953).
25. S.-S. Chern, On a generalization of Kähler geometry, Algebraic Geometry and Topology, A Symposium in Honour of S. Lefschetz, pp. 103-121 (1957).
26. S.-S. Chern, Complex manifolds, University of Chicago, Mimeographed Notes (1956).
27. C. Chevalley, "Lie Groups," Princeton University Press, Princeton, New Jersey, 1946.
28. B. Eckmann and A. Frölicher, Sur l'intégrabilité des structures presques complexes, *Compt. rend. acad. sci.* **232**, 2284-2286 (1951).
29. B. Eckmann and H. Guggenheimer, Sur les variétés closes à métrique hermitiennes sans torsion, *Compt. rend. acad. sci.* **229**, 503-505 (1949).
30. C. Ehresmann, Sur les variétés presques complexes, *Proc. Intern. Congr. Mathematicians, Harvard*, pp. 412-419 (1950).
31. M. P. Gaffney, Hilbert space methods in the theory of harmonic integrals, *Trans. Am. Math. Soc.* **78**, 426-444 (1955).
32. S. I. Goldberg, Tensorfields and curvature in Hermitian manifolds with torsion, *Ann. Math.* **63**, 64-76 (1956).
33. S. I. Goldberg, Note on projectively Euclidean Hermitian manifolds, *Proc. Natl. Acad. Sci. U. S.* **42**, 128-130 (1956).
34. S. I. Goldberg, On pseudo-harmonic and pseudo-Killing vectors in metric manifolds with torsion, *Ann. Math.* **64**, 364-373 (1956).
35. S. I. Goldberg, Conformal transformations of Kaehler manifolds, *Bull. Am. Math. Soc.* **66**, 54-58 (1960).
36. S. I. Goldberg, Groups of automorphisms of almost Kaehler manifolds, *Bull. Am. Math. Soc.* **66**, 180-183 (1960).
37. S. I. Goldberg, Groups of transformations of Kaehler and almost Kaehler manifolds, *Comm. Math. Helv.* **35**, 35-46 (1961).
38. S. I. Goldberg, Conformal maps of almost Kaehler manifolds, *Tôhoku Math. J.* **13**, 120-131 (1961).
39. W. V. D. Hodge, "The Theory and Applications of Harmonic Integrals," Cambridge University Press, London and New York, 1952.
40. E. Hopf, Elementare Bemerkungen über die Lösungen partieller Differentialgleichungen zweiter Ordnung vom elliptischen Typus, *Sitzber. preuss. Akad. Wiss. Physik.-math. Kl.* **19**, 147-152 (1927).
41. H. Hopf and H. Samelson, Ein Satz über die Wirkungsräume geschlossener Lieshcher Gruppen, *Comm. Math. Helv.* **13**, 240-251 (1941).
42. S. Ishihara and M. Obata, Affine transformations in a Riemannian manifold, *Tôhoku Math. J.* **7**, 146-150 (1955).
43. E. Kaehler, Ueber eine bemerkenswerte Hermitische Metrik, *Abhandl. Math. Seminar Hamburgischen Univ.* **9**, 173-186 (1933).
44. M. Kervaire, A manifold which does not admit any differentiable structure, *Comm. Math. Helv.* **34**, 257-270 (1960).
45. S. Kobayashi, A theorem on the affine transformation group of a Riemannian manifold, *Nagoya Math. J.* **9**, 39-41 (1955).
46. K. Kodaira, Harmonic fields in Riemannian manifolds (generalized potential theory), *Ann. Math.* **50**, 587-665 (1949).
47. K. Kodaira, On a differential-geometric method in the theory of analytic stacks, *Proc. Natl. Acad. Sci. U. S.* **39**, 1268-1273 (1953).
48 J.-L. Koszul, Homologie et cohomologie des algèbres de Lie, *Bull. soc. math. France*, **78**, 65-127 (1950).

49. J.-L. Koszul, Sur la forme hermitienne canonique des espaces homogènes complexes, *Can. J. Math.* **7**, 562-576 (1955).
50. J.-L. Koszul, Variétés kählériennes, Inst. Mat. Pura e Apl. C. N. Pq., São Paulo, Mimeographed Notes (1957).
51. A. Lichnerowicz, Courbure et nombres de Betti d'une variété riemannienne compacte, *Compt. rend. acad. sci.* **226**, 1678-1680 (1948).
52. A. Lichnerowicz, Courbure, nombres de Betti et espaces symétriques, *Proc. Intern. Congr. Mathematicians, Harvard*, pp. 216-223 (1950).
53. A. Lichnerowicz, Espaces homogènes kähleriens, *Colloq. Intern. géométrie différentielle Strasbourg*, pp. 171-184 (1953).
54. A. Lichnerowicz, Sur les groupes d'automorphismes de certaines variétés kähleriennes, *Compt. rend. acad. sci.* **239**, 1344-1347 (1954).
55. A. Lichnerowicz, Transformations infinitésimales conformes de certaines variétés riemanniennes compactes, *Compt. rend. acad. sci.* **241**, 726-729 (1955).
56. A. Lichnerowicz, "Théorie globale des connexions et des groupes d'holonomie," Travaux et Recherches Mathématique, Dunod, Paris (1955).
57. A. Lichnerowicz, Sur les transformations analytiques des variétés kähleriennes compactes, *Compt. rend. acad. sci.* **244**, 3011-3014 (1957).
58. A. Lichnerowicz, "Géométrie des groupes de Transformations," Travaux et Recherches Mathématique, Dunod, Paris (1958).
59. Y. Matsushima, Sur la structure du groupe d'homéomorphismes analytique d'une certaine variété kaehlérinne, *Nagoya Math. J.* **11**, 145-150 (1957).
60. J. Milnor, On manifolds homeomorphic to the 7-sphere, *Ann. Math.* **64**, 399-405 (1956).
61. D. Montgomery, Simply connected homogeneous spaces, *Proc. Am. Math. Soc.* **1**, 467-469 (1950).
62. S. Myers, Riemannian manifolds with positive mean curvature, *Duke Math. J.* **8**, 401-404 (1941).
63. K. Nomizu, Lie groups and differential geometry, Publications of the Mathematical Society of Japan, No. 2, Mathematical Society of Japan (1956).
64. H. E. Rauch, A contribution to differential geometry in the large, *Ann. Math.* **54**, 38-55 (1951).
65. G. De Rham, "Variétés différentiables," Actualités scientifiques et industrielles, No. 1222. Hermann, Paris, 1955.
66. G. De Rham and K. Kodaira, Harmonic integrals, Inst. for Adv. Study, Mimeographed Notes (1950).
67. L. Schwartz, Lectures on complex analytic manifolds, Tata Inst. of Fund. Research, Bombay, Mimeographed notes (1955).
68. S. Tachibana, On almost-analytic vectors in almost-Kählerian manifolds, *Tôhoku Math. J.* **11**, 247-265 (1959).
69. H. C. Wang, Complex parallisable manifolds, *Proc. Am. Math. Soc.* **5**, 771-776 (1954).
70. A. Weil, Sur la théorie des formes différentielles attachées à une variété analytique complexe, *Comm. Math. Helv.* **20**, 110-116 (1947).
71. A. Weil, Sur les théorèmes de De Rham, *Comm. Math. Helv.* **26**, 119-145 (1952).
72. A. Weil, "Introduction à l'étude des variétés kählériennes," Actualités scientifiques et industrielles, No. 1267. Hermann, Paris, 1958.
73. K. Yano, On harmonic and Killing vector fields, *Ann. Math.* **55**, 38-45 (1952).
74. K. Yano, Some remarks on tensor fields and curvature, *Ann. Math.* **55**, 328-347 (1952).

75. K. Yano and S. Bochner, "Curvature and Betti Numbers," Annals of Mathematics Studies, No. 32, Princeton University Press, Princeton, New Jersey, 1953.
76. K. Yano and I. Mogi, On real representations of Kaehlerian manifolds, *Ann. Math.* **61**, 170-189 (1955).
77. K. Yano and T. Nagano, Einstein spaces admitting a one-parameter group of conformal transformations, *Ann. Math.* **69**, 451-461 (1959).
78. S. I. Goldberg and S. Kobayashi, The conformal transformation group of a compact Riemannian manifold, *Proc. Natl. Acad. Sci. U.S.* **48**, 25-26 (1962).
78′. S. I. Goldberg and S. Kobayashi, The conformal transformation group of a compact Riemannian manifold, *Am. J. Math.* **84**, 170-174 (1962).
79. S. I. Goldberg and S. Kobayashi, The conformal transformation group of a compact homogeneous Riemannian manifold, *Bull. Am. Math. Soc.* **68**, 378-381 (1962).
80. F. Hirzebruch, "Neue topologische Methoden in der algebraischen Geometrie," Ergebnisse, p. 2 (1956).
81. S. Kobayashi, On compact Kaehler manifolds with positive definite Ricci tensor, *Ann. Math.* **74**, 570-574 (1961).
82. B. Kostant, On invariant skew tensors, *Proc. Natl. Acad. Sci. U.S.* **42**, 148-151 (1956).
83. N. H. Kuiper, On conformally-flat spaces in the large, *Ann. Math.* **50**, 916-924 (1949).
84. A. Newlander and L. Nirenberg, Complex analytic coordinates in almost complex manifolds, *Ann. Math.* **65**, 391-404 (1957).
85. A. Nijenhuis and W. B. Woolf, Some integration problems in almost complex and complex manifolds, *Ann. Math.* **77**, 424-489 (1963).
86. K. Nomizu and H. Ozeki, The existence of complete Riemannian metrics, *Proc. Am. Math. Soc.* 12, 889-891 (1961).
87. R. L. Bishop and S. I. Goldberg, On curvature and Euler-Poincaré characteristic, *Proc. Natl. Acad. Sci. U.S.* **49**, 814-817 (1963).
88. R. L. Bishop and S. I. Goldberg, On the topology of positively curved Kaehler manifolds, *Tôhoku Math. J.* **15**, 359-364 (1963).
89. R. L. Bishop and S. I. Goldberg, Some implications of the generalized Gauss-Bonnet theorem, *Trans. Am. Math. Soc.* **112**, 508-535 (1964).
90. R. L. Bishop and S. I. Goldberg, On the second cohomology group of a Kaehler manifold of positive sectional curvature, *Proc. Am. Math. Soc.* **16**, 119-122 (1965).
91. R. L. Bishop and S. I. Goldberg, Rigidity of positively curved Kaehler manifolds, *Proc. Natl. Acad. Sci. U.S.* **54**, 1037-1041 (1965).
92. R. L. Bishop and S. I. Goldberg, On the topology of positively curved Kaehler manifolds II, *Tôhoku Math. J.* **17**, 310-318 (1965).
93. R. L. Bishop and S. I. Goldberg, A characterization of the Euclidean sphere, *Bull. Am. Math. Soc.* **72**, 122-124 (1966).
94. S. I. Goldberg and S. Kobayashi, Holomorphic bisectional curvature, *J. Differential Geometry* **1**, 225-233 (1967).
95. S. I. Goldberg, Integrability of almost Kaehler manifolds, *Proc. Am. Math. Soc.* **21**, 96-100 (1969).
96. W. C. Weber and S. I. Goldberg, "Conformal deformations of Riemannian manifolds," Queen's Papers on Pure and Applied Mathematics, No. 16, Queen's University, Kingston, Ontario, 1969.
97. J. Morrow and K. Kodaira, "Complex Manifolds," Holt, Rinehart and Winston, New York, 1971.
98. S. I. Goldberg and I. Vaisman, On compact locally conformal Kaehler manifolds with non-negative sectional curvature, *Ann. Fac. Sci. Univ. Toulouse* **2**, 117-123 (1980).

AUTHOR INDEX

Numbers in parentheses are reference numbers and are included to assist in locating references when the authors' names are not mentioned at the point of reference in the text. Numbers in italic indicate the page on which the full reference is listed.

SUBJECT INDEX

D

A CATALOG OF SELECTED

DOVER BOOKS

IN SCIENCE AND MATHEMATICS

A CATALOG OF SELECTED

DOVER BOOKS

IN SCIENCE AND MATHEMATICS

QUALITATIVE THEORY OF DIFFERENTIAL EQUATIONS, V.V. Nemytskii and V.V. Stepanov. Classic graduate-level text by two prominent Soviet mathematicians covers classical differential equations as well as topological dynamics and erqodic theory. Bibliographies. 523pp. 5⅜ × 8½. 65954-2 Pa. $10.95

MATRICES AND LINEAR ALGEBRA, Hans Schneider and George Phillip Barker. Basic textbook covers theory of matrices and its applications to systems of linear equations and related topics such as determinants, eigenvalues and differential equations. Numerous exercises. 432pp. 5⅜ × 8½. 66014-1 Pa. $8.95

QUANTUM THEORY, David Bohm. This advanced undergraduate-level text presents the quantum theory in terms of qualitative and imaginative concepts, followed by specific applications worked out in mathematical detail. Preface. Index. 655pp. 5⅜ × 8½. 65969-0 Pa. $10.95

ATOMIC PHYSICS (8th edition), Max Born. Nobel laureate's lucid treatment of kinetic theory of gases, elementary particles, nuclear atom, wave-corpuscles, atomic structure and spectral lines, much more. Over 40 appendices, bibliography. 495pp. 5⅜ × 8½. 65984-4 Pa. $11.95

ELECTRONIC STRUCTURE AND THE PROPERTIES OF SOLIDS: The Physics of the Chemical Bond, Walter A. Harrison. Innovative text offers basic understanding of the electronic structure of covalent and ionic solids, simple metals, transition metals and their compounds. Problems. 1980 edition. 582pp. 6⅛ × 9¼. 66021-4 Pa. $14.95

BOUNDARY VALUE PROBLEMS OF HEAT CONDUCTION, M. Necati Özisik. Systematic, comprehensive treatment of modern mathematical methods of solving problems in heat conduction and diffusion. Numerous examples and problems. Selected references. Appendices. 505pp. 5⅜ × 8½. 65990-9 Pa. $11.95

A SHORT HISTORY OF CHEMISTRY (3rd edition), J.R. Partington. Classic exposition explores origins of chemistry, alchemy, early medical chemistry, nature of atmosphere, theory of valency, laws and structure of atomic theory, much more. 428pp. 5⅜ × 8½. (Available in U.S. only) 65977-1 Pa. $10.95

A HISTORY OF ASTRONOMY, A. Pannekoek. Well-balanced, carefully reasoned study covers such topics as Ptolemaic theory, work of Copernicus, Kepler, Newton, Eddington's work on stars, much more. Illustrated. References. 521pp. 5⅜ × 8½. 65994-1 Pa. $11.95

PRINCIPLES OF METEOROLOGICAL ANALYSIS, Walter J. Saucier. Highly respected, abundantly illustrated classic reviews atmospheric variables, hydrostatics, static stability, various analyses (scalar, cross-section, isobaric, isentropic, more). For intermediate meteorology students. 454pp. 6⅛ × 9¼. 65979-8 Pa. $12.95

RELATIVITY, THERMODYNAMICS AND COSMOLOGY, Richard C. Tolman. Landmark study extends thermodynamics to special, general relativity; also applications of relativistic mechanics, thermodynamics to cosmological models. 501pp. 5⅜ × 8½. 65383-8 Pa. $11.95

APPLIED ANALYSIS, Cornelius Lanczos. Classic work on analysis and design of finite processes for approximating solution of analytical problems. Algebraic equations, matrices, harmonic analysis, quadrature methods, much more. 559pp. 5⅜ × 8½. 65656-X Pa. $11.95

SPECIAL RELATIVITY FOR PHYSICISTS, G. Stephenson and C.W. Kilmister. Concise elegant account for nonspecialists. Lorentz transformation, optical and dynamical applications, more. Bibliography. 108pp. 5⅜ × 8½. 65519-9 Pa. $3.95

INTRODUCTION TO ANALYSIS, Maxwell Rosenlicht. Unusually clear, accessible coverage of set theory, real number system, metric spaces, continuous functions, Riemann integration, multiple integrals, more. Wide range of problems. Undergraduate level. Bibliography. 254pp. 5⅜ × 8½. 65038-3 Pa. $7.00

INTRODUCTION TO QUANTUM MECHANICS With Applications to Chemistry, Linus Pauling & E. Bright Wilson, Jr. Classic undergraduate text by Nobel Prize winner applies quantum mechanics to chemical and physical problems. Numerous tables and figures enhance the text. Chapter bibliographies. Appendices. Index. 468pp. 5⅜ × 8½. 64871-0 Pa. $9.95

ASYMPTOTIC EXPANSIONS OF INTEGRALS, Norman Bleistein & Richard A. Handelsman. Best introduction to important field with applications in a variety of scientific disciplines. New preface. Problems. Diagrams. Tables. Bibliography. Index. 448pp. 5⅜ × 8½. 65082-0 Pa. $10.95

MATHEMATICS APPLIED TO CONTINUUM MECHANICS, Lee A. Segel. Analyzes models of fluid flow and solid deformation. For upper-level math, science and engineering students. 608pp. 5⅜ × 8½. 65369-2 Pa. $12.95

ELEMENTS OF REAL ANALYSIS, David A. Sprecher. Classic text covers fundamental concepts, real number system, point sets, functions of a real variable, Fourier series, much more. Over 500 exercises. 352pp. 5⅜ × 8½. 65385-4 Pa. $8.95

PHYSICAL PRINCIPLES OF THE QUANTUM THEORY, Werner Heisenberg. Nobel Laureate discusses quantum theory, uncertainty, wave mechanics, work of Dirac, Schroedinger, Compton, Wilson, Einstein, etc. 184pp. 5⅜ × 8½. 60113-7 Pa. $4.95

INTRODUCTORY REAL ANALYSIS, A.N. Kolmogorov, S.V. Fomin. Translated by Richard A. Silverman. Self-contained, evenly paced introduction to real and functional analysis. Some 350 problems. 403pp. 5⅜ × 8½. 61226-0 Pa. $7.95

PROBLEMS AND SOLUTIONS IN QUANTUM CHEMISTRY AND PHYSICS, Charles S. Johnson, Jr. and Lee G. Pedersen. Unusually varied problems, detailed solutions in coverage of quantum mechanics, wave mechanics, angular momentum, molecular spectroscopy, scattering theory, more. 280 problems plus 139 supplementary exercises. 430pp. 6½ × 9¼. 65236-X Pa. $10.95

ASYMPTOTIC METHODS IN ANALYSIS, N.G. de Bruijn. An inexpensive, comprehensive guide to asymptotic methods—the pioneering work that teaches by explaining worked examples in detail. Index. 224pp. 5⅜ × 8½. 64221-6 Pa. $5.95

OPTICAL RESONANCE AND TWO-LEVEL ATOMS, L. Allen and J.H. Eberly. Clear, comprehensive introduction to basic principles behind all quantum optical resonance phenomena. 53 illustrations. Preface. Index. 256pp. 5⅜ × 8½. 65533-4 Pa. $6.95

COMPLEX VARIABLES, Francis J. Flanigan. Unusual approach, delaying complex algebra till harmonic functions have been analyzed from real variable viewpoint. Includes problems with answers. 364pp. 5⅜ × 8½. 61388-7 Pa. $7.95

ATOMIC SPECTRA AND ATOMIC STRUCTURE, Gerhard Herzberg. One of best introductions; especially for specialist in other fields. Treatment is physical rather than mathematical. 80 illustrations. 257pp. 5⅜ × 8½. 60115-3 Pa. $4.95

APPLIED COMPLEX VARIABLES, John W. Dettman. Step-by-step coverage of fundamentals of analytic function theory—plus lucid exposition of 5 important applications: Potential Theory; Ordinary Differential Equations; Fourier Transforms; Laplace Transforms; Asymptotic Expansions. 66 figures. Exercises at chapter ends. 512pp. 5⅜ × 8½. 64670-X Pa. $10.95

ULTRASONIC ABSORPTION: An Introduction to the Theory of Sound Absorption and Dispersion in Gases, Liquids and Solids, A.B. Bhatia. Standard reference in the field provides a clear, systematically organized introductory review of fundamental concepts for advanced graduate students, research workers. Numerous diagrams. Bibliography. 440pp. 5⅜ × 8½. 64917-2 Pa. $8.95

UNBOUNDED LINEAR OPERATORS: Theory and Applications, Seymour Goldberg. Classic presents systematic treatment of the theory of unbounded linear operators in normed linear spaces with applications to differential equations. Bibliography. 199pp. 5⅜ × 8½. 64830-3 Pa. $7.00

LIGHT SCATTERING BY SMALL PARTICLES, H.C. van de Hulst. Comprehensive treatment including full range of useful approximation methods for researchers in chemistry, meteorology and astronomy. 44 illustrations. 470pp. 5⅜ × 8½. 64228-3 Pa. $9.95

CONFORMAL MAPPING ON RIEMANN SURFACES, Harvey Cohn. Lucid, insightful book presents ideal coverage of subject. 334 exercises make book perfect for self-study. 55 figures. 352pp. 5⅜ × 8¼. 64025-6 Pa. $8.95

OPTICKS, Sir Isaac Newton. Newton's own experiments with spectroscopy, colors, lenses, reflection, refraction, etc., in language the layman can follow. Foreword by Albert Einstein. 532pp. 5⅜ × 8½. 60205-2 Pa. $8.95

GENERALIZED INTEGRAL TRANSFORMATIONS, A.H. Zemanian. Graduate-level study of recent generalizations of the Laplace, Mellin, Hankel, K. Weierstrass, convolution and other simple transformations. Bibliography. 320pp. 5⅜ × 8½. 65375-7 Pa. $7.95

GEOMETRY OF COMPLEX NUMBERS, Hans Schwerdtfeger. Illuminating, widely praised book on analytic geometry of circles, the Moebius transformation, and two-dimensional non-Euclidean geometries. 200pp. 5⅝ × 8¼.
63830-8 Pa. $6.95

MECHANICS, J.P. Den Hartog. A classic introductory text or refresher. Hundreds of applications and design problems illuminate fundamentals of trusses, loaded beams and cables, etc. 334 answered problems. 462pp. 5⅜ × 8½. 60754-2 Pa. $8.95

TOPOLOGY, John G. Hocking and Gail S. Young. Superb one-year course in classical topology. Topological spaces and functions, point-set topology, much more. Examples and problems. Bibliography. Index. 384pp. 5⅝ × 8¼.
65676-4 Pa. $7.95

STRENGTH OF MATERIALS, J.P. Den Hartog. Full, clear treatment of basic material (tension, torsion, bending, etc.) plus advanced material on engineering methods, applications. 350 answered problems. 323pp. 5⅜ × 8½. 60755-0 Pa. $7.50

ELEMENTARY CONCEPTS OF TOPOLOGY, Paul Alexandroff. Elegant, intuitive approach to topology from set-theoretic topology to Betti groups; how concepts of topology are useful in math and physics. 25 figures. 57pp. 5⅜ × 8½.
60747-X Pa. $2.95

ADVANCED STRENGTH OF MATERIALS, J.P. Den Hartog. Superbly written advanced text covers torsion, rotating disks, membrane stresses in shells, much more. Many problems and answers. 388pp. 5⅜ × 8½. 65407-9 Pa. $8.95

COMPUTABILITY AND UNSOLVABILITY, Martin Davis. Classic graduate-level introduction to theory of computability, usually referred to as theory of recurrent functions. New preface and appendix. 288pp. 5⅜ × 8½. 61471-9 Pa. $6.95

GENERAL CHEMISTRY, Linus Pauling. Revised 3rd edition of classic first-year text by Nobel laureate. Atomic and molecular structure, quantum mechanics, statistical mechanics, thermodynamics correlated with descriptive chemistry. Problems. 992pp. 5⅜ × 8½. 65622-5 Pa. $18.95

AN INTRODUCTION TO MATRICES, SETS AND GROUPS FOR SCIENCE STUDENTS, G. Stephenson. Concise, readable text introduces sets, groups, and most importantly, matrices to undergraduate students of physics, chemistry, and engineering. Problems. 164pp. 5⅜ × 8½. 65077-4 Pa. $5.95

THE HISTORICAL BACKGROUND OF CHEMISTRY, Henry M. Leicester. Evolution of ideas, not individual biography. Concentrates on formulation of a coherent set of chemical laws. 260pp. 5⅜ × 8½. 61053-5 Pa. $6.00

THE PHILOSOPHY OF MATHEMATICS: An Introductory Essay, Stephan Körner. Surveys the views of Plato, Aristotle, Leibniz & Kant concerning propositions and theories of applied and pure mathematics. Introduction. Two appendices. Index. 198pp. 5⅜ × 8½. 25048-2 Pa. $5.95

THE DEVELOPMENT OF MODERN CHEMISTRY, Aaron J. Ihde. Authoritative history of chemistry from ancient Greek theory to 20th-century innovation. Covers major chemists and their discoveries. 209 illustrations. 14 tables. Bibliographies. Indices. Appendices. 851pp. 5⅜ × 8½. 64235-6 Pa. $15.95

THE FOUR-COLOR PROBLEM: Assaults and Conquest, Thomas L. Saaty and Paul G. Kainen. Engrossing, comprehensive account of the century-old combinatorial topological problem, its history and solution. Bibliographies. Index. 110 figures. 228pp. 5⅜ × 8½. 65092-8 Pa. $6.00

CATALYSIS IN CHEMISTRY AND ENZYMOLOGY, William P. Jencks. Exceptionally clear coverage of mechanisms for catalysis, forces in aqueous solution, carbonyl- and acyl-group reactions, practical kinetics, more. 864pp. 5⅜ × 8½. 65460-5 Pa. $18.95

PROBABILITY: An Introduction, Samuel Goldberg. Excellent basic text covers set theory, probability theory for finite sample spaces, binomial theorem, much more. 360 problems. Bibliographies. 322pp. 5⅜ × 8½. 65252-1 Pa. $7.95

LIGHTNING, Martin A. Uman. Revised, updated edition of classic work on the physics of lightning. Phenomena, terminology, measurement, photography, spectroscopy, thunder, more. Reviews recent research. Bibliography. Indices. 320pp. 5⅜ × 8¼. 64575-4 Pa. $7.95

PROBABILITY THEORY: A Concise Course, Y.A. Rozanov. Highly readable, self-contained introduction covers combination of events, dependent events, Bernoulli trials, etc. Translation by Richard Silverman. 148pp. 5⅜ × 8¼. 63544-9 Pa. $4.50

THE CEASELESS WIND: An Introduction to the Theory of Atmospheric Motion, John A. Dutton. Acclaimed text integrates disciplines of mathematics and physics for full understanding of dynamics of atmospheric motion. Over 400 problems. Index. 97 illustrations. 640pp. 6 × 9. 65096-0 Pa. $16.95

STATISTICS MANUAL, Edwin L. Crow, et al. Comprehensive, practical collection of classical and modern methods prepared by U.S. Naval Ordnance Test Station. Stress on use. Basics of statistics assumed. 288pp. 5⅜ × 8½. 60599-X Pa. $6.00

WIND WAVES: Their Generation and Propagation on the Ocean Surface, Blair Kinsman. Classic of oceanography offers detailed discussion of stochastic processes and power spectral analysis that revolutionized ocean wave theory. Rigorous, lucid. 676pp. 5⅜ × 8½. 64652-1 Pa. $14.95

STATISTICAL METHOD FROM THE VIEWPOINT OF QUALITY CONTROL, Walter A. Shewhart. Important text explains regulation of variables, uses of statistical control to achieve quality control in industry, agriculture, other areas. 192pp. 5⅜ × 8½. 65232-7 Pa. $6.00

THE INTERPRETATION OF GEOLOGICAL PHASE DIAGRAMS, Ernest G. Ehlers. Clear, concise text emphasizes diagrams of systems under fluid or containing pressure; also coverage of complex binary systems, hydrothermal melting, more. 288pp. 6½ × 9¼. 65389-7 Pa. $8.95

STATISTICAL ADJUSTMENT OF DATA, W. Edwards Deming. Introduction to basic concepts of statistics, curve fitting, least squares solution, conditions without parameter, conditions containing parameters. 26 exercises worked out. 271pp. 5⅜ × 8½. 64685-8 Pa. $7.95